# CATERPILLARS IN THE FIELD AND GARDEN

**Field Guide Series**

edited by Jeffrey Glassberg

*Butterflies through Binoculars: The Boston–New York–Washington Region*
by Jeffrey Glassberg

*Butterflies through Binoculars: The East*
by Jeffrey Glassberg

*Butterflies through Binoculars: Florida*
by Jeffrey Glassberg, Marc C. Minno, and John V. Calhoun

*Dragonflies through Binoculars: A Field Guide to Dragonflies of North America, north of Mexico*
by Sidney W. Dunkle

*Butterflies through Binoculars: The West*
by Jeffrey Glassberg

*Caterpillars in the Field and Garden: A Field Guide to the Butterfly Caterpillars of North America*
by Thomas J. Allen, Jim P. Brock, and Jeffrey Glassberg

# CATERPILLARS IN THE FIELD AND GARDEN

## A FIELD GUIDE TO THE BUTTERFLY CATERPILLARS OF NORTH AMERICA

**Thomas J. Allen,**
**Jim P. Brock, and**
**Jeffrey Glassberg**

OXFORD
UNIVERSITY PRESS
2005

OXFORD
UNIVERSITY PRESS

Oxford University Press, Inc., publishes works
that further Oxford University's objective of excellence
in research, scholarship, and education.

Oxford New York
Auckland Cape Town Dar es Salaam Hong Kong Karachi
Kuala Lumpur Madrid Melbourne Mexico City Nairobi
New Delhi Shanghai Taipei Toronto

With offices in
Argentina Austria Brazil Chile Czech Republic France Greece
Guatemala Hungary Italy Japan Poland Portugal Singapore
South Korea Switzerland Thailand Turkey Ukraine Vietnam



Published by Oxford University Press, Inc.
198 Madison Avenue, New York, New York 10016
www.oup.com



Library of Congress Cataloging-in-Publication Data

Allen, Thomas, J., 1940–
Caterpillars in the field and garden : a field guide to butterfly caterpillars of North America /
by Thomas J. Allen, Jim P. Brock, and Jeffrey Glassberg.
p. cm. — (Butterflies [and others] through binoculars field guide series)
Includes bibliographical references and index.

ISBN-13: 978-0-19-518371-9 (cl)
ISBN-10: 0-19-518371-1 (cl)
ISBN-13: 978-0-19-514987-6 (pbk)
ISBN-10: 0-19-514987-4 (pbk)

1. Caterpillars—North America—Identification.
2. Caterpillars—North America—Pictorial works.
I. Brock, James P. II. Glassberg, Jeffrey. III. Title. IV. Butterflies through binoculars series.
QL548.A44 2004 595.78'139'097—dc22 2003057958

9 8 7 6 5 4

Printed in China
on acid-free paper

# Contents

# Acknowledgments

THIS GUIDE is not just the work of its authors. Many people played a major role in its development, and without their efforts this book would not be as comprehensive. Some helped with the loan of their personal images of various species to complete the major genera. Others provided live material in the form of eggs, caterpillars, or adults to be reared or photographed. Still others reviewed sections of the manuscript, adding valuable comments and information to help make the guide more complete and accurate. We are indebted to all these folks and thank each one individually for their contribution.

We are grateful to the following people for the loan of caterpillar images: Greg Ballmer, Susan S. Borkin, Rick and Nora Bowers, Thomas Emmel, John Emmel, Jeffery Fengler, Eve and Rob Gill, Jack Harry, Richard Heitzman, Gabriela Jimenez, Marc Minno, Diane Pierce, Steven Prchal, John Rawlins, Jane Ruffin, Jorge Soberon, Steve Spomer, Jennifer Szymanski, Dave Wagner, Keith Wolfe, Jacque Wolfe, and Dave Wright.

Many people contributed their time and homes to assist us in finding caterpillars in the field, while others provided various species in the form of eggs, caterpillars, and on occasion adults for us to photograph. Without their assistance, we could not have obtained all the species included in this guide. Warm thanks go to: Andrew Brand, Tom Carr, Chris Conlin, Michael Delasantro, Pete Haggard, Ken Hansen, Jack Harry, Paulette Haywood, Steve Kohler, Merv Larson, Douglas Mullins, Mike Nelson, Troy Pabst, Shawn and Joe Patterson, Steven Prchal, Mike Quinn, James Reiser, Kilian Roever, Andres and Pilar Sada, Dale Schweitzer, Jeffery Slotten, the late Mike Smith, Noel Snyder, Steve Sommerfeld, Todd Stout, Mike Thomas, Monty Volouski, Dave Wagner, Reginald Webster, Keith Wolfe, Jacque Wolfe, and Dave Wright. A special thanks to Richard Boscoe, who helped locate many eastern species for photography, while Ken Hansen led us to many western and arctic species.

We owe a special debt to those who reviewed sections of the guide, making valuable additions and corrections, especially to Keith Wolfe for his review of the Swallowtails, Whites and Sulphurs and the Brushfoots.

Lastly, we wish to thank our wives for their patience and understanding while we chased after caterpillars and filled our houses with rearing containers over a period of a few years.

A complete credit listing of species, locality, and contributors, can be found in the appendix.

# Introduction

WHEN ALICE VISITED WONDERLAND, she spoke with the hookah-smoking Caterpillar. "When you have to turn into a chrysalis—you will some day, you know—and then after that into a butterfly, I should think you'll feel it a little queer." "Not a bit" said the Caterpillar, who then told Alice how she could adjust to the new experiences of Wonderland by controlling her size. Alice used the Caterpillar's advice to navigate the incredible world of Wonderland. Now, you too can enter a magical world that takes place at a different scale and that follows rules different from those of the everyday world in which you live—the world of caterpillars!

Most people love adult butterflies. Butterflies have been admired, praised, and revered for centuries. Moreover, their popularity continues to rise as ever greater numbers of people turn to butterfly watching and butterfly gardening as ways of interacting with the natural world and relieving some of the stresses of modern life.

But many people's attitude toward caterpillars—the early stages of butterflies (and moths)—is ambivalent, at best. "Aren't caterpillars those worm-like creatures that are eating all the plants in my garden?" they might say. Well, the answers are yes and yes. Yes, although not related at all, caterpillars are shaped somewhat like worms, but most of them are not so long and slender, and many are quite beautiful. And yes, they are eating all the plants in your garden. But here we need to make a subtle distinction. Overwhelmingly, the caterpillars that are eating your everyday garden plants and making pests of themselves are caterpillars that become adult moths. In contrast, butterfly caterpillars will turn up their noses at your roses, your rosemary, and your rhododendrons. If you have caterpillars tasting your tomatoes or your corn, they are moth caterpillars, not butterflies. Few butterfly caterpillars eat standard garden plants and even fewer become garden pests. Unlike moths, most butterfly caterpillars do not strip foodplants so bare that it harms them (well, occasionally they do become over-exuberant). Most of the readers of this book—people who are, or want to be, butterfly gardeners—will welcome the sight of caterpillars in their garden. They are as varied, as fascinating, and often as colorful as the adult butterflies they become. Think about this. For every adult butterfly in North America there is a caterpillar. For every caterpillar there is often more than one color form. For nearly all species, the younger stages differ from the older stages. So many possibilities for enjoyment!

Beyond your own garden, the world of caterpillars offers wonderful opportunities for both pure recreation and for adding to existing knowledge. Because knowledge of caterpillars is still quite limited—this book is the first guide to the butterfly caterpillars of North America—time spent in the field searching for and observing caterpillars is almost certain to yield new and important information.

It is easy to see why most people have, up until now, ignored this small but fascinating world. For many people, caterpillars are often just plain hard to find and too small to see! But, we believe that if you will take the time to search for, find, and get acquainted with caterpillars, then it is almost certain that your attitude will change. You will be intrigued by the aphid-eating Harvester caterpillars. You will marvel at the gregarious, brilliant red, slug-like Atala caterpillars as they crawl around in unison on their coontie foodplant. You will find yourself lurking near scrub oaks, just after dusk (flashlight in hand), hoping to see Edwards' Hairstreak caterpillars crawling up the small trees, attended by ants.

Caterpillars provide fascinating glimpses into a world that most people have chosen to ignore. It is hoped the following pages will inspire you to seek and identify new caterpillars, to explore the enchanting life-styles of both common and rare species and, finally, to boldly enter into new habitats and solve some of the remaining mysteries of North American caterpillars.

## Distinguishing Butterfly Caterpillars from Moth Caterpillars

Moth species of North America outnumber butterflies from ten to perhaps twenty to one. New species of moths are still being discovered in regions where butterflies have not yielded a new species in decades. Is it any wonder that moth caterpillars can and are easily confused with butterfly caterpillars?

Some of the larger species of moths have familiar caterpillars and thus are easily identified; the same holds true for large, common, and well-known butterflies. Unfortunately, there are thousands of other species, both moths and butterflies, to provide confusion for caterpillar seekers in all regions and all habitats.

The illustrations and the text in this book are meant to provide a clear idea of what might be a butterfly versus what might be a moth. Still, in many cases it will be difficult to determine whether a caterpillar is a butterfly or moth without the use of a dissecting microscope, a technique outside the scope of this guide. This is especially true when dealing with early-stage caterpillars and the hundreds of very small species. To help you recognize the kinds of caterpillars that are *not* butterfly caterpillars, we have included a small potpourri of moth caterpillars on pages 152–54.

## How To Find Butterfly Caterpillars

Searching for butterfly caterpillars is an undeveloped art. Relatively few people have attempted this exercise, and fewer still have succeeded to any great degree. Finding caterpillars is more difficult than finding adult butterflies. In fact, some of North America's caterpillars have never been seen in their natural habitats. Most closely related butterfly caterpillars have a similar appearance. As an aid to identification, we have included a color illustration of at least one species from almost every resident butterfly genus north of Mexico, omitting only a few of the extremely rare genera.

So, how do you get started? First, unlike the search for adult butterflies, a sunny day or great weather is not required. Caterpillars most often rest on some part of their favored foodplant, regardless of the weather. Thus caterpillar searches are the perfect butterflying activity for inclement weather!

Second, although binoculars will prove very useful in searching for stands of suspected caterpillar foodplants, they will be less helpful in finding the caterpillars on the foodplants because most caterpillars hide under leaves or in leaf nests. Instead, your eyes will need to be fine-tuned to caterpillar activities, many of which are disguises and strategies they use to conceal themselves. The most difficult caterpillars to find are those that mimic the substrate upon which they rest. Most of the time they rest quietly, preferring to move only while feeding or when they are ready to find a pupation site. Also, most caterpillars are much smaller than the adult butterflies you are familiar with. When you do find one, you can use the opposite end of your binoculars for close-up magnification to examine your find.

Before proceeding with your search, there are some important things to know. To be successful at finding a particular caterpillar you must first know the type of food that it might be feeding on, when it will be feeding, and what it might look like.

### Caterpillar Foodplants

All butterfly caterpillars need food, and for all but one of our species (Harvester), that food is plant(s). Although some species of butterfly caterpillars eat only one species of plant, most will use a number of related plant species as foodplants and some caterpillars eat a wide variety of plants from different plant families.

Learning to recognize butterfly foodplants is a gigantic aid in finding (and identifying) butterfly caterpillars. Unfortunately, for most people this is not easy without assistance. This difficulty is magnified for the absolute beginner. Plant guides with color photos, botanical gardens, well signed nature trails, and native plant nurseries provide some help. What works even better is going into the field with someone who has extensive knowledge of the local flora.

Foodplants can also be determined by observing where female butterflies lay their eggs. This activity can be time-consuming, but in the end you will be rewarded with not only a foodplant but caterpillars, too. Previously unreported foodplants are often discovered by this method.

### When To Look

When you know the foodplant for a particular butterfly it is critical to know the caterpillar's feeding period. This varies greatly from species to species. Butterflies with many broods may have caterpillars present for most of the season, whereas those species that fly just once or twice a year may have a very short feeding period, often only lasting a month or less. In addition, it is important to know if the feeding period occurs before or after the flight of the adults. You will also want to know when you are most likely to find late instar caterpillars, because early instar caterpillars are smaller and less likely to be found.

### Other Helpful Hints

Once you know the foodplant and the feeding period, it is worthwhile to have a search image for the caterpillar. Will it be smooth, hairy, or spiny? Is its color variable or is there only one form for the species? The illustrations in this book are intended to be your best guide. However, keep in mind that the early instars of most species often look very different from the later instars. It is beyond the scope of this book to include more than just a few examples of early instars.

It is useful to know what part of the plant the caterpillar eats. Many species eat only the flowering parts and ignore the leaves. Some species eat only tender new leaves, bypassing older foliage. Some species burrow into their foodplants. Clues like these can guide your search and save time, especially when foodplants are large shrubs or trees.

Knowing a little about behavior can assist in finding some species. For example, most skipper caterpillars and some brushfoot species rest inside nests made of leaves from their foodplant. It is often easier to scan for these nests than for the caterpillars. Most blues, some hairstreaks, and a few coppers are tended by ants. Following a trail of ants will often lead to caterpillars.

Sometimes caterpillars can be located by examining leaves for damage. Most early instar caterpillars eat the top layers of the leaves, whereas older caterpillars eat entire leaves or flowers. Remember, what goes in must come out. A caterpillar's droppings (frass) may reveal its hiding place, especially when the foodplant is a low-growing shrub. Many species feed only at night in the later instars. These may be located by flashlight just after dark.

### Starting Your Search

For the novice, a butterfly garden is a great place to begin to gain familiarity with butterfly caterpillars. Ideally, there will be a variety of foodplants so that you can practice looking for different species. If you have some experience, you may want to try your favorite butterfly spots. After you've had some success with common species you may be ready to search for the more difficult varieties. In any event, it is better to go with knowledgeable friends than to go alone. The more eyes the better chances of success and, consequently, the more fun!

### Reaching Your Goal

Even though finding caterpillars is difficult and often time-consuming, don't give up. There are great rewards for success. Locating a particular butterfly caterpillar can provide a great sense of accomplishment. More important, such discoveries give a valuable insight into the life history of a particular species Also, you may discover a new foodplant or some previously unknown behavior. A butterfly's life history is critical to understanding its requirements, and to conservation. Now, at the start of a new century with emerging environmental challenges, more and more butterfly species will require conservation measures for survival.

## Identifying Caterpillars

### Swallowtails (family Papilionidae)

Swallowtails are the largest and most conspicuous of North American butterfly caterpillars. They are often encountered resting on twigs or on the top of leaves in gardens, parks, and other urban settings. Most are smooth and brightly colored; some are cryptically colored.

All have a fleshy, retractable organ above the head, the osmeterium, which gives off a pungent odor when everted (see Giant Swallowtail photo, page 33). For most species, the early instars are darkly colored, with a white or cream-colored saddle, giving them the appearance of a bird dropping. As they sit motionless on a leaf, this subterfuge often allows them to go undetected. Winter or dry seasons are passed as a chrysalis in all except the parnassians.

The parnassians, a distinctive subgroup of the swallowtail family, are mostly black, covered with very short hairs that give them a velvety appearance, but without spines or fleshy tubercles. The osmeterium is present, but seldom used. These caterpillars are sun-loving; they often rest on the ground when not feeding, apparently mimicking toxic millipedes. The eggs are somewhat flattened and dimpled. Early instars are similar to the last instar and do not resemble a bird dropping. Unlike true swallowtails, pupation of parnassian caterpillars takes place on the ground by connecting surrounding debris with silk to form a wispy shelter. Eggs overwinter. Parnassian caterpillars feed close to the ground and can easily hide among plants, debris, or rocks, and often leave the hostplant to do so.

Spicebush, Tiger, Two-tailed, and Palamedes swallowtails all have large eyespots giving them a ferocious head-on appearance that could scare predators.

These species also live in folded leaf shelters lined with silk and go unnoticed unless the fold is opened. To find them, remember that they always fold the leaf on the upper surface, never under as is done by many moth caterpillars.

Pipevine-feeding species are equipped with fleshy knobs and filaments called "tubercles." The caterpillars accumulate chemicals from the pipevines that are toxic to many potential predators.

### **Whites and Yellows** (family Pieridae)

The Whites and Yellows family contains many species that are cryptically colored—green and blending with the hostplant leaves, or with colors and resting postures that resemble the flowers or seedpods of the hostplant—and many that are showy. Most are rather smooth but some have short colorless hairs or small raised bumps. Most are solitary feeders. Many secrete oily and sticky droplets at the tips of short hairs which tend to repel ants and other predators. Some of them are very sensitive to disturbances and will either wriggle violently or drop on a fine thread of silk to the ground, or nearly so to escape a predator. Like swallowtails many of these caterpillars rest in the open on leaves or stems of their hostplant. Many feed on specific parts of the plant. Some grow very rapidly and are in the caterpillar stage for only about 3 weeks.

### **Whites** (subfamily Pierinae)

Hostplants for the Whites are primarily mustards, but capers and pines, among others, are also used. A few species use annuals. The eggs of most whites are spindle-shaped, blue-green, yellow, or cream turning orange before hatching and laid one at a time. Early instars are generally dull or yellow-green, developing patterns and other colors as they mature. Seeds and flowering parts of the hostplant are the main food source (orangetips, marbles). Leaves are preferred in only a few species (genus *Neophasia*). Most overwinter as a chrysalis.

### **Sulphurs** (subfamily Coliadinae)

Most Coliadinae are green with a yellow, cream, or orange lateral stripe, but some of our species with tropical affinities have beautiful banded patterns dorsally. Sulphurs are primarily legume-feeders but heaths and other plants are also used. Eggs are spindle-shaped and cream- to orange-colored. In general, early instars resemble last instars, to some degree. Leaves are the preferred diet for most, but flowers and seeds are also eaten. Pupation is similar to swallowtails. Dry seasons and winters are spent as adults in most *Eurema, Phoebis,* partially grown caterpillars in *Colias,* and perhaps as chrysalids in other species.

### **Gossamer-wings** (family Lycaenidae)

The Gossamer-wings are a large family of small, slug-shaped caterpillars that use a variety of hostplants. Many species exhibit a wide range of color variation, presumably because as a protective measure they adopt the color of that portion of the host on which they are feeding. Most feed on the new buds, flowers, or fruit of their hostplant, but occasionally they feed on leaves as well. These characteristics make these caterpillars very difficult to locate.

Many of these species are myrmecophilus (ant-loving). When ants stroke special glands located on the abdominal segments of the caterpillars, the glands secrete a sugary solution. This "honeydew" substance is avidly sought by some ants, who tend the caterpillars, protecting them from predators and parasites. When searching for these small caterpillars, first try searching the buds and flowers of small plants for the evidence of frass or the presence of ants.

Some tree-feeding hairstreaks (for example, Coral Hairstreak) often climb down a tree to its base to rest, and may even descend into the leaf litter at the tree-base to take shelter in a silk-lined nest (for example, Edwards' Hairstreak). They may also be tended by ants. To find these species, one should search around small saplings near the ground, or in the litter, for they feed at night and are resting during the day.

Often caterpillars that feed on fruits bore into the fruit and are hidden from view. For these one must look for evidence of frass. These include Early Hairstreak and others. By feeding inside the fruit, the caterpillar is protected from predators and parasites, at least while it is in the fruit. Some of these caterpillars adopt different strategies on different hostplants or at different times. For example, endangered Miami Blues, when feeding on ballonvines, eat the seeds within the balloon-like seedpod; however, when feeding on nickerbeans, they feed exposed on tender new leaf growth.

Eggs of these species are flattened and dimpled or sculptured. Pupation usually occurs under debris on a flat surface such as a leaf or bark. Chrysalids range in color from tan to dark brown and resemble a small mummy case or perhaps a rodent dropping.

### Harvesters (subfamily Meletinae)

Worldwide, Harvesters comprise a small group of 50 species, one of which is found in North America. All feed on aphids (Homoptera) and are found in association with wooly aphid colonies on various trees. The Harvester is our only carnivorous butterfly.

### Coppers (subfamily Lycaeninae)

Sixteen species of Coppers are found in temperate regions of North America. All are slug-like and similar to other Gossamer-winged species, but often with narrower, more elongated bodies. They generally lack chevron markings dorsally, although chevrons are present on early instars of some species. Most overwinter as eggs and therefore caterpillars are best found 2–4 weeks prior to flights. Copper caterpillars are primarily leaf-eaters and can usually be found resting on the undersides of the host leaves. All lack honeydew glands, but a few species are tended by ants. For some species we have not seen enough examples to determine the extent of variation.

### Hairstreaks (subfamily Theclinae)

A very large group worldwide, there are about 80 species of Hairstreaks in North America, mainly in southern regions. A variety of plants are used as hosts. Some species eat only flowers or fruits, and others are strictly leaf-eaters. Red-banded Hairstreak prefers rotting leaves as a food source. Early season fliers (elfins, Bramble Hairstreaks) spend the winter as chrysalids. Later season fliers (for example, *Satyrium*), overwinter as eggs, feeding prior to their flights. The more tropical species may breed continually.

For identification purposes the ultimate disclaimer belongs to the Gray Hairstreak. This species occurs almost everywhere and its caterpillars eat just about anything so even in those instances where one has found a caterpillar on a supposedly unique hostplant, a Gray Hairstreak can also be a possibility (see page 63).

### Blues (subfamily Polyommatinae)

The blues are small and difficult to identify because of their similarities in color and pattern and because similar species often share the same hostplants. Blues are found mainly in temperate regions, with more than 30 species in North America. Eggs are usually laid on flower buds or fresh growth of hostplants; the flowers or young leaves are then fed upon by the caterpillars. Blue caterpillars are cryptic, blending perfectly with the plants they feed on and most are tended by ants. Caterpillars can often be found easily by following foraging ants on the host. They will eventually lead one to the flowers or seeds, where most caterpillars feed. Blues generally hibernate as chrysalids, but a few spend the winter as eggs or partially grown caterpillars. Early instars are especially difficult to identify to species.

### **Metalmarks** (family Riodinidae)

The Metalmarks are mostly tropical, but 24 species occur in North America. They are similar to Gossamer-wings, except that Metalmark caterpillars have broader heads and are either covered with long, fine hairs or have numerous tufts of long hairs. Eggs are round and flattened, similar to those of the Gossamer-wings. The eggs of scintillant metalmarks and greenmarks (genus *Caria*) are sculptured so as to look like tiny flying saucers. Chrysalids are similar to those of Gossamer-wings but with more elongated abdomens. Most species overwinter as partially grown caterpillars.

### **Brushfoots** (family Nymphalidae)

Brushfoot caterpillars are best characterized as being spiny. Many species are protected from predation by having rows of branched spines on their bodies. This gives them a fearsome appearance, and for that reason are left alone by many predators. The spines may also serve as a physical impediment to being eaten by predators or having eggs laid on them by parasitic wasps. In the early growth stages some of the brushfoots, especially the admirals, resemble bird droppings and, like the swallowtails, go unnoticed by predators. Others are either cryptically patterned, blending into their surroundings, or are marked with a blotched pattern that breaks up their shape and makes them difficult to see.

The genus *Danaus*, which includes the Monarch, Queen, and Soldier, are brightly marked with colorful stripes. Because these species feed on toxic plants and store the toxins in their tissues, the bright markings serve as a sign of caution to predators. As there is no need to hide, most can be easily found on their hostplant.

Other species garner protection from predators and parasites by feeding gregariously—remaining together until the later instars. Many of these species drop quickly from the hostplant when disturbed, especially the checkerspots. Species such as Crimson Patch may jerk their bodies violently when disturbed or drop on a fine thread of silk to the ground. The *Chlosyne* and crescents feed within webs of silk constructed around parts of the hostplant. These webs protect the young caterpillars from predators and parasites. Young caterpillars winter in these webs, and they can be found in the fall in webs near the tops of hostplants.

### **Skippers** (family Hesperiidae)

Skippers comprise almost a third of our butterfly fauna. Skipper caterpillars have smooth bodies covered with short barely noticeable hairs. Heads are large relative to the thorax, creating a constriction between the head and the first segment. Facial markings are important for identification in many cases.

From the Arctic Skipper in a rolled leaf blade in Washington State to the Mangrove Skipper in a tied leaf shelter on a red mangrove in the Florida Keys, skippers depend more on shelters than on coloration for protection. Almost all skipper caterpillars conceal themselves in a nest of leaves created by folding the leaf over and tying it with silk or by silking together more than one leaf. These nests (unlike those that some moth caterpillars make) are devoid of droppings, expelled by caterpillars while feeding outside the nest. (For simplicity's sake, in the species accounts we make note only of those few species that do not make nests).

Eggs are usually laid singly and on or very near the hostplant, but there are exceptions (see Giant Skippers, p. 174). In most cases partially grown or full-grown caterpillars overwinter. Pupation may occur on the hostplant in a nest, or on the ground in a cocoon-like enclosure or nest of silken debris. In Skippers the stage after the caterpillar stage is referred to as the "pupal stage," and the small case is a pupa, not a chrysalis.

### **Firetips** (subfamily Pyrrhopyginae)

Firetips are tropical skippers with brilliantly colored caterpillars. There is one species in our area.

### **Spread-wing Skippers** (subfamily Pyrginae)

Caterpillars in this large subfamily generally feed on the leaves of legumes, mallows, oaks or other trees, and shrubs. To our knowledge, all live in a nest made of leaves until they are full-grown. Early instars make their first nest immediately after hatching from the egg. This nest is made simply by cutting and folding over a leaf flap. Subsequent nests may involve silking together more than one leaf, especially in larger species. Some species outgrow the size of the host leaves and either nest on adjacent plants or on the ground at the base of the hostplant. In most cases partially grown or full-grown caterpillars overwinter, although a few species do so as pupas, and others feed through the colder months and develop at a slower rate.

### **Grass-Skippers** (subfamily Hesperiinae)

The group is aptly named since most caterpillars feed specifically on grasses, although some also use sedges. Almost all make nests on their hostplant. These nests can be aerial (high up on the host) or hidden down in the clump of grass, and even extending slightly below ground level. The early instars of at least one species, Southern Broken-Dash, construct a shelter and then cut the shelter free and carry it with them as they move from grass blade to blade!

### **Giant Skippers** (subfamily Megathyminae)

This sub-family is characterized by grub-like caterpillars that bore into leaves, stems, and rootstocks of agaves, yuccas, and related plants.

## Butterfly Biology

Each butterfly goes through four distinct stages in its life. Egg, caterpillar, pupa, and adult. Let's start with a female adult butterfly and follow the next generation.

### Eggs

Assuming that the adult female butterfly has mated (and almost all do), she will be carrying fertilized eggs. She will spend a considerable part of her day searching for an appropriate plant on which to lay an egg. An appropriate plant is one that the caterpillar that will hatch from the egg will be able to use as food. Later, we will learn more about caterpillar foodplants.

There is much to enjoy about eggs, especially if you have a magnifying lens that allows you to see them better. Their size, shape, and color vary from species to species and, most dramatically, among the various families of butterflies. Many eggs have a beautiful and intricate architecture.

How and where the female places her eggs is also interesting. Some species lay only one egg per plant. Others place a mass of eggs together. Some species lay their eggs mainly on flower buds; others place them on the undersides of leaves, and still others lay their eggs at the base of a tree. How many eggs a particular female lays varies greatly from species to species. Over the course of their lives some butterflies lay only a few dozen eggs; most probably lay a few hundred, and some, such as the Regal Fritillary, lay more than a thousand.

### Caterpillars

When the egg hatches, usually after less than a week, a tiny caterpillar emerges. This voracious eating machine is the star of our show! As it rapidly increases in size, the young caterpillar outgrows its skin (the exoskeleton), which splits and is shed, revealing a new, larger, and baggier skin below. As the caterpillar grows, this process is repeated a number of times (usually 4 or 5) over the course of about 2 or 3 weeks. Each stage between skin shedding is termed an "instar." Different instars of the same species may have a dramatically different appearance. For example, last instar caterpillars of Guava Skippers are milky white, whereas the penultimate instar caterpillars are brilliant red with cream bands (see page 115). Caterpillars of related species often look different but have similarities. For example, Zebra Heliconian caterpillars are white with black dots, and Gulf Fritillaries are orange with black dots, yet both have similar black spines.

Many people use the terms *caterpillar(s)* and *larva(e)* interchangeably. Actually, *caterpillar* is not only much more widely understood, it is also more precise. The word *larva* refers to the immature stages of all insects that undergoes complete metamorphosis, and it is often used even more generally, referring to the immature stages of other invertebrates. The word *caterpillar* refers only to the immature stages of butterflies and moths.

### Pupas (Chrysalids)

When the caterpillar has grown to its full size, it attaches itself to a support and pupates. Sometimes this happens on the foodplant itself, but more often the caterpillar wanders away from the foodplant and attaches itself to a twig or a blade of grass. The caterpillar, now encased in a hard outer shell, becomes a chrysalis—seemingly lifeless and inert. But inside that hard shell, an amazing transformation is taking place. The tissues and structures of the caterpillar are being broken down and replaced with the tissues and structures of the adult butterfly. If development is proceeding straight through, this process usually takes 1 to 2 weeks. In some species, the chrysalis enters a resting state for a few months—or for the winter—and then development will proceed.

### Adults

When the adult is fully formed, the chrysalis splits open and the adult butterfly emerges (ecloses). Before this occurs, inside the chrysalis the wings are wrapped tightly around the butterfly's body. Upon emergence, fluid is pumped through the "veins" in the wings, and they are unfurled. This is a very vulnerable time in a butterfly's life, as it basks in the sunshine to harden and set its wings. Once the adult butterfly emerges from the chrysalis it no longer grows larger.

### Broods

As we've seen, the butterfly life cycle goes: adult—egg—caterpillar—pupa—adult. Some species of butterflies go through this cycle only once a year. Over much of the northern part of its range, the Mourning Cloak fits this pattern. Adult butterflies spend the winter in shelters, such as hollows in trees. In the springtime they become active and the females lay eggs. The eggs hatch, the caterpillars eat, and then they pupate in early summer. Toward the end of June, the new adults emerge and begin a new cycle.

Most butterflies have more than one cycle, or brood, each year. Many species have a definite number of broods that fly at specific times of the year. For example in the northeastern United States, Juniper Hairstreak adults fly in April–May. The females lay their eggs; caterpillars hatch, eat, and pupate; and new butterflies emerge and fly in July–August. The offspring of the July–August brood spend the winter in the pupal stage, then the adults emerge and fly the following April–May.

Some species are more or less continuously brooded throughout the year. The Cabbage White is a good example. A succession of individuals of this species fly from the first warm weather in spring until hard frosts in the fall.

### Predators, Parasites, and Diseases

Most people realize that caterpillars are avidly eaten by a host of creatures, such as birds, mammals, spiders, and other predatory insects. We're amazed when we stop to consider that most female butterflies are capable of laying from 100 to 300 eggs (some of the Fritillaries lay as many as 1500 eggs), but that individuals from only 1 or 2 of these eggs will survive to become an adult butterfly. The mortality rate of a butterfly population is phenomenal, and most are lost in the caterpillar stage. For many predators, caterpillars would be sitting ducks, if they weren't butterflies.

Few people realize that diseases and parasites devastate butterfly populations. When people think of disease and parasites they often think of fleas and ticks as parasites and the diseases they may carry and inflict upon us or our pets. But, caterpillars are also exposed to many diseases. These include fungal diseases, especially if the weather is too moist, bacterial diseases, and viral diseases. Such diseases can have a large and rapid impact on caterpillar numbers, especially if the species is a communal feeder. Scientists have used this information to control gypsy moth populations using viral and fungal disease organisms.

Perhaps even more important than disease organisms are the numerous parasites that affect eggs, caterpillars, and pupas. Wasp (Hymenoptera) parasites are by far the most significant. They may parasitize any life stage, but especially the caterpillar, by laying their eggs inside body of the host butterfly species. The parasitic larva then feeds on the tissues of the host butterfly species, eventually consuming it or killing it. Often these parasites are almost microscopic in size, but a quick look at a caterpillar can reveal whether it has been parasitized (see, Silver-banded Hairstreak, page 53). Caterpillars that have been parasitized have a black dot or several where the skin has been punctured by the ovipositor (egg-laying organ) of the wasp. There are also a few flies (Diptera) that parasitize caterpillars. These generally adhere their eggs to the body of the caterpillar, and the newly hatched fly larva then bores into the body.

### Behavior

Caterpillars exhibit a wide behavioral repertoire, much of which helps them avoid predators and parasites and increases their chances of survival. Many patterns of behavior are designed to conceal the caterpillar, which makes caterpillar hunting a challenge.

Probably the most common form of concealment is the use of nests. Many caterpillars construct leaf shelters on the foodplant and hide in these nests when they are not feeding. Other caterpillars conceal themselves in the leaf litter beneath a foodplant or construct a silken nest of various ground debris. Still others, such as Early Hairstreaks, Miami Blues, and Giant Skippers, bore into parts of the foodplant such as the fruit, flowers, stem, leaves, or roots and feed within the plant, hidden from view. Many caterpillars are camouflaged, as their colors and markings mimic an inedible object. Most often caterpillars mimic the foodplant itself, but numbers of species, including many of the swallowtails and some of the brushfoots, also resemble bird droppings.

Many species feed at night which prevents them from being eaten or parasitized by day-flying predators and wasps. They are, however, vulnerable to nocturnal predators such as spiders and mice. If these protective tactics don't work, some caterpillars use the element of surprise to ward off danger and frighten predators. Spicebush and Palamedes Swallowtails, among others, have large false eyespots that frighten predators. Other species, such as Crimson Patch, tortoiseshells, and emperors, have a habit of twitching when threatened by danger. Some butterfly caterpillars, such as the checkerspots, drop from the plant when disturbed, either on a thread of silk or directly to the ground, and become lost in the vegetation.

Swallowtails have a gland just behind the head which can be everted when they are disturbed. This fleshy gland (see Giant Swallowtail, page 33) gives off a very offensive and pungent odor that is intended to frighten away the intruder. Other warning signs such as the bright colors of Monarch caterpillars say "don't eat me, I don't taste good." The milkweed-feeding caterpillars and the pipevine-feeding swallowtails contain bad-tasting, toxic chemicals (specifically cardioglycosides) produced by the foodplants. These caterpillars store the chemicals in their tissues rendering them unpalatable.

A few butterflies, such as Julia Heliconian and Malachite can extrude an acidic fluid from the tips of their spines which can cause a skin rash when touched or handled by humans. Many of the brushfoot caterpillars, such as Baltimore Checkerspot, have spines, some of which are very long, and prevent parasites, especially the smaller ones, from being able to reach the body surface with their egg-laying organ (ovipositor). The use of toxic foodplants or foodplants with spines, such as nettles and thistles by some species, deters many predators from making a meal of the caterpillars.

Finally, some species are communal feeders and aggregate in large numbers during a portion of their caterpillar stage. This insures better survival because many of these species live within silken shelters they construct on the foodplant. The checkerspots, which do this, also winter in these shelters, further improving their chances for survival. Mourning Cloaks feed as a group throughout the caterpillar stage.

## Butterfly Gardening

A rapidly growing number of people are combining the satisfaction that comes from gardening with the excitement and drama that butterflies provide. So, how do you actually create a butterfly garden? Plant some pretty flowers? Plant flowers that "attract" butterflies? Well, yes. In part. But if you want to create a real butterfly garden you need to provide the plants that the caterpillars eat. Remember that without caterpillars there are no adult butterflies. Flowers are the easy part because there are many at which butterflies love to nectar. And because very few species of butterflies have particular likes when it comes to flowers, there are many flowers you can choose that will be attractive to a broad array of butterfly species, even ones in different families. But the caterpillars are more persnickety! Some eat only a single group of related plants, and others eat only one particular plant species. For example, Yucca Giant-Skipper caterpillars will eat only yuccas, but they will use a number of different species, whereas caterpillars of Harris' Checkerspot will eat only flat-topped white aster! Species that are especially likely to be in your garden (assuming that you live within their range and plant the caterpillar foodplant) are indicated by an asterisk in the species accounts that accompany the illustrations.

Because caterpillars are so particular, what plants are useful in a butterfly garden will vary tremendously from place to place, depending upon what species of butterflies are nearby. Practically speaking, the would-be butterfly gardener needs not only continent-wide information but also site-specific information and local experience. The broad picture we paint here should be supplemented with local color and detail. Ultimately, the best sources of local information will be the butterfly gardening brochures available at the North American Butterfly Association's Web site, www.naba.org, and from local NABA chapters. Even with all available information in hand, the best procedure is to test various plants in your garden to see how well they do.

General suggestions for groups of plants useful for butterfly gardening are listed later in this section. Many other suggestions are made throughout this book, in the "Garden Tips" included in many of the species accounts. Part of the fun is the experimentation. Of course, actually getting to see those caterpillars grow in your garden is pretty exciting also!

Although the list of plants that are potentially useful in the butterfly garden is extensive, some groups of plants top the list as must-haves for most parts of the United States. Very high on the list are the milkweeds. Not only are milkweeds the required caterpillar foodplants for Monarchs and relatives, they are important native nectar sources and beautiful to boot! Asters are another useful group that serve two functions—as the caterpillar foodplants for a number of Crescents and as very useful fall nectar sources. Especially in the West, every butterfly garden should have buckwheats, sensational plants that succor caterpillars and adults of many butterfly species. In the eastern two-thirds of the United States, if you have the space, seriously consider planting a hackberry tree. Caterpillars of three species of emperors (two of them widespread and relatively common), American Snout, and Question Mark all use these trees for food. In addition, their fruits are used by other wildlife (just because you're focused on butterflies doesn't mean that you have to ignore other animals). In southern climes, passion vines and sennas are *de rigueur* if you want heliconians and yellows in your garden.

For nectar, there is a broad array of possibilities. Try, of course, to have a variety of plants that will provide nectar throughout the butterfly season, taking into account the blooming seasons of individual plants. One plant that is particularly attractive to almost all butterflies, and that grows over most of the United States, is butterfly bush. Most butterfly gardens outside of the hottest areas include this non-native plant, but there is growing concern that it may become invasive and a problem. At least in the Northeast, there are a few areas where this alien is now naturalized, forming dense thickets along highway borders.

Plants that are widely available through local nurseries are attractive because they are easy to obtain, have been bred to grow easily in gardens, and are more consistent than are wild plants. But, ultimately, using plants that are native to your local area will often prove to more effective for the butterflies, and more satisfying to you.

In addition to providing caterpillar foodplants and adult nectar sources, you should consider providing windbreaks to shelter butterflies, flat stones on which adult butterflies can bask and warm up, and damp sand or gravel at which they can obtain salts. Fermenting fruit is also welcome to many adults, especially to many of the brushfoots. Keep in mind, however, that it sometimes attracts unwanted insects in addition to butterflies.

And, of course, avoid the use of pesticides in your garden. They will kill your caterpillars and butterflies.

Many garden plants are not eaten by any butterfly caterpillars. So, if you find a caterpillar eating your tomatoes, you can be certain that it is a moth, not a butterfly. Very few plants commonly used for landscaping are affected by butterflies. Here is a short list of some common garden plants not normally used by butterflies as caterpillar foodplants.

andromedas (*Pieris japonica*)
apples
astilbes
azaleas
basils
butterfly bush (*Buddleia davidii*)
camellias
chrysanthemums
corn
crocus
daffodils
dahlias
daphnes
day lilies
delphiniums
forsythias
fuschsias
geraniums
gladioluses
hostas
hydrangeas
impatiens
irises
lettuces
lilacs
lilies
magnolias
marigolds
maples
oleanders
pentas
peonies
peppers
petunias
phlox
potatoes
primroses
pyracanthas
rhododendrons
roses
sage
squashes
thyme
tomatoes
tulips
vinca
wisteria
yews
zinnias

## Some Stellar Butterfly Garden Plants

| PLANT | GOOD NECTAR | USED BY THESE CATERPILLARS |
|---|---|---|
| Alfalfa | + | Orange Sulphur, Clouded Sulphur |
| Asters | + | Pearl Crescent, Northern Crescent, Field Crescent |
| Blazing-stars | + | |
| Buckwheats (West) | + | Some blues, Mormon Metalmark |
| Butterfly bush | + | |
| Ceanothus | + | California Tortoiseshell, some duskywings |
| Clovers, red & white | + | Orange Sulphur, Clouded Sulphur, Eastern Tailed-Blue |
| Coneflowers | + | |
| Dandelions | + | |
| Deerweed (West) | | Orange Sulphur, Bramble Hairstreak, Marine Blue |
| Fennel | | Anise Swallowtail, Black Swallowtail |
| Fogfruits (South) | + | Phaon Crescent, Common Buckeye, White Peacock |
| Goldenrods | + | |
| Hackberry trees | | American Snout, Question Mark, emperors |
| Heliotrope | + | |
| Kidneywood (South) | + | Southern Dogface |
| Lantanas (South) | + | |
| Lupines | | Boisduval's Blue, Melissa Blue, Silvery Blue |
| Mallows | + | Gray Hairstreak, Mallow Scrub-Hairstreak, Painted Lady, West Coast Lady, checkered-skippers |
| Mexican sunflower | + | Bordered Patch |
| Milkweeds | + | Monarch, Queen |
| Mistflowers | + | Some scintillant metalmarks |
| Passionvines (South) | | Gulf Fritillary, heliconians, Variegated Fritillary |
| Penta (South) | + | |
| Pipevines | | Pipevine Swallowtail |
| Rabbitbrush (West) | + | |
| Rock-cresses (North) | + | Marbles and orangetips |
| Sennas (South) | | Little Yellow, Sleepy Orange, Cloudless Sulphur, Orange-barred Sulphur |
| Sunflowers | + | Bordered Patch, checkerspots |
| Thyme | + | |
| Verbenas (South) | + | |
| Violets (North and West) | ± | Greater fritillaries |
| Wild black cherry(East) | | Eastern Tiger Swallowtail, Coral Hairstreak, Spring Azure, Red-spotted Purple |
| Willows (shrubs) | | Sylvan Hairstreak, Mourning Cloak, Viceroy, Lorquin's Admiral |
| Zinnias | + | |

## Raising Butterfly Caterpillars

Once you have located a caterpillar or some eggs, you may want to raise them. Raising caterpillars and watching them grow into butterflies is fun and immensely satisfying. Common species are relatively easy to raise if you provide them with appropriate food and shelter.

Start by providing your caterpillar with leaves from the plant upon which it was found. You may want to collect some extra foliage because caterpillars consume ever more leaves as they grow. Be sure to collect leaves of the correct plant, as more often than not your caterpillar will not eat other types of plants. When using cuttings, they should be placed in a container of water or refrigerated, otherwise the leaves will dry out and become inedible. To keep your caterpillar from falling into the water cover the mouth of the jar with some sort of wrap through which you can push the plant stems.

Next you will want to protect your caterpillar from predators, parasites, and other dangerous elements. An aquarium or even a large glass jar will do. Make sure that your caterpillar has air by punching a hole in the lid of the jar. A screen top will do for an aquarium. The point is to make sure there is frequent air exchange. Keep the container out of direct sun and away from other potentially harmful elements. Try to simulate as natural a condition for your caterpillar as possible. Raising your caterpillar in a jar will require more upkeep than if it is kept in an aquarium.

Your caterpillar, simply put, is an eating machine, it is necessary to clean out the droppings frequently. Make sure that you provide fresh food. Provide it with twigs or branches so that it has something to climb on prior to pupation. Caterpillars often pupate on the lid of the container, and they can sometimes be deterred from this by giving them lots of twigs or branches.

An alternative to raising your caterpillar under glass is to sleeve it out on a potted foodplant or even a branch of the foodplant in the backyard. This reduces your daily chores and is more natural for the caterpillar. It is advisable to surround a sleeved plant with fine screen to further ward off predators. The drawback to this method is that it seriously hinders the ease of observation and that, in turn, reduces the fun. We do most of our rearing on plants either grown from seed or bought from a nursery.

## Conservation

Happily, butterflies are still plentiful throughout much of the United States. However, in the areas where most people live, the urban and suburban population centers, both the diversity of butterfly species and the numbers of individual butterflies have been greatly diminished. The major reason for the reduced numbers of butterflies is that people have destroyed their habitats. Each time a new subdivision is created, or a new parking lot paved, the butterflies that had lived on these plots of land are gone. Of course, gone with them are all the other plants and animals that had lived there. Because most butterflies are small and do not need vast vistas of wilderness to survive, we can do much to ameliorate the effect of human population growth on butterflies.

Throughout much of the United States there is a basic design to the suburban land, using a limited palette of non-native shrubs, flowers, and grasses, such as rhododendrons, roses, and lawn grasses. Although these gardens and lawns look attractive to many people, they might as well be artificial movie sets as far as butterflies are concerned. Once one realizes that the typical home gardens are sterile environments, they begin to look much less attractive. As ever more people landscape using native plants that support butterflies, there is no reason why we cannot have healthy populations of many species of butterflies coexisting with people.

A second factor causing butterfly populations to decline is the use of pesticides. Especially worrisome in this regard is the broadcast spraying of anti–gypsy moth and anti-mosquito toxic chemicals onto our forests and homes, respectively. Both programs kill butterflies wholesale and are probably responsible for causing the endangered situation affecting some butterfly species. Both programs are misguided, in that they will not attain their goals and cause more harm than good, not only to butterflies, but, in the case of the anti-mosquito sprays, to people as well.

Some might say "Butterflies are okay, but why save them? Let's focus on people." Butterflies are early warning indicators of the deterioration of the environment and are significant actors on the ecological stage, serving as food for other animals and as pollinators of many plants. But the importance of butterflies goes well beyond these functional roles, and includes the symbolic role that butterflies play for people. Butterflies represent beauty, freedom, and the human soul, concepts all worth cherishing. Preserving a healthy environment of course has a direct positive effect on future generations. An often overlooked fact is that butterflies are important to large numbers of people, and will be so to even more in the future. People who care passionately about butterflies are hurt when butterflies are harmed and helped when butterflies are helped. Beautiful paintings, such as the *Mona Lisa*, are protected and cared for only because they have great meaning for many people. The growing number of people interested in all facets of the world of butterflies will help to conserve them not only by being a direct constituency for butterfly conservation but also by becoming a highly visible group of people whose needs are seen as legitimate even by people who themselves do not really care about butterflies. So by helping to communicate your passion for butterflies to others, you will be helping to improve the world.

## About the Species Accounts

In the pages that follow, the order in which species are treated more or less follows the taxonomic order of the North American Butterfly Association's (NABA) *Checklist and English Names of North American Butterflies*. Occasionally a slightly different order is followed to group similar-appearing species or to facilitate a smooth and pleasing layout. The accounts are organized as follows:

*An asterisk preceding the name of a species means that, if you live within the range of this species and have the appropriate caterpillar foodplant in your garden, there is a good chance that you will find the caterpillar in your garden.

**Name** English and scientific names follow the North American Butterfly Association's (NABA) *Checklist and English Names of North American Butterflies*. Until recently, each author of a book about butterflies used whatever set of names struck his or her fancy. The result has been a confusing plethora of names that has bewildered the uninitiated and made it more difficult for the public to become involved with butterflies. With the publication of NABA's list in 1995 (and the second edition of the Checklist in 2001) we are now on the road toward standardization, although this process will take years to be completed.

**Identification** Each species section begins with an account of how to identify that species. The most important identification clues are presented in boldface type.

**We must emphasize that our knowledge of how to identify butterfly caterpillars is still far from complete. For a very large number of species, the images in this book will be the first published photographs of that species of caterpillar. For many species, especially many of the gossamer-wings and skippers, we are not yet sure if certain markings or colorations have strong value in identification because we have seen so few individual caterpillars of those species and their relatives.**

Many caterpillars have stripes or bands of colors. We have used the word "stripe" to mean a line of color (of varying thickness) that runs from the head to the rear of the caterpillar—the entire distance or only a portion of the distance. We have used the word "band" to mean a line of color (of varying thickness) that runs from the dorsal (back) side of the caterpillar to the ventral (belly) side of the caterpillar:—the entire distance or only a portion of the distance. *Lateral* refers to the side of the caterpillar.

The size of an adult butterfly can be a useful clue to its identification, but the size of caterpillars is, in general, less useful. This is because the size of individual caterpillars can vary tremendously, both because of variation caused by genetics and environment and because the caterpillar grows very rapidly as it matures.

**Habitat** This section describes the types of areas where this caterpillar might normally be found.

**Found** Here we try to provide a rough guide to when to search for the caterpillar in question, usually giving information about whether it is most fruitful to search before or after the flights of the adult butterflies.

Flight dates can vary tremendously depending upon the weather pattern of the year. Thus, finding a caterpillar species at a particular time in a particular year may depend on the vagaries of that year's brood sequence and abundance level. In desert areas especially, timing of rains may dramatically affect flight times and abundance. Although optimal search time may be given as, for example, May–June, in most cases this will refer to much of the range of the butterfly. You've probably noticed that much of the West is quite hilly. If you missed finding that caterpillar in May and now its June, just go higher up the hill! At any one spot the optimal search time will generally be shorter than the time period given, and the best search time will vary between localities depending in large part, but not entirely, on the latitude and the elevation of the locality.

**Host** Here we indicate the **major** hostplant(s), or group of plants that are eaten by the caterpillars. If a species commonly will be found on unlisted plants, we say "and others." For most caterpillars, the easiest way to find them is to search areas where the foodplant is common. We use the terms hostplant and caterpillar foodplant interchangeably.

**Garden Tips** For certain species that are easily attracted to gardens we suggest particular garden-worthy plants that are especially useful.

**Comments** Here we include assorted information and/or thoughts that didn't easily fit into one of the above-listed categories.

## Parts of a Caterpillar

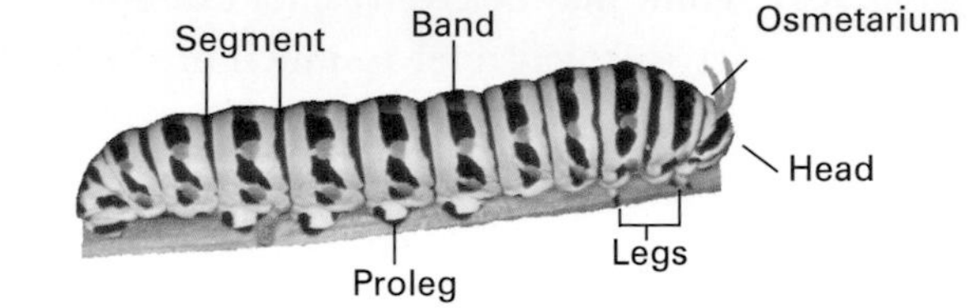


**Baird's Swallowtail**

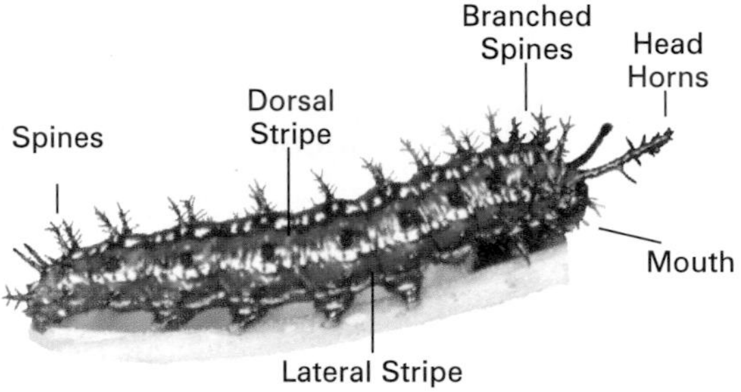


**Variegated Fritillary**

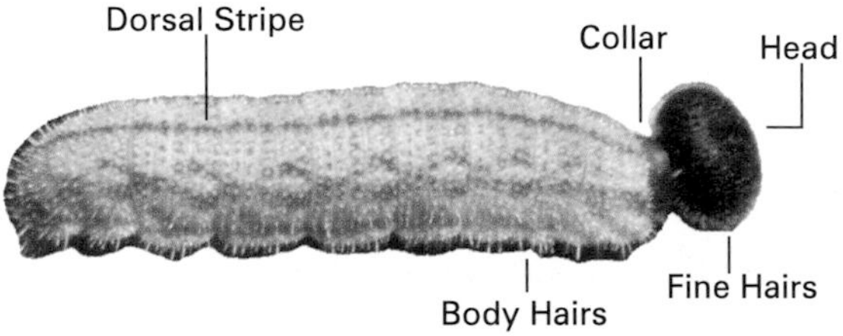


**Dorantes Longtail**

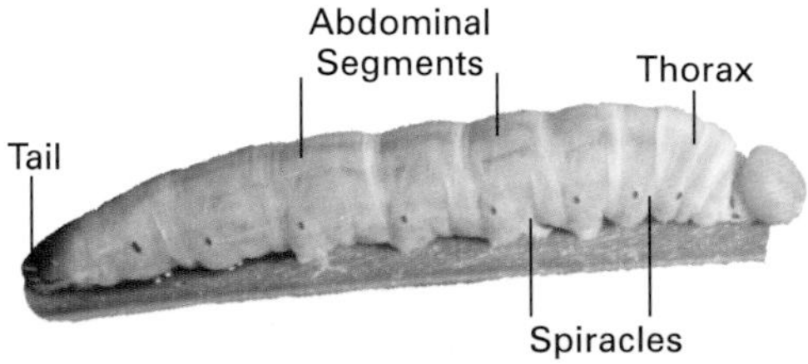


**Monk Skipper**

## Range Map Colors and Information

Two types of range maps accompany the species accounts. Most range maps show the range of only one species. For such maps the following color legend applies:

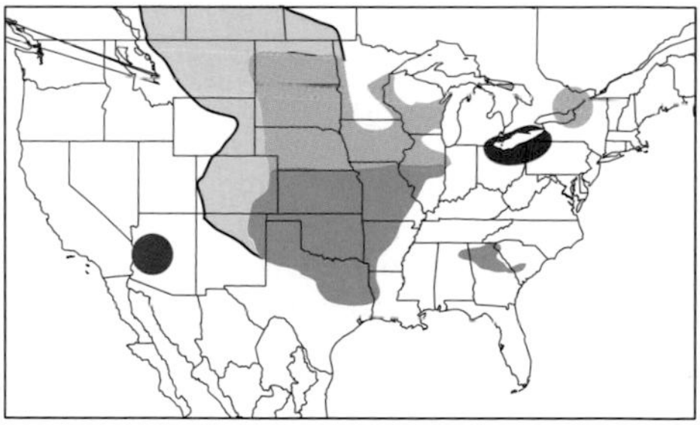

One Brood

Two Broods

Three Broods

No Longer Found in This Region

Range Enlarged for Visibility

Some maps show the range of two species, and a few show the range of three species. For such maps broods are not included, and the following color legend applies:

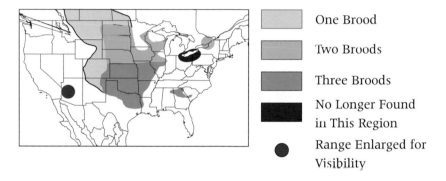

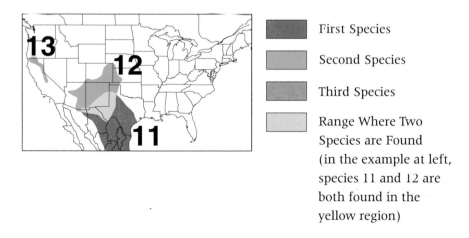

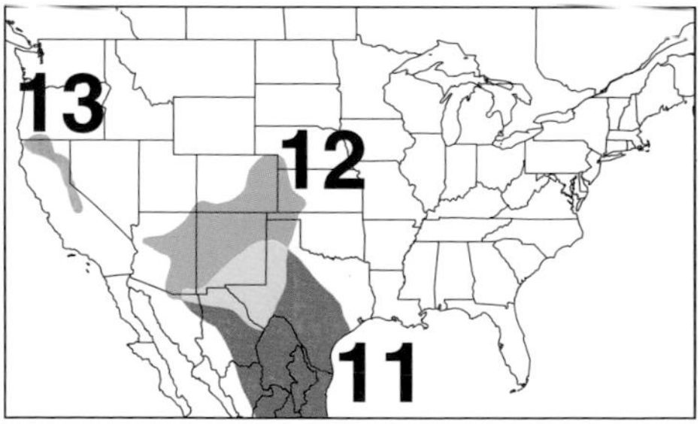


First Species

Second Species

Third Species

Range Where Two Species are Found (in the example at left, species 11 and 12 are both found in the yellow region)

See page 6 for general information about parnassians and true swallowtails.

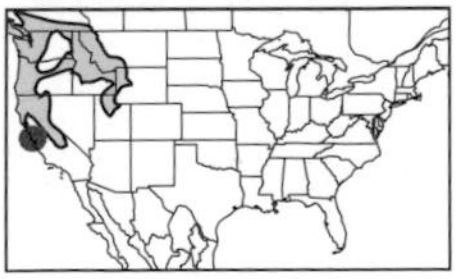

**Clodius Parnassian** *Parnassius clodius*
**Two rows of yellow spots** or streaks, one on either side of its back. Sometimes with dorsal row of black chevrons. Body usually black or dark brown. *Habitat* Mountain meadows and moist, open mountain woodland down to sea level in the Northwest. *Found* Prior to single flight, mainly May–June, July at higher altitudes. *Host* Bleeding hearts. *Comments* Often rests on ground when not feeding.

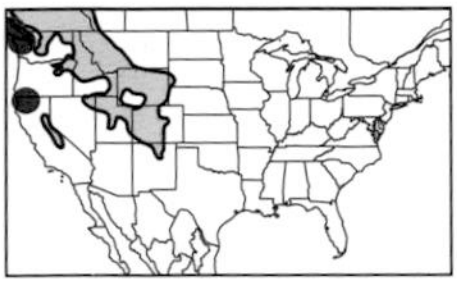

**Phoebus Parnassian** *Parnassius phoebus*
**Four rows of yellow spots** on its back, two on either side. Outer row of spots sometimes nearly connect to form a line. Body usually black. *Habitat* Open montane habitats, from about 4500′ to above treeline. Restricted to alpine areas especially rocky summits in Sierra Nevada Mtns. of California. *Found* Prior to single flight, May–July, depending on elevation. *Host* Stonecrops.

***Polydamus Swallowtail** *Battus polydamus*
Pair of **long fleshy tubercles (filaments) just behind head** with raised, short fleshy orange-red tubercles down either side of back extending to posterior. Body usually black or reddish-black but may also be brownish, greenish, or yellowish-black. *Early Instars* Similar to last instar. *Habitat* Vacant lots and edges of woodlands. *Found* May–Oct. *Host* Pipevines.

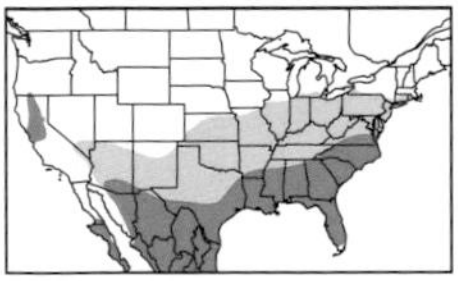

***Pipevine Swallowtail** *Battus philenor*
Pair of **very long fleshy tubercles just behind head** with raised, short fleshy orange or red tubercles down either side of back and three pairs of long fleshy tubercles on the posterior. Body usually purplish, red (especially in West), or reddish-black. *Early Instars* Similar to last instar. *Habitat* Deciduous woodlands, dry or moist, or grasslands where its host occurs. *Found* March–Oct in the south; May–Sept northward. *Host* Pipevines.

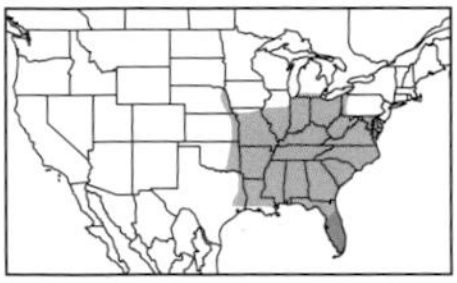

**Zebra Swallowtail** *Eurytides marcellus*
**Prominent blue, black, and yellow bands**. Body thickest at thorax, usually green (see inset to photo 7), but also brown or black, usually with thin bands (yellow, orange, or white) on each segment, no eyespots. *Habitat* Open, brushy fields and woodlands, especially along watercourses. *Found* March–Oct in south, April–June, Aug–Sept northward. *Host* Pawpaws and squirrel-bananas.

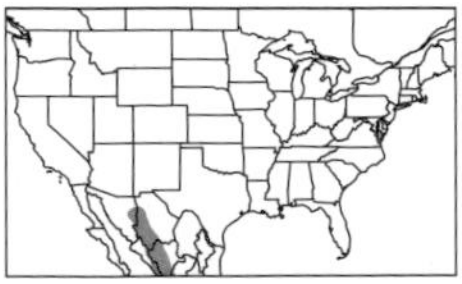

**White-dotted Cattleheart** *Parides alopius*
Distinctive. Cream-colored bands between segments. Body reddish-black with creamy diagonal stripe extending from dorsal to ventral midway down back. *Habitat* Mid-elevation tropical forest. *Found* May–Oct in NW Mexico, after flights. One U.S. record. *Host* Pipevines (*A. watsoni* in NW Mexico).

1. Clodius Parnassian

2. Pipevine Swallowtail adult

3. Phoebus Parnassian

4. Polydamus Swallowtail

5. Pipevine Swallowtail

6. Pipevine Swallowtail

7. Zebra Swallowtail

8. White-dotted Cattleheart

## "Black" Swallowtail Group

As a group, these caterpillars are distinctive, but identification to species can be difficult. There is much individual variation including mostly green or mostly black forms and everything in between. Early instar caterpillars are black with rows of raised tubercles often white, cream, yellow, or orange. With each successive instar a white patch or blotch (often called a saddle) in the center of the back becomes better developed and the caterpillar resembles a bird dropping (see inset photo 3). Last instar caterpillars lack fleshy tubercles.

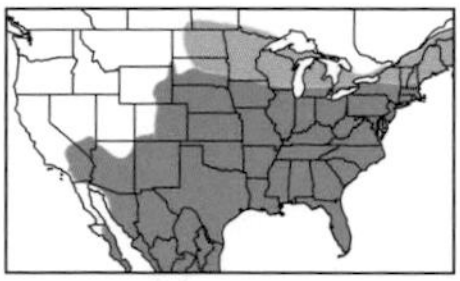

### *Black Swallowtail *Papilio polyxenes*

Caterpillars found in most gardens on carrot family plants will probably be this species. *Habitat* Open areas, especially disturbed habitats; fields, meadows, tidal marshes, suburban lawns and gardens. Deserts for 'Desert' Black Swallowtail *Found* All year in the South; April–June, Aug–Sept northward. *Host* Dill, parsley, fennel, carrot and similar plants (cultivated and wild) in the carrot family. Turpentine broom for "Desert" Black Swallowtail. *Garden Tips* Fennel, parsley, and other garden variety plants in the carrot family.

### Short-tailed Swallowtail *Papilio brevicauda*

Slightly overlaps range of Black Swallowtail and probably is not distinguishable from it in these areas. The individual shown (inset to photo 3) is an early instar caterpillar. *Habitat* Heathland, grassy sea cliffs, tidal marshes and dunes, gardens, and glades. *Found* May–June, August–Sept, after flights. *Host* Scotchman's lovage, cow parsnip, and angelica.

### Ozark Swallowtail *Papilio joanae*

Identical to Black Swallowtail but reportedly occupies different habitat and prefers different hosts. *Habitat* Cedar glades and woodlands. *Found* May–June, Aug–Sept, after flights. *Host* Meadow parsnip, yellow pimpernel, golden alexander, and other members of the carrot family.

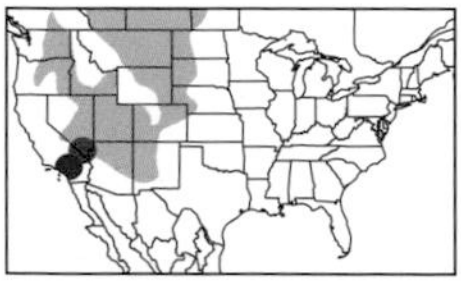

### Old World Swallowtail *Papilio machaon*

Usually found on wild tarragon. *Habitat* Arid and semi-arid hills, prairie, sagebrush steppes, to open woodland. High elevation mountains and mountain meadows in the Southwest for "Baird's" Old World Swallowtail. *Found* April–Sept, after flights. *Host* Wild tarragon or dragonwort, rarely, members of the carrot family in nature.

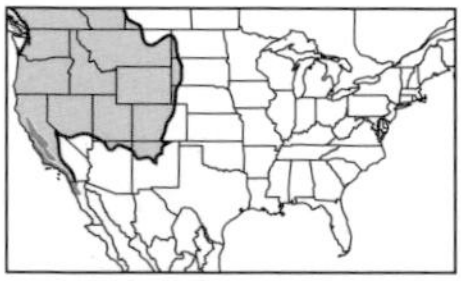

### Anise Swallowtail *Papilio zelicaon*

Caterpillars found in western gardens on carrot family plants will probably be this species. *Habitat* Deciduous and mixed mountainous woodlands from sea level to timberline, citrus groves, suburban backyards, and vacant city lots. *Found* March–Nov, after flights. Varies with region and habitat. *Host* Wild parsley, wild anise, or fennel (introduced), also cultivated citrus. *Garden Tips* Wild anise.

1. Black Swallowtail

2. Black Swallowtail adult

3. Black Swallowtail (inset, Short-tailed Sw.)

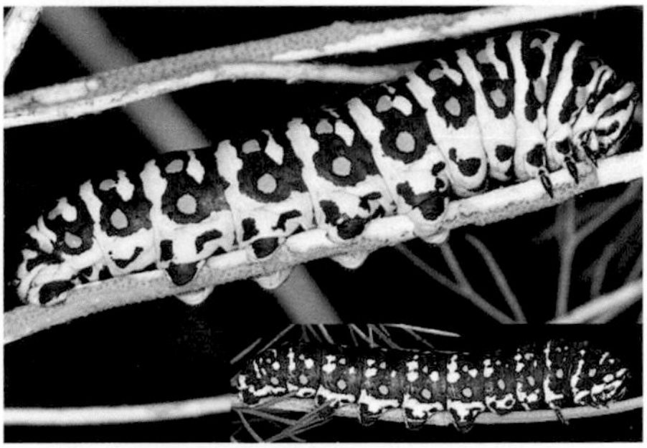

4. 'Desert' Black Swallowtail

5. Ozark Swallowtail

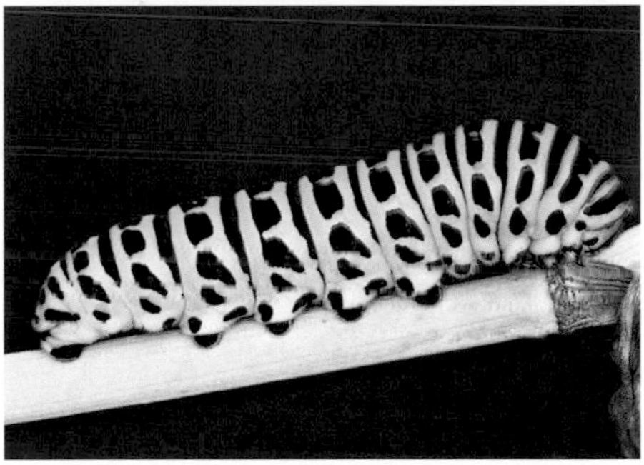

6. Old World Swallowtail

7. 'Baird's' Old World Swallowtail

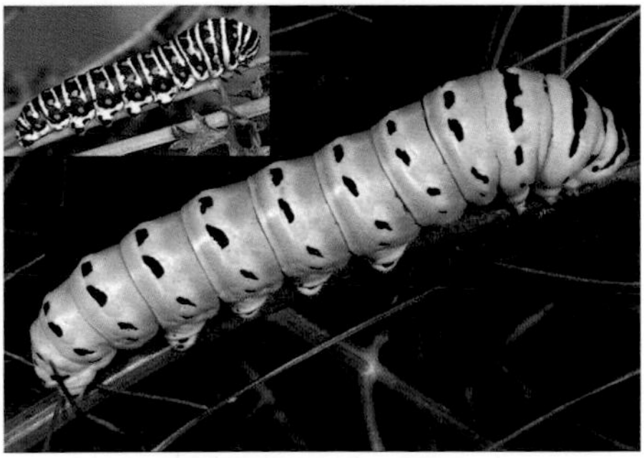

8. Anise Swallowtail

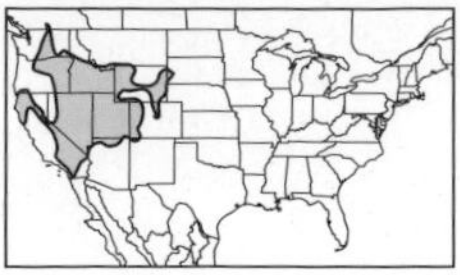

## Indra Swallowtail *Papilio indra*

**Alternately banded black and cream or pink** (sometimes yellow or salmon, or bands sometimes lacking altogether) with yellow or orange dots. *Early Instars* Resembles a bird dropping (photo 1). *Habitat* Dry, rocky slopes, steep canyons and hilltops, especially in arid regions but also open woodlands. *Found* After flights, April–June in deserts, June–Aug in mountains; later in double brood areas. *Host* Wild parsleys and relatives. *Comments* Caterpillars often located in rocky canyon bottoms at the base of steep cliffs.

## Ruby-spotted Swallowtail *Papilio anchisiades*

Finely mottled brown, without white patches. *Early Instars* Yellow-orange to yellow-green with yellow heads. Does not resemble bird a dropping. Gregarious. *Habitat* Subtropical woodlands and citrus groves. *Found* May–Dec. *Host* Citrus. *Comments* Sibling caterpillars remain together in clusters during development.

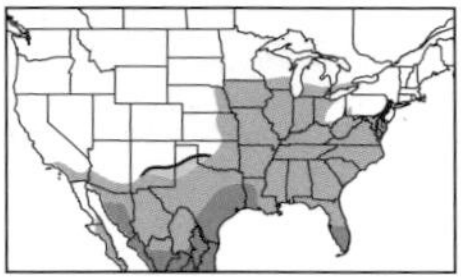

## *Giant Swallowtail *Papilio cresphontes*

Finely mottled brown with **white saddle** as well as **white blotches on thorax and rear.** Osmeteria are orange. *Early Instars* Shiny yellow-brown with white saddle. Resembles a bird dropping. *Habitat* Open woodlands and abandoned fields and hillsides near woodlands; also citrus groves. *Found* All year in South; 2 broods northward; June–July, Aug–Oct. *Host* Hop tree, wild lime, torchwood, cultivated citrus, and others in citrus family. *Garden Tips* Prickly ash, grapefruit, lemon, orange, and hop tree. *Comments* Most likely of all citrus swallowtail caterpillars to occur in the garden.

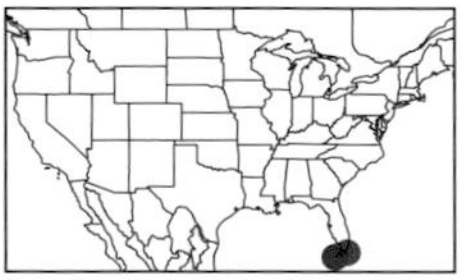

## Schaus' Swallowtail *Papilio aristodemus*

Mottled maroon-brown with **white blotches along side**. **No white saddle on back**. *Early Instars* Somewhat resembles a bird dropping but lacks white saddle. *Habitat* Tropical hardwood hammocks. *Found* June–Aug. *Host* Wild lime and torchwood. *Comments* Federally listed as an endangered species, Schaus' Swallowtail is the subject of recovery and reintroduction efforts in the Florida Keys.

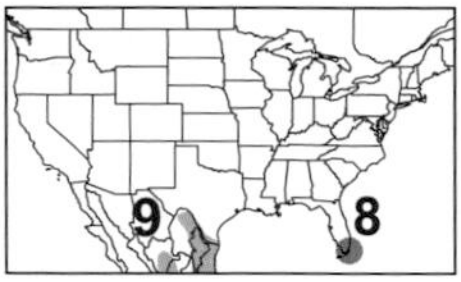


## 8. Bahamian Swallowtail *Papilio andraemon*
## 9. Ornythion Swallowtail *Papilio ornythion*

Not shown.

Similar to Giant Swallowtail. Ornythion with more yellow. *Early Instars* Shiny blackish-brown with white blotch behind head, along side, and on rear. Resembles a bird dropping. *Habitat* Tropical hardwood hammocks and woodland. *Found* May/June–Oct–Nov. *Host* Torchwood and other citrus relatives. *Comments* Bahamian was found for a number of years in extreme southern Florida but has not been seen in recent years. Ornythion Swallowtail is a regular stray to southern Texas.

1. Indra Swallowtail

2. Giant Swallowtail adult

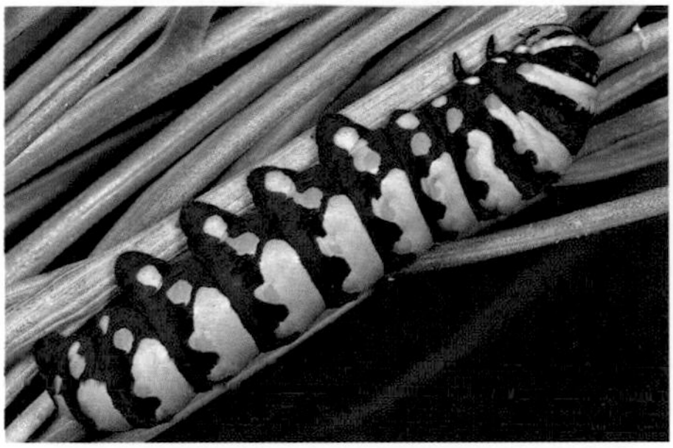

3. Indra Swallowtail

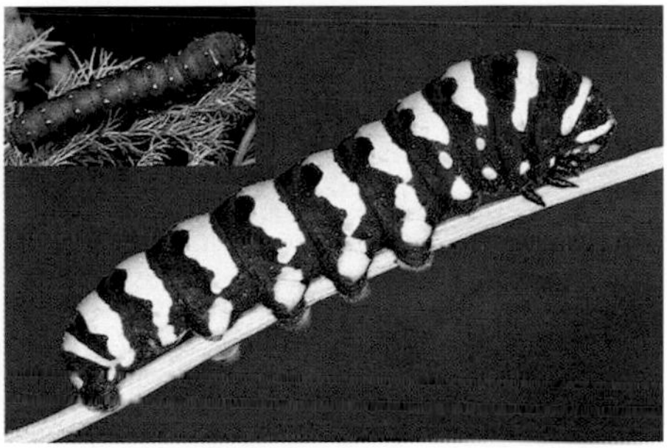

4. Indra Swallowtail

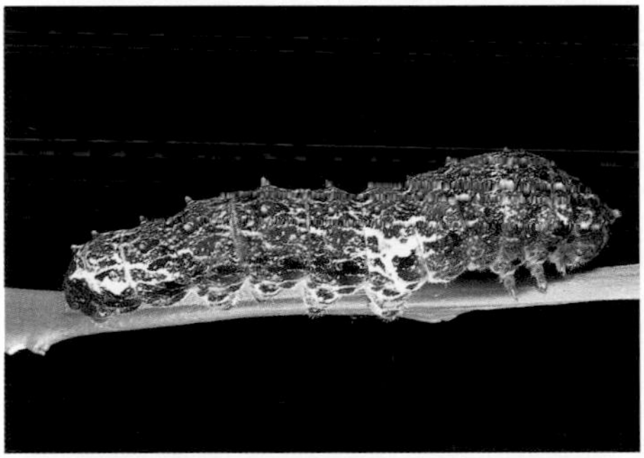

5. Ruby-spotted Swallowtail

6. Giant Swallowtail

7. Schaus' Swallowtail

8. Bahamian Swallowtail

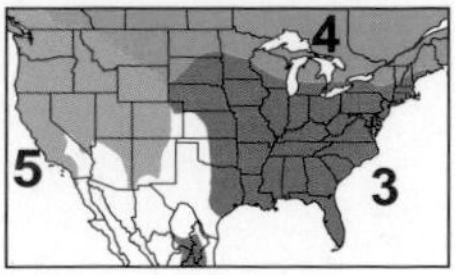


### *3. Eastern Tiger Swallowtail *Papilio glaucus*

### 4. Canadian Tiger Swallowtail *Papilio canadensis*

Not shown.

### 5. Western Tiger Swallowtail *Papilio rutulus*

Two yellow eyespots on thorax **centered with white or blue** spot **thickly ringed with black**. The yellow eyespots on both Pale and Two-tailed Swallowtail are often more elongated and slightly extend over the side of the thorax more than on tiger swallowtails. *Early Instars* Resembles a bird dropping (photo 5 inset). *Habitat* Deciduous woodlands, including suburban areas, especially woodland edges and wooded watercourses. *Found* April–Sept/Oct. *Host* Yellow poplar or tulip tree, sweet bay, and black cherry; birch, aspens, and black cherry for Canadian; willows, sycamore, cherry, aspen and others for Western. *Comments* Turns from green to brown prior to pupating.

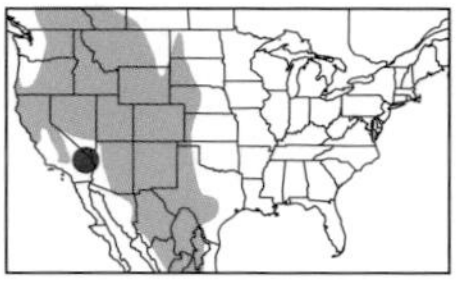

### Two-tailed Swallowtail *Papilio multicaudata*

Yellow spots on thorax **centered in green and/or blue thinly ringed with black**. Pale Swallowtail uses different hosts. *Early Instars* Resembles a bird dropping. *Habitat* Wooded areas, especially canyons and ravines near watercourses, but also towns and parks. *Found* April–Oct after flights. *Host* Cherry, ash, and hop tree. *Comments* At just over 2 inches fully grown, this is one of our largest butterfly caterpillars.

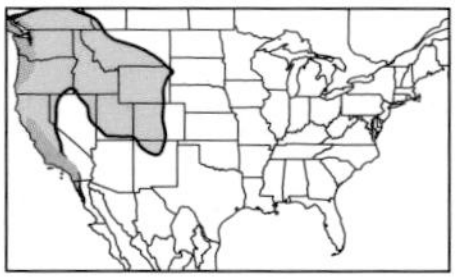

### Pale Swallowtail *Papilio eurymedon*

Host often differs for this species. The lower edge of the yellow spots on the thorax may be narrower than on other species. *Habitat* Deciduous-coniferous woodlands and chaparral in hilly or mountainous areas, often near moist canyons with permanent water. *Found* April–Nov in lowlands (2 broods), May–Oct in single-brood areas. *Host* Buckthorns, buckbrush, and wild plum.

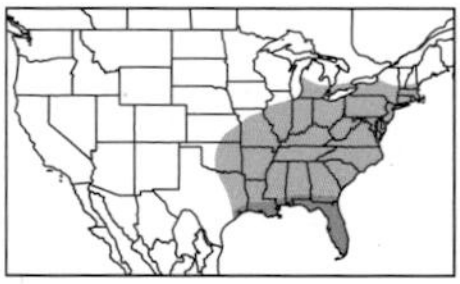

### *Spicebush Swallowtail *Papilio troilus*

**Lives in folded leaf shelters** on the host. Not closely related to the tiger swallowtail complex and easily separated from them by host and the larger eyespots on the thorax. *Early Instars* Resembles a bird dropping (photo 8, top left inset). Shiny brown with white saddle beyond thorax separated down middle and white patch on rear interrupted by brown. *Habitat* Open woodlands and woodland edges. *Found* March–Oct. *Host* Sassafras, spicebush, red bay, and swamp bay. *Comments* Turns golden yellow prior to pupating (see photo 8, top right inset).

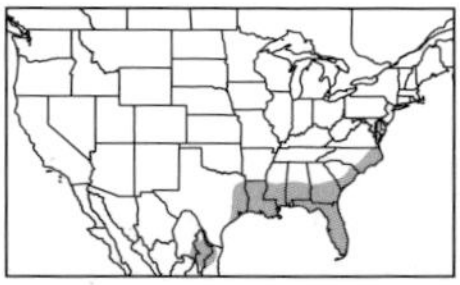

### *Palamedes Swallowtail *Papilio palamedes*

Nearly identical to Spicebush Swallowtail but **does not make leaf shelters**. The lateral yellow band often is thinner than that on Spicebush Swallowtail, and the orange spots on the abdomen are usually smaller than those on Spicebush Swallowtail. *Early Instars* Like Spicebush Swallowtail but white splotch on rear not interrupted by brown. *Habitat* Coastal marshes and swampy areas; open woodlands and edges. *Found* March–Dec. *Host* Red bay, swamp bay, pond spice, and perhaps others.

1. Eastern Tiger Swallowtail chrysalis

2. Eastern Tiger Swallowtail adult

3. Eastern Tiger Swallowtail

4. Western Tiger Swallowtail

6. Two-tailed Swallowtail

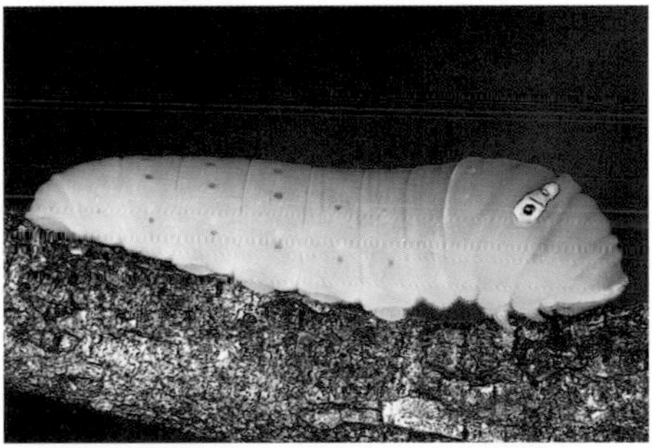

7. Pale Swallowtail

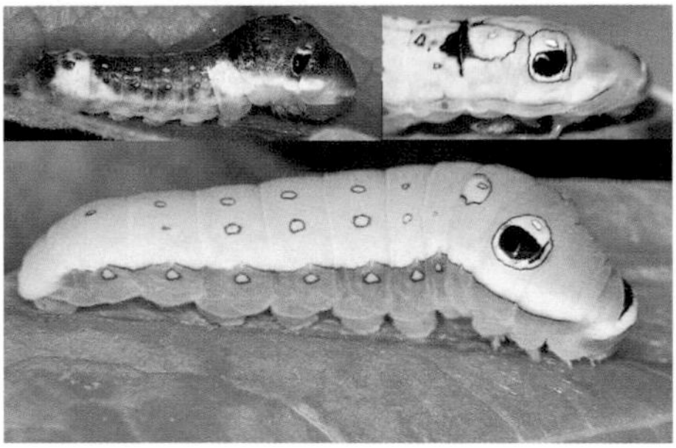

8. Spicebush Swallowtail

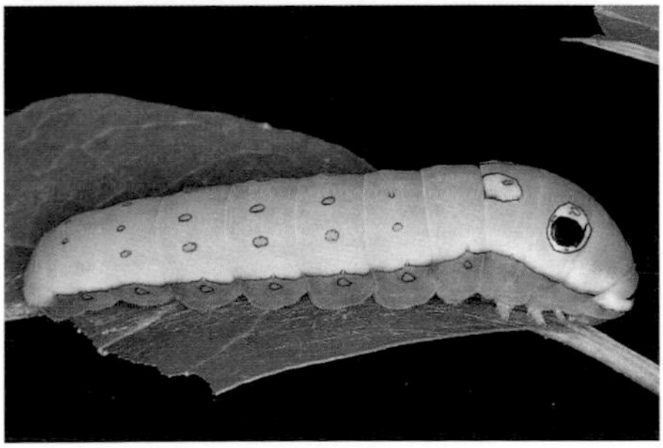

9. Palamedes Swallowtail

See page 7 for general information about group.

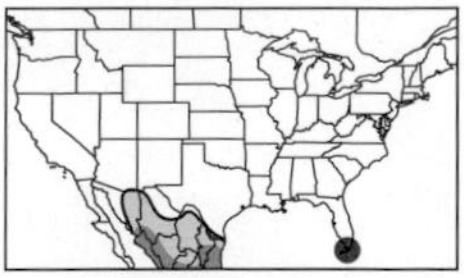

## Florida White *Appias drusilla*

Granular green body with short blue and black tubercles and small yellow tubercles at front and rear. Two short tails at rear. No similar butterfly caterpillars use capers as hosts. *Habitat* Tropical hardwood hammocks. *Found* All year. *Host* Capers. *Garden Tips* Jamaican caper.

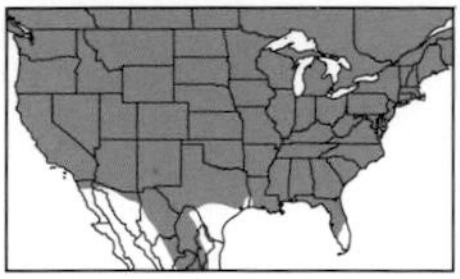

## *Cabbage White *Pieris rapae*

Considered a pest by some; on cabbage, broccoli, etc. Note **yellow sub-lateral dashes**. Mustard White is usually found in more natural settings; otherwise indistinguishable. *Habitat* Any type of open or slightly wooded terrain, especially gardens, roadsides, and agricultural lands. In the East, open woodlands are readily used during spring. *Found* Most of year; all year in warmest habitats, primarily on leaves. *Host* Cabbage, broccoli, water cress and other mustards, nasturtium; rarely capers. *Garden Tips* Cabbage, broccoli, cauliflower, Brussels sprouts. *Comments* The clear, oily fluid that collects at the tips of glandular hairs repels ants and possibly other predators.

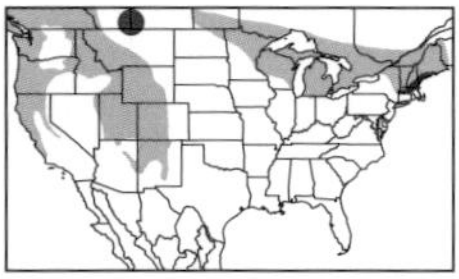

## Mustard White *Pieris napi*

Nearly identical to Cabbage White but lacks yellow dashes. Prefers more natural areas. *Habitat* In the West, openings in moist forest, usually coniferous. Elsewhere, northern deciduous forests. *Found* Feb–Sept, after flights. *Host* Mustards. *Comments* When provoked, a bright green fluid is disgorged.

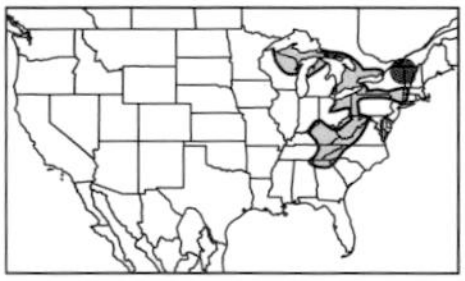

## West Virginia White *Pieris virginiensis*

Like Cabbage White without yellow dashes. *Habitat* Rich, deciduous woodlands usually in mountains with good stands of the host plant (usually near streams). *Found* March to mid-May, after flight. *Host* Toothworts.

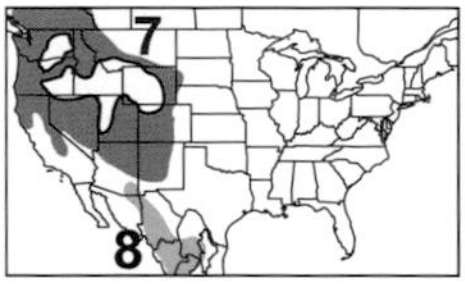


## 7. Pine White *Neophasia menapia*

## 8. Chiricahua White *Neophasia terlootii* Not shown.

Green with four bold lateral white stripes. *Habitat* Pine forests. *Found* Prior to flight, May–June for Pine White; May–Oct for Chiricahua White; on needles. *Host* Ponderosa pines and other pinaceae. *Garden Tips* Have ponderosa pines in your yard and live in the mountain West! *Comments* Gregarious in early instars. Spraying the caterpillars with a fine water mist elicits violent gyrations and thus, they can be located on small pines.

## Great Southern White *Ascia monuste*

Laterally striped with raised black dots and sparse long hairs and **yellow face**. Stripe width varies. See Checkered White (page 38). *Habitat* Open situations, especially along the coast including dunes, salt marshes, fields, and gardens. *Found* All year in South. *Host* Saltwort, mustards, and capers. *Garden Tips* Arugula, nasturtium, and sea rocket.

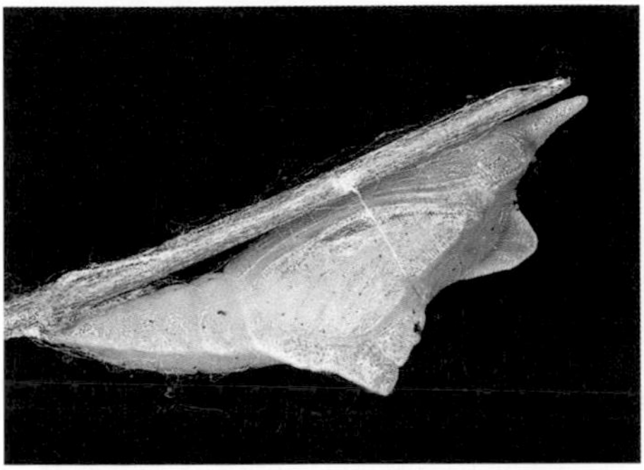

1. West Virginia White chrysalis

2. Cabbage White adult

3. Florida White

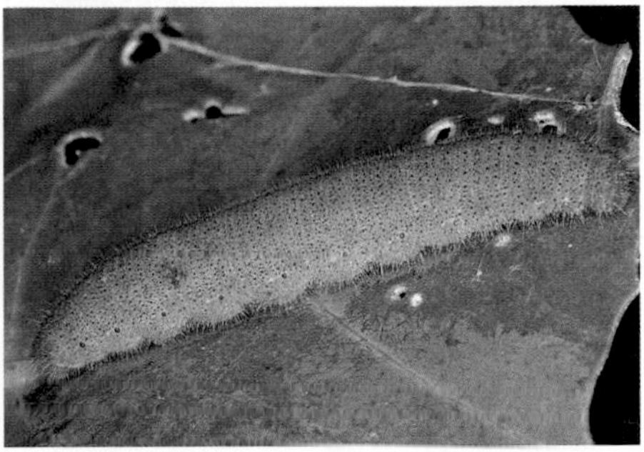

4. Cabbage White

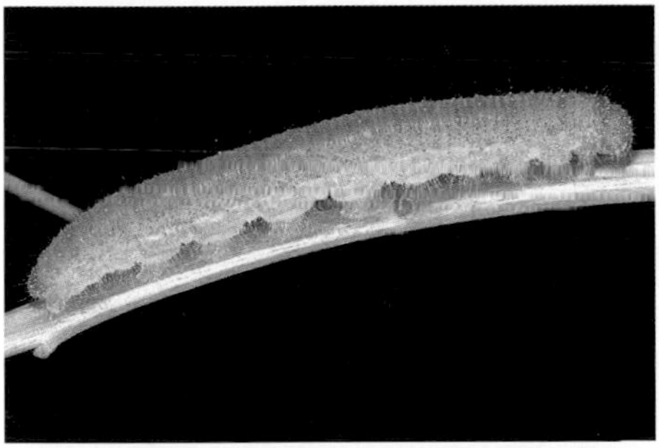

5. Mustard White

6. West Virginia White

7. Pine White

8. Great Southern White

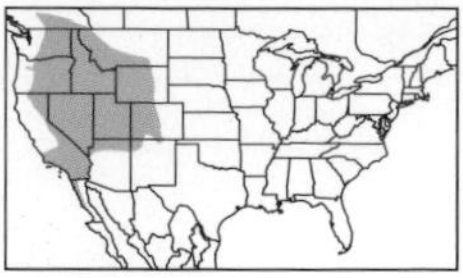

## Becker's White *Pontia beckerii*

Sparsely **covered with long hairs**. *Habitat* Desert, juniper hills, roadsides, sagebrush steppes and other arid regions. *Found* April–Sept, after flights, mainly on flowers and seeds. *Host* Prince's plume, bladderpod, and other mustards. *Comments* Growth stages rapid, as little as 23 days.

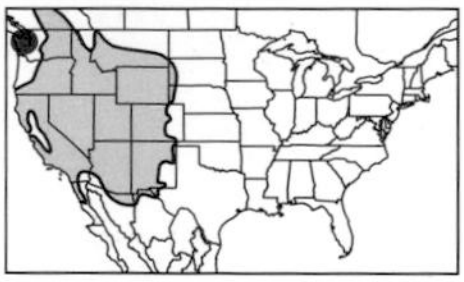

## Spring White *Pontia sisymbrii*

Distinctively banded. *Habitat* High deserts, sagebrush, dry rocky areas, and coniferous woodlands. *Found* Varies with region, March–April in south; May–Aug northward, on flowers and seeds. *Host* Mustards, especially rock cresses and lacepod. *Comments* A distinctive butterfly caterpillar for sure, but one may encounter moth caterpillars with similar pattern and colors.

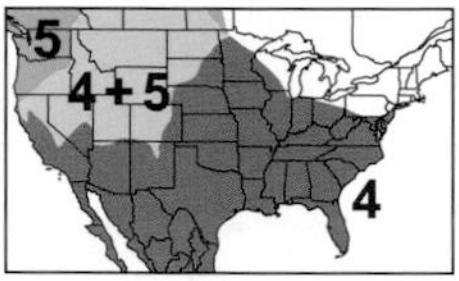


## 4. Checkered White *Pontia protodice* Not shown.

## 5. Western White *Pontia occidentalis*

Essentially identical, both have gray-purple and yellow dorsal and lateral stripes and are sparsely covered with **short fine hair**. *Habitat* Widespread in disturbed open habitats; also lowlands and high peaks. *Found* March–Nov, after flights, on flowers and seeds. *Host* A wide variety of mustards from native to exotic. *Garden Tips* Peppergrass for Checkered White.

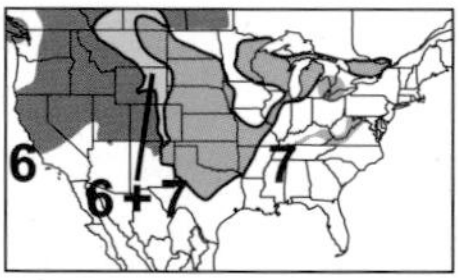


## 6. Large Marble *Euchloe ausonides*

## 7. Olympia Marble *Euchloe olympia*

Mostly gray-purple with yellow dorsal and yellow and **white lateral stripe**s. **Not covered with hair**. *Habitat* Large Marble in a wide variety of open habitats including meadows, roadsides, farmland, and forest openings. Olympia in poor soil areas such as barrens and grasslands. *Found* 6. March–June in lowlands; July–Sept at high elevations, after flights, on flowers and seeds. 7. May–June. *Host* Large uses tower mustard, tansy mustard and other mustards. Olympia uses rock cresses.

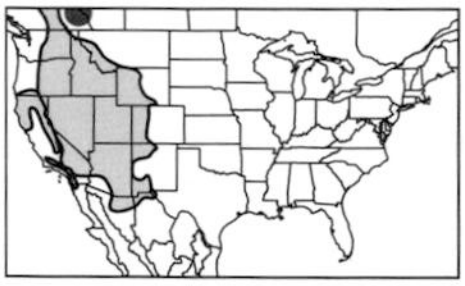

## Pearly Marble *Euchloe hyantis*

White lateral stripe edged above by a purple stripe of varying width and intensity. Sara Orangetip has white lateral stripe not edged with purple. Pentultimate instar Desert Orangetip similar but usually with thinner white lateral stripe. *Habitat* Open arid regions including desert, juniper-pinyon, pine, and sagebrush; also, forests and chaparral. *Found* April–June in arid regions, June–Aug at higher elevations, after single flight, on flowers and seeds. *Host* Tansy mustard, twist flower, and other mustards. *Comments* Turns purple prior to pupating.

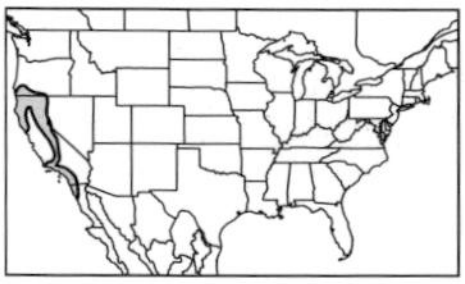

## Gray Marble *Anthocharis lanceolata*

**White lateral stripe edged above by broken yellow stripe**. Sara Orangetip also has a white lateral stripe but without yellow. Colias sulphurs do not feed on mustards. *Habitat* Rocky slopes and road cuts in woodlands, from near sea level to 7500′. *Found* March–July, depending on elevation, after single flight, on flowers and seeds. *Host* Rock cresses and other mustards.

1. Becker's White

2. Large Marble adult

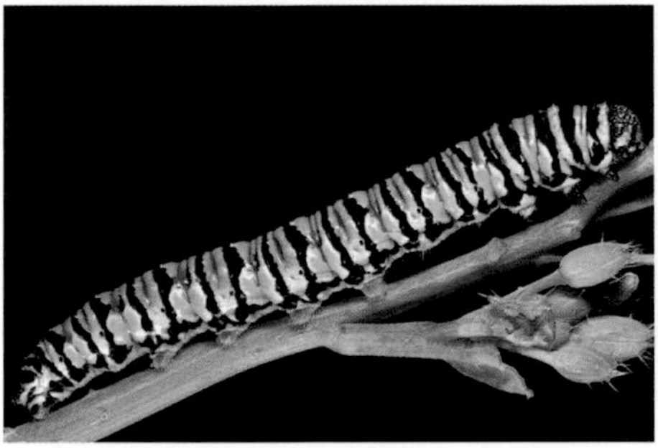
3. Spring White

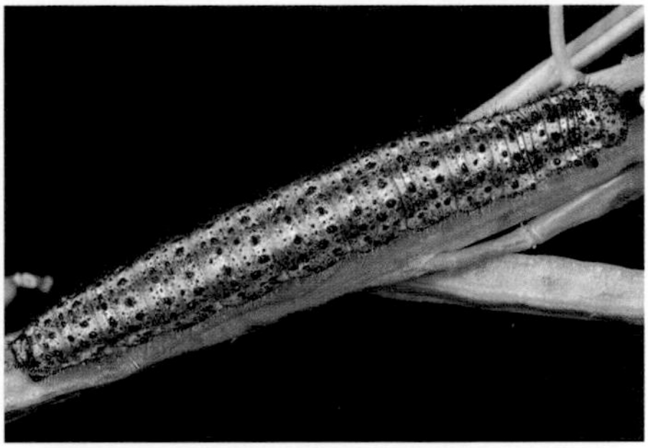
5. Western White

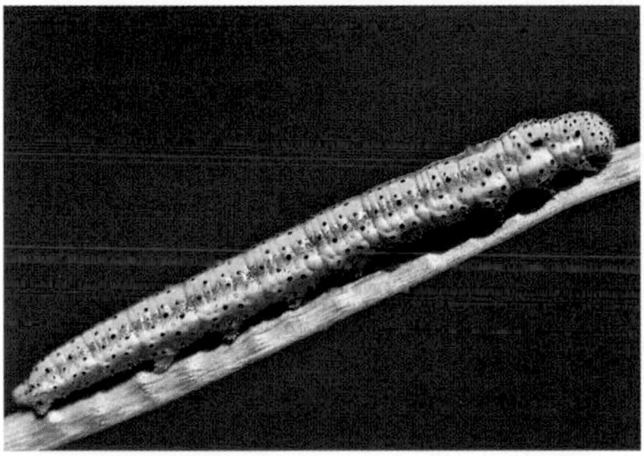
6. Large Marble

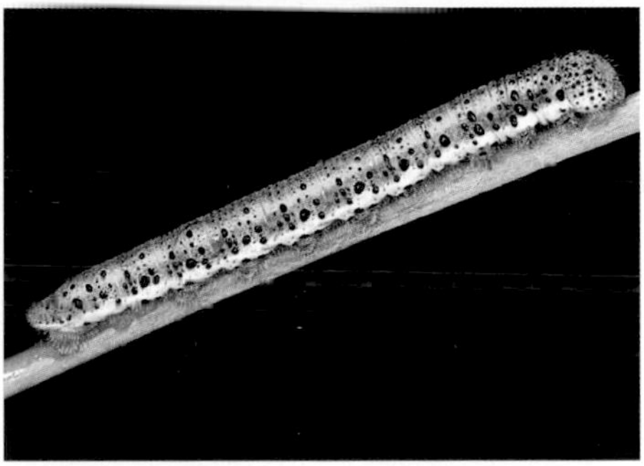
7. Olympia Marble

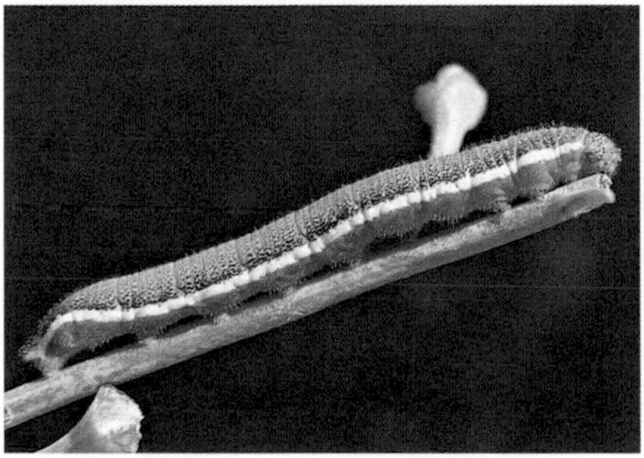
8. Pearly Marble

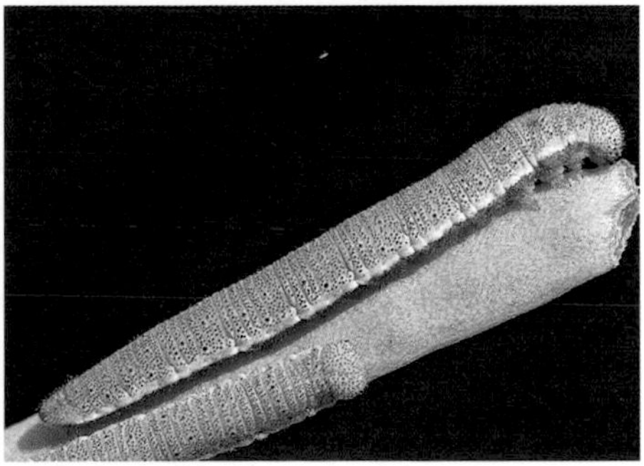
9. Gray Marble

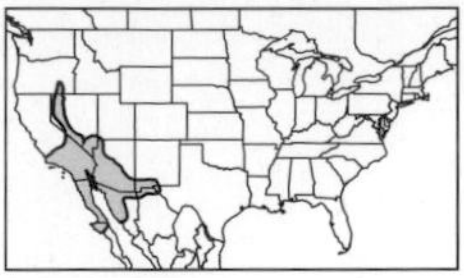

## Desert Orangetip *Anthocharis cethura*

Distinctive crossbanding. Pentultimate instar (photo 1 inset) is similar to last instar Pearly Marble. *Habitat* Open arid regions including desert, chaparral, and juniper covered hills. *Found* March–June after single flight; earlier in southern deserts, later in western Great Basin, on flowers and seeds. *Host* Twist flower, lacepod, and other mustards, including exotics.

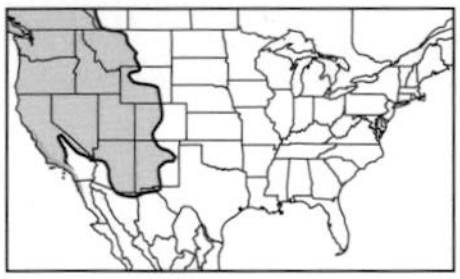

## Sara Orangetip *Anthocharis sara*

Green with well-defined white lateral stripe. Pearly Marble has a white lateral stripe edged above with purple. *Habitat* Wide variety of habitats, from high desert hills to coniferous forests. *Found* March–Aug, depending on elevation, after 1–2 flights, on flowers and seeds. *Host* Mainly rock cresses but many other mustards, native and exotic are used. *Garden Tips* Rock cresses. *Comments* Chrysalis may be either tan or green.

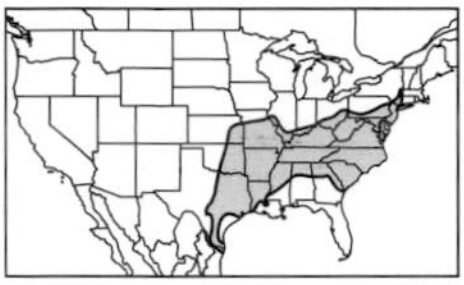

## Falcate Orangetip *Anthocharis midea*

Green or blue-green with yellow or orange stripe on top and a white lateral stripe. *Habitat* Open woodlands and shale barrens. *Found* April–June after the single flight. *Host* Mustards, especially rock-cresses and bitter cresses. *Comments* Similar to marbles, caterpillars are very aggressive toward each other.

## Suphurs (subfamily Coliadinae).

See page 7 for general information

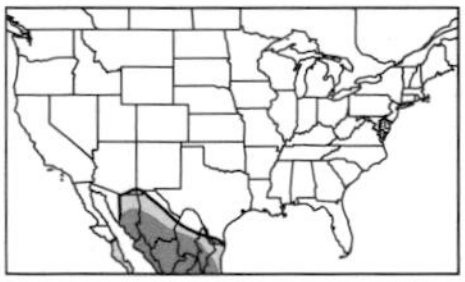

## White Angled-Sulphur *Anteos clorinde*

**White or cream lateral stripe edged above with varying amounts of black speckles** (sometimes absent). Reportedly has other forms including a yellow one that feeds primarily on flowers. Green forms feed on leaves. Green form of Cloudless Sulphur (page 45) is similar, but lateral stripe is yellow. *Habitat* Open areas close to tropical thorn forest and thorn scrub. *Found* All year, mostly on leaves. *Host* Tree senna, probably shrubby sennas in U.S. *Comments* Caterpillars not yet reported from U.S.

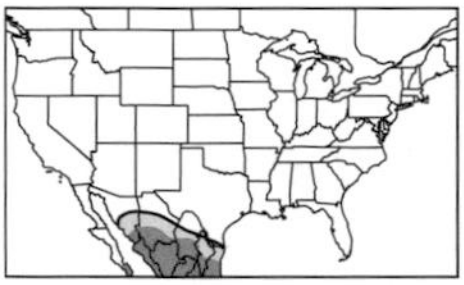

## Yellow Angled-Sulphur *Anteos maerula*

Yellow-green with raised blue color on thorax and thin blue banding. *Habitat* Open areas close to tropical thorn forest and thorn scrub. *Found* Most of year in tropical regions. *Host* Tree senna. *Comments* Caterpillars not yet reported from U.S.

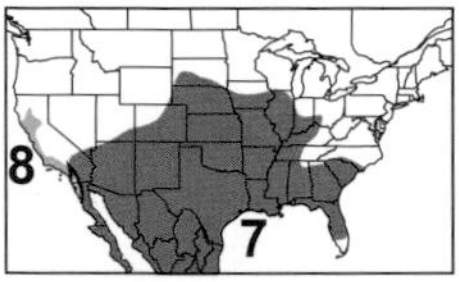


## 7. *Southern Dogface *Colias cesonia*

## 8. California Dogface *Colias eurydice*

Both have two forms: **Yellow bands edged with black** and a yellow or cream lateral stripe; or without the bands. For Southern Dogface, band form is most common; For California, band form is rare. *Habitat* Open areas, usually dry, including thorn scrub and agricultural lands. Foothill canyons and meadows for California. *Found* March–Nov, on leaves. *Host* False indigo for both, plus indigo bushes, leadplant, and other legumes for Southern. *Garden Tips* Indigo bushes (in Southwest), leadplant.

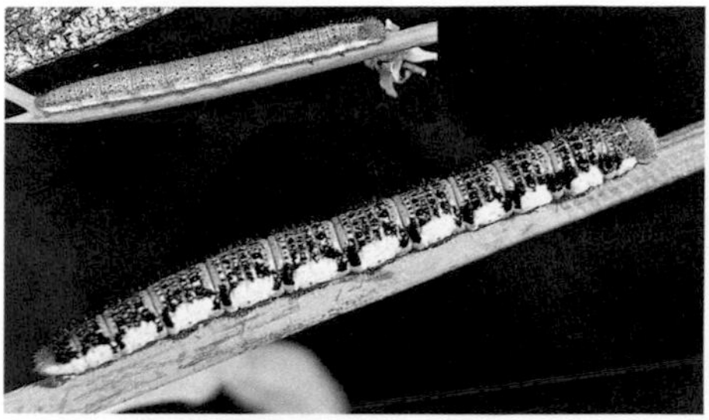

1. Desert Orangetip

2. Sara Orangetip adult

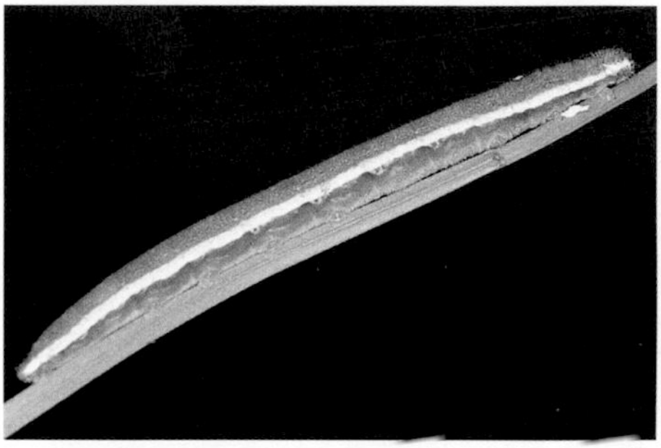

3. Sara Orangetip

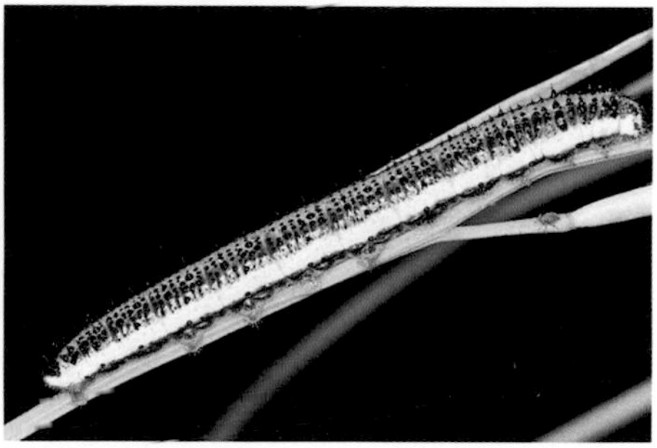

4. Falcate Orangetip

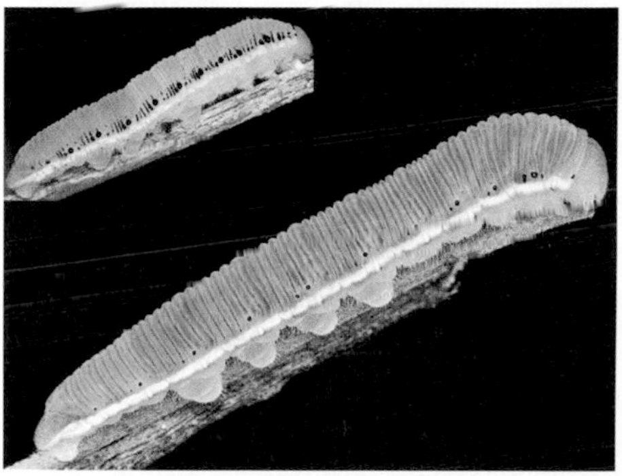

5. White Angled-Sulphur

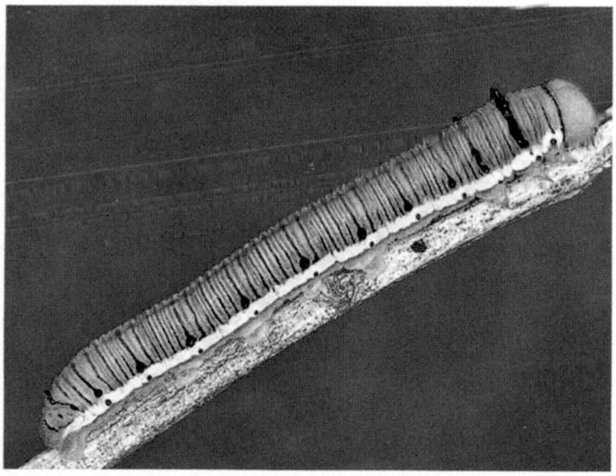

6. Yellow Angled-Sulphur

7. Southern Dogface

8. Southern Dogface (inset, Calif. Dogface)

See page 155 for more information about *Colias* Sulphurs

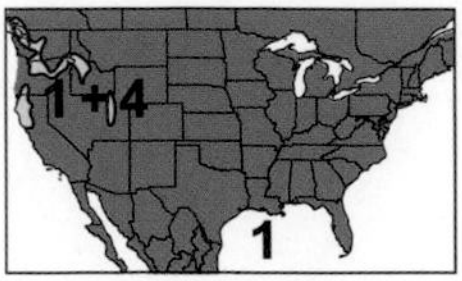


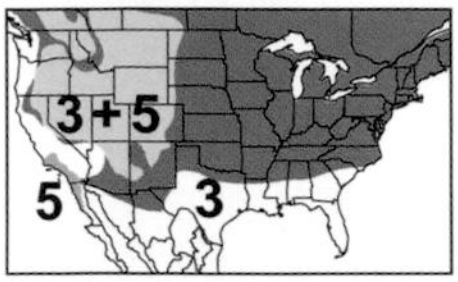


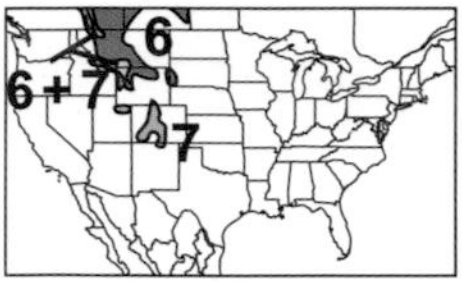


## *Colias* sulphurs that feed on legumes

### 1. Orange Sulphur *Colias eurytheme*

### 3. Clouded Sulphur *Colias philodice*

### 4. Western Sulphur *Colias occidentalis*

### 5. Queen Alexandra's Sulphur *Colias alexandra*

### 6. Christina's Sulphur *Colias christina* Not shown.

### 7. Mead's Sulphur *Colias meadii*

Species probably not separable from one another except for Mead's, which has a white lateral stripe and creamy subdorsal stripe edged with black. Orange and Clouded are often common to abundant. *Habitat* Almost any open habitat for Orange and Clouded, especially agricultural lands and roadsides. Western in coniferous forests and included meadows. Queen Alexandra's and Christina's mainly in prairie, foothill grassland and wet meadows in woodland. Mead's in alpine and subalpine meadows. *Found* Mainly prior to flight. Most of the year for Orange and Clouded; May–June for Western, Queen Alexandra's and Christina's; June–July for Mead's. *Host* A wide variety of legumes, but Western especially on *Lathyrus*, Queen Alexandra's and Western especially on rattleweeds and clovers. *Garden Tips* White sweet clover and alfalfa for the widespread Orange and Clouded Sulphurs.

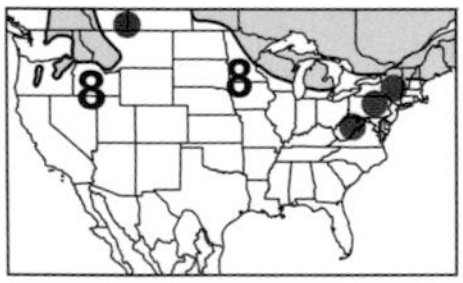


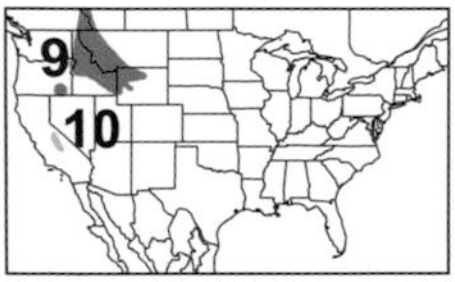


## *Colias* sulphurs that feed on blueberries & relatives

### 8. Pink-edged Sulphur *Colias interior*

### 9. Pelidne Sulphur *Colias pelidne* Not shown.

### 10. Sierra Sulphur *Colias behrii*

Pink-edged Sulphur is similar to other sulphurs. Pelidne caterpillar is unknown. Sierra Sulphur has white lateral and subdorsal stripes. *Habitat* Pink-edged and Pelidne in forest openings with heaths (Pelidne in subalpine). Sierra in subalpine moist meadows. *Found* Mainly prior to flight. Pink-edged in April–May in East, later to the north and mountain West. Pelidne and Sierra in June–July. Early instars following flight. For Pink-edged, third stage caterpillars overwinter underneath hostplant leaves. *Host* Blueberries for Pink-edged and Pelidne (which also uses wintergreen); dwarf bilberry for Sierra.

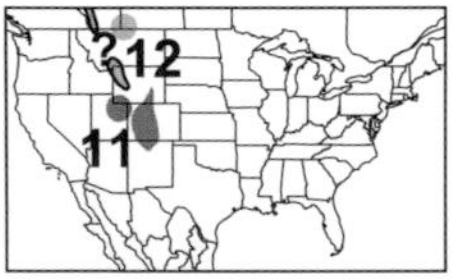


## *Colias* sulphurs that feed on willows

### 11. Scudder's Sulphur *Colias scudderi* Not shown.

### 12. Giant Sulphur *Colias gigantea* Not shown.

Scudder's Sulphur with cream to yellow lateral stripe and no dorsal stripe. Giant Sulphur caterpillar is unknown. *Habitat* Willow bogs. Scudder's in alpine from 9 to 12,000′. Giant in boreal and mixed forest. *Found* Probably prior to single flight, June–July. *Host* Willows.

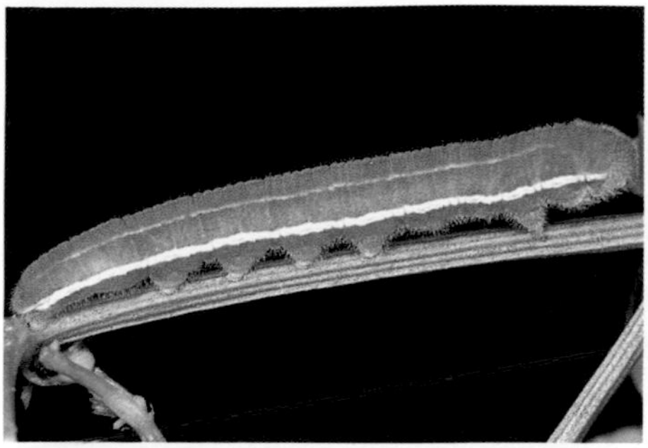
1. Orange Sulphur

2. Orange Sulphur adult

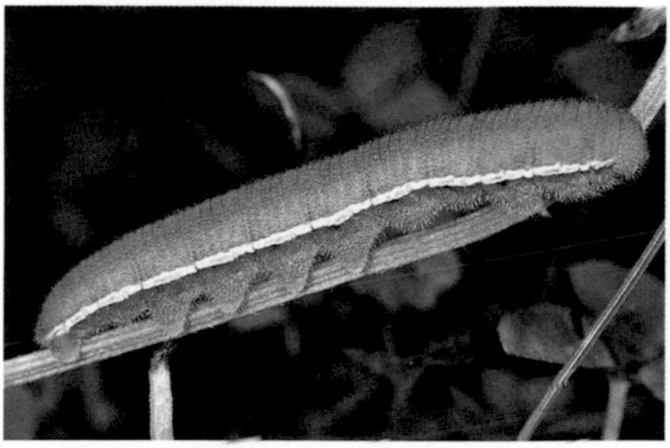
3. Clouded Sulphur

4. Western Sulphur

5. Queen Alexandra's Sulphur

6. Mead's Sulphur

8. Pink-edged Sulphur

10. Sierra Sulphur

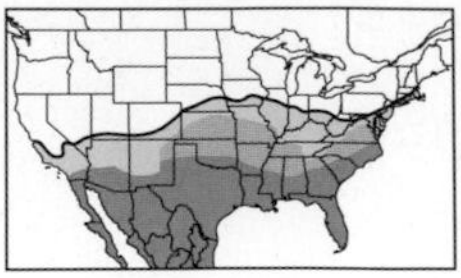

***Cloudless Sulphur** *Phoebis sennae*

Has both green and yellow forms. The **green form with a yellow lateral stripe** and often with characteristic **tripled blue lateral dashes**. Large Orange Sulphur is less boldly marked, lacks blue lateral dashes, and uses different hosts. Yellow form is distinctive. *Habitat* A wide variety of open situations. *Found* All year in the South; northward Aug–Oct, on both leaves and flowers. *Host* Sennas (formerly Cassia). *Garden Tips* Partridge pea, slim-pod senna (in Arizona). *Comments* An easy butterfly to attract to the garden in warm southern areas.

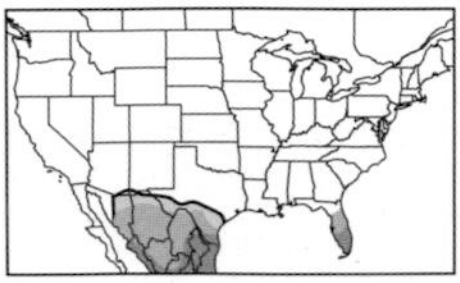

***Orange-barred Sulphur** *Phoebis philea*

Green to yellowish with a dark blue speckled sublateral line creating a **lateral line of yellow crescents**. *Habitat* Gardens and open woodlands in subtropical areas. *Found* All year in the South, Aug–Sept in the Southwest. *Host* Sennas. *Garden Tips* Christmas senna, candle plant (in Florida).

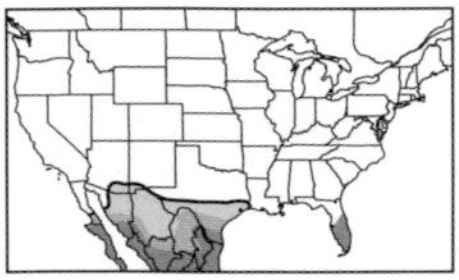

***Large Orange Sulphur** *Phoebis agarithe*

Green with **yellow sublateral line edged above with blue stripe**. Not banded or as boldly marked as forms of Cloudless Sulphur. *Habitat* General in open tropical and subtropical situations, including gardens and woodland edges. *Found* Most of year in southern areas, on tender leaves. *Host* Blackbead, wild tamarind, Texas ebony, guaymuchil, feather tree, and others. *Garden Tips* Blackbead (in Florida), Texas ebony (in Texas), feather tree (in Arizona). *Comments* Has a much more compact chrysalis than either Cloudless and Orange-barred Sulphurs.

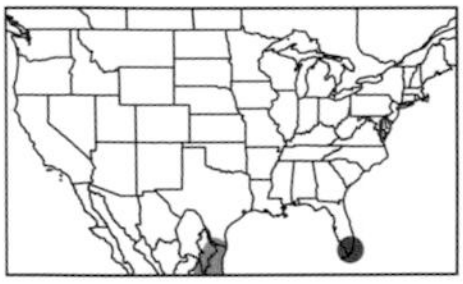

**Statira Sulphur** *Phoebis statira* Not shown.

Green with yellow lateral stripe. Other sulphurs use different hosts. *Habitat* Open areas near salt marshes or mangroves—near stands of coinvine; occasionally inland residential areas. *Found* All year. *Host* Coinvine.

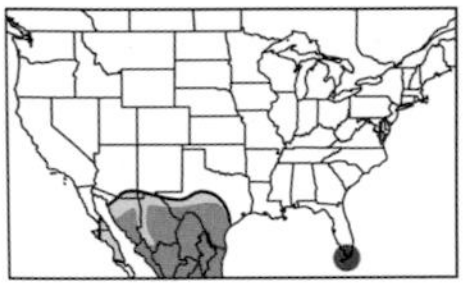

***Lyside Sulphur** *Kricogonia lyside*

Variable. May be boldly marked or unmarked. **No other butterfly uses lignum vitae.** *Habitat* Tropical and subtropical scrub. *Found* Most of year. Best following rain-triggered flights, mainly on fresh leaves. *Host* **Lignum vitae**. *Garden Tips* Lignum vitae. *Comments* Able to go from egg to adult in 13 days in the heat of summer!

1. Cloudless Sulphur chrysalis

2. Orange-barred Sulphur adult

3. Cloudless Sulphur

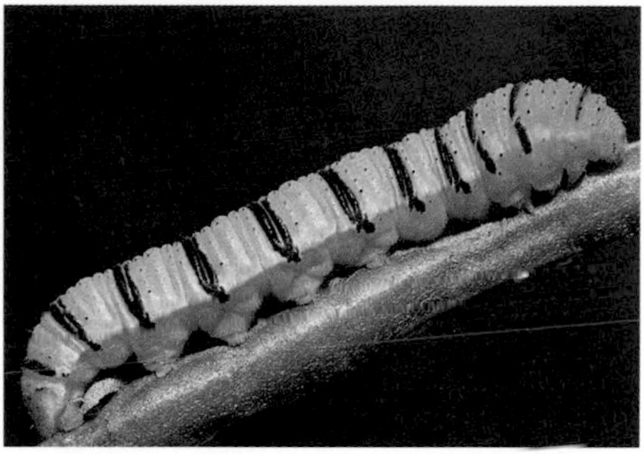

4. Cloudless Sulphur

5. Orange-barred Sulphur

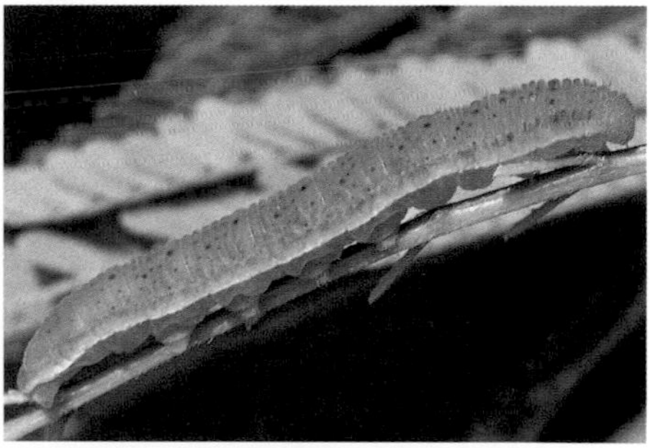

6. Large Orange Sulphur

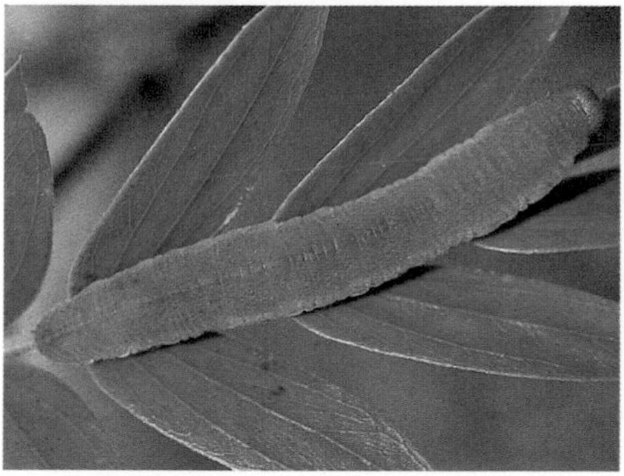

8. Lyside Sulphur

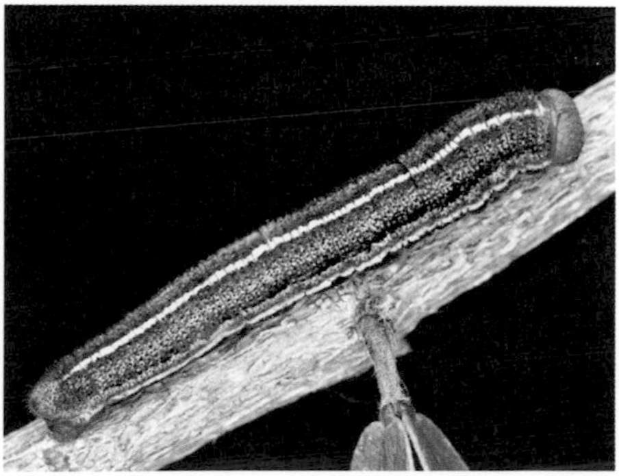

9. Lyside Sulphur

**Yellows** (*Eurema*) are green or blue-green with white, cream, or yellow lateral stripes. Only Mexican is reliably distinguishable from the others based upon appearance.

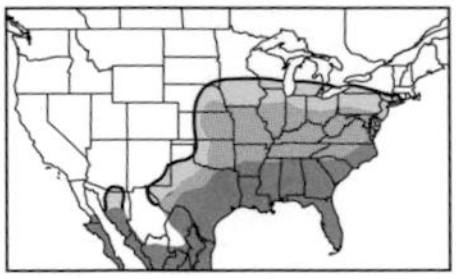

***Little Yellow** *Eurema lisa*

*Habitat* Disturbed open areas, especially in dry, sandy, grassy fields and roadsides. *Found* Most of year in the South, July–Sept northward. *Host* Sennas. *Garden Tips* Partridge pea, Bahama senna, Christmas senna.

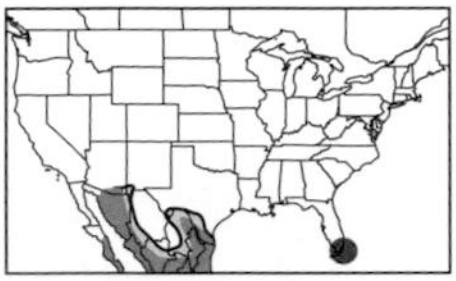

**Mimosa Yellow** *Eurema nise*

*Habitat* Subtropical brushy areas. *Found* All year in southern Texas; May–Dec in the Southwest. Rare in southern Florida. *Host* Acacias, mimosas (not known to use fern acacia).

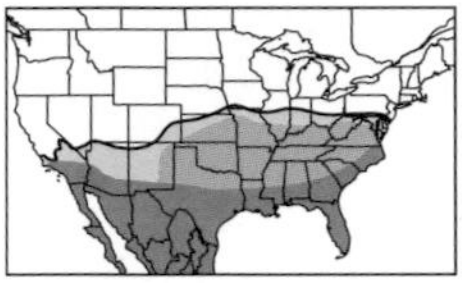

***Sleepy Orange** *Eurema nicippe*

*Habitat* A wide variety of open areas. *Found* Most of year in the South, July–Sept northward, mainly on leaves. *Host* Sennas. *Garden Tips* Sicklepod (Florida), Christmas senna, desert senna, slim-pod senna (Southwest). *Comments* Perhaps the clear drops of fluid on the ends of hairs on the body of yellows (see photo 4) serve the same purpose as those found on Cabbage Whites.

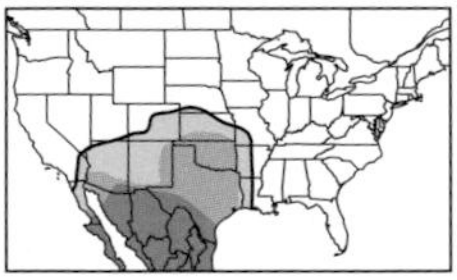

**Mexican Yellow** *Eurema mexicana*

**Double creamy or yellow dorsal stripe**, cream or yellow lateral stripe with **light crossbanding**. *Habitat* Open habitats from wooded edges to desert grasslands. *Found* Most of the year, after flights, on leaves. *Host* Fern (or prairie) acacia, locust. *Garden Tips* Fern acacia.

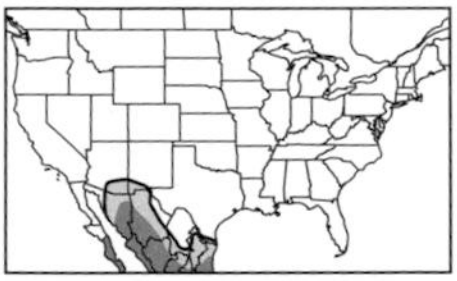

**Tailed Orange** *Eurema proterpia*

Usually brighter yellow-green than Sleepy Orange. Other yellows in range have not yet been found on this species' hostplant. *Habitat* Tropical woodlands and thornscrub. *Found* July–Oct, after flights. *Host* Plants in the genus *Chamaecrista,* a summer annual in the Southwest.

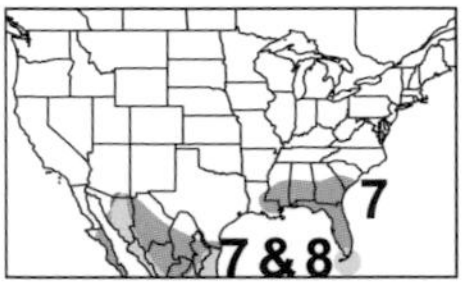


**7. Barred Yellow** *Eurema daira*

**8. DinaYellow** *Eurema dina* Not shown.

*Host* Barred Yellow on pencil flowers and joint vetches. Dina Yellow on Mexican alvaradoa and bitterbush. See page 155 for more information.

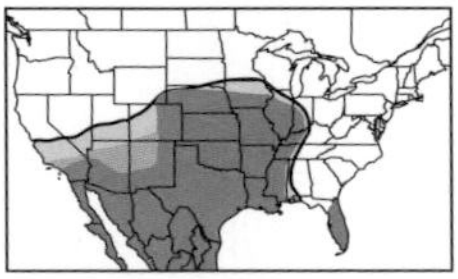

***Dainty Sulphur** *Nathalis iole*

With or without maroon stripes. **Two red bumps above the head**. Our only sulphur that feeds on members of the sunflower family. *Habitat* Wide variety of open habitats, from deserts to prairie to open woodland. *Found* Most of the year in southern areas; July–Sept northward. *Host* Dogweed, beggar ticks, greenthread, Spanish needles, and other sunflowers. *Garden Tips* Spanish needles (Florida), dogweed (Southwest).

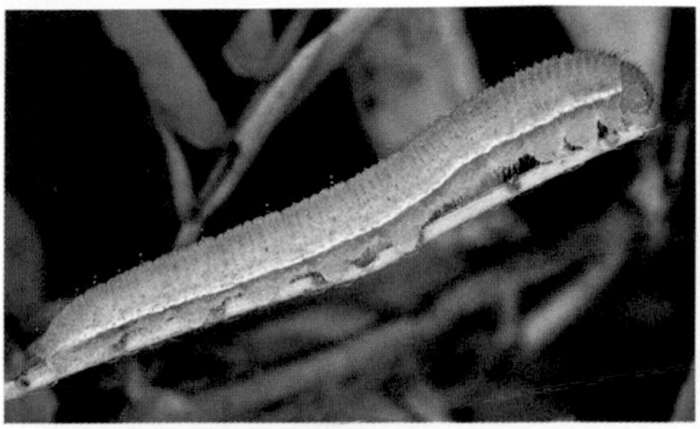

1. Little Yellow

2. Little Yellow adult

3. Mimosa Yellow

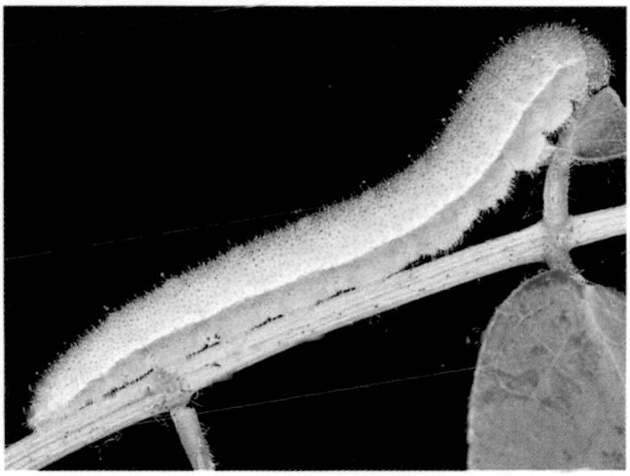

4. Sleepy Orange

5. Mexican Yellow

6. Tailed Orange

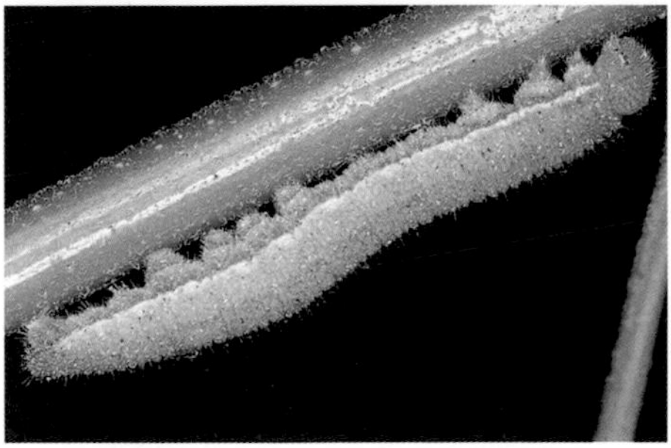

7. Barred Yellow

9. Dainty Sulphur

## Gossamer-wings (Family Lycaenidae).

See page 8 for general information about gossamer-wings.

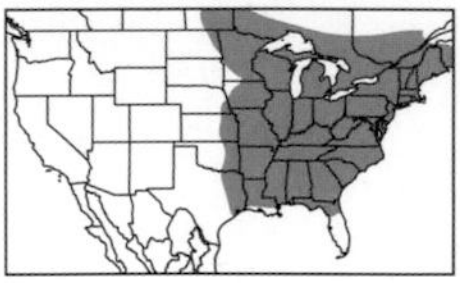

### Harvester *Feniseca tarquinius*

Distinctive, as shown, but usually covered with white powder when feeding on aphids. *Habitat* Woodlands, especially near watercourses or wet areas with alders. *Found* May–Oct. *Host* Our only carniverous caterpillar eats wooly aphids, usually on alder or beech. *Comments* Development is rapid and caterpillars are usually found under the aphids. Chrysalis resembles a monkey's head.

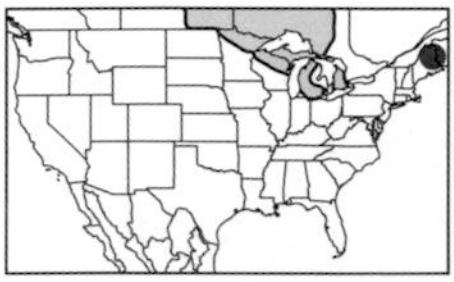

### Dorcas Copper *Lycaena dorcas*

Green with faint wavy dorsal and lateral lines. Uses different hosts than similar coppers. *Habitat* Northern bogs and marshes where shrubby cinquefoil occurs. *Found* May–June. *Host* Shrubby cinquefoil.

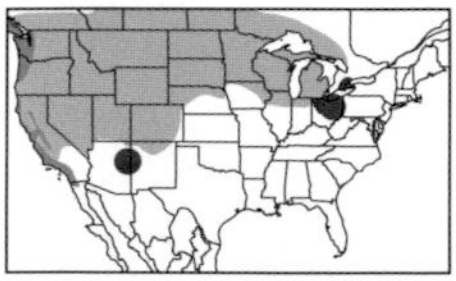

### Purplish Copper *Lycaena helloides*

Green with faint wavy lateral lines. Great and Gray Coppers usually have maroon dorsal stripe. *Habitat* Open moist situations, often disturbed. *Found* April–May, and June–Sept, 2–3 broods. *Host* Docks and knotweeds.

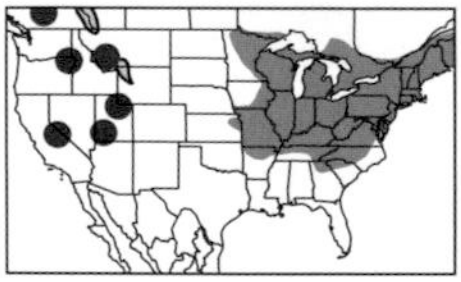

### American Copper *Lycaena phlaeas*

Green, sometimes with red or maroon lateral stripe or entire sides. In West, Lustrous Copper may not be distinguishable. *Habitat* Disturbed open areas, fields, sandy prairies, power-line cuts in the East, rocky alpine slopes in the West. *Found* May–Sept, 2–3 broods in the East, June–Aug prior to single brood in the West. *Host* Sheep sorrel and docks.

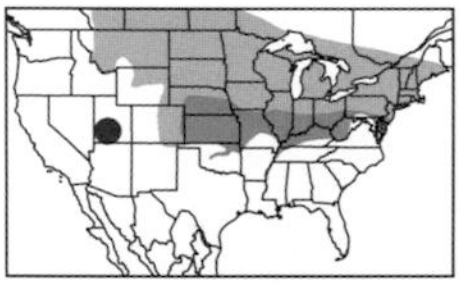

### Bronze Copper *Lycaena hyllus*

**Yellow-green**, more slender than similar coppers. *Habitat* Low wet meadows and marshes, especially in river flood plains. *Found* Late May–Sept, 2–3 broods. *Host* Water dock and curly dock.

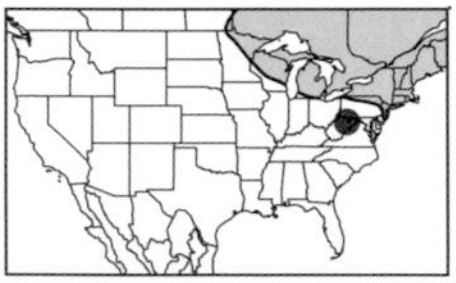

### Bog Copper *Lycaena epixanthe*

Blue-green. Hostplant is unique for coppers. *Habitat* Acid bogs with cranberries. *Found* May–June, to mid-July on occasion. *Host* Cranberries.

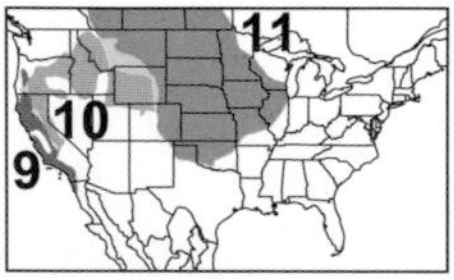


### 9. Great Copper *Lycaena xanthoides* Not shown.

### 10. Edith's Copper *Lycaena editha* Not shown.

### 11. Gray Copper *Lycaena dione* Not shown.

Green (usually) or maroon, usually with a maroon stripe. *Habitat* Varied. Mainly wet meadows and prairie for Gray; moist areas in sagebrush steppes or mountain forest for Edith's; chaparral for Great. *Found* April–July. *Host* Docks. See pg. 155 for more information. *Comments* Tended by ants.

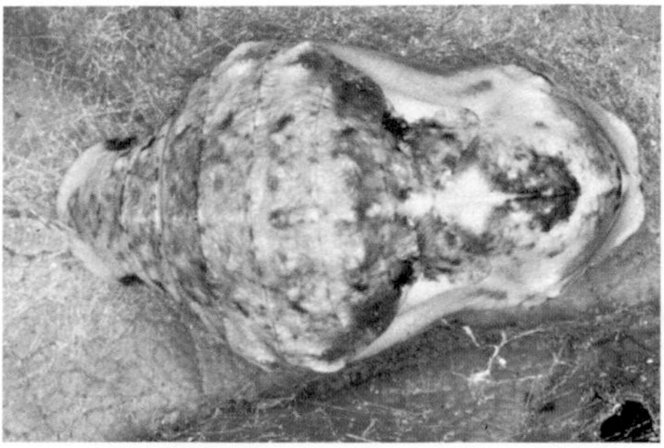

1. Harvester chrysalis

2. American Copper adult

3. Harvester

4. Dorcas Copper

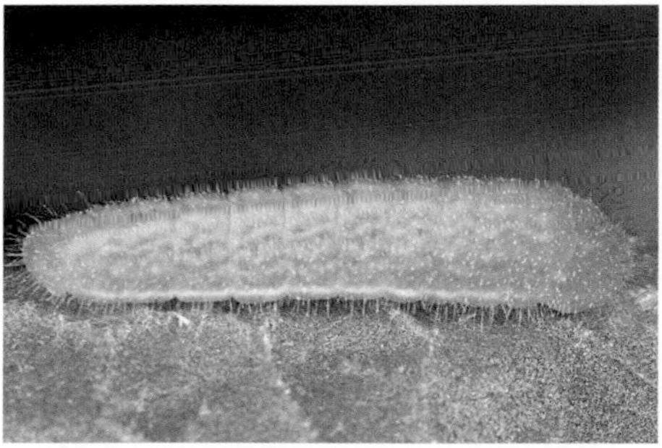

5. Purplish Copper

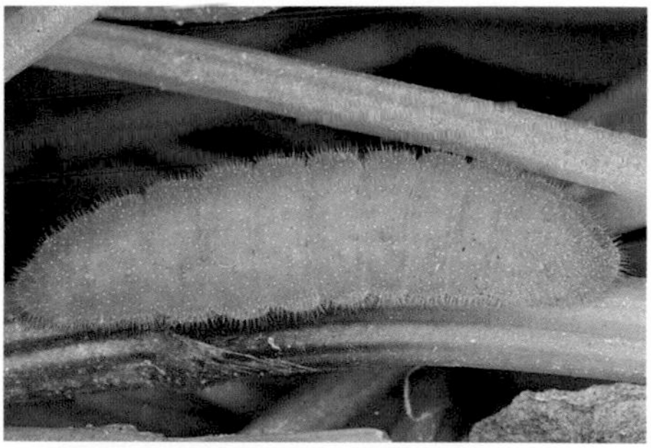

6. American Copper

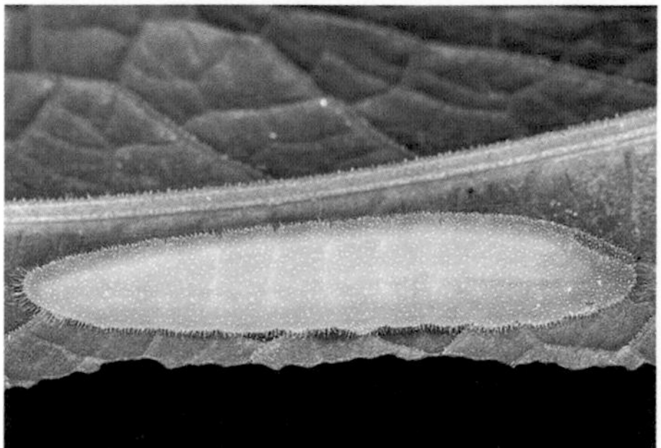

7. Bronze Copper

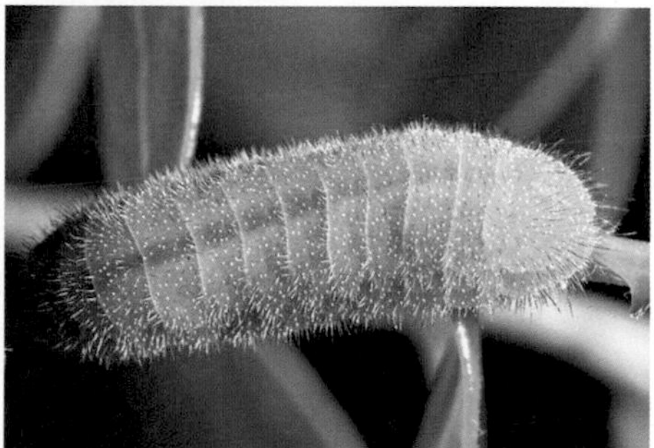

8. Bog Copper

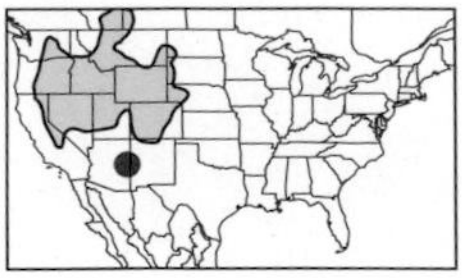

## Ruddy Copper *Lycaena rubidus*

Green, **covered with fine orange hair**. *Early Instars* Pink with red chevrons *Habitat* Moist meadows, streamsides, and other open moist areas in arid country. *Found* May–July. *Host* Docks *Comments* Tended by ants and usually rests at base of host.

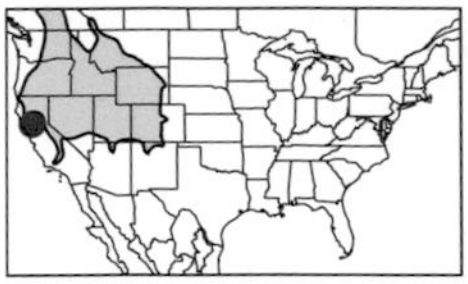

## Blue Copper *Lycaena heteronea*

Green to bluish-green, unmarked with covering of **fine white hair**. *Habitat* Sagebrush steppes, dry hillsides, barren rock outcrops, and mountain meadows. *Found* May–July. *Host* Sulphur buckwheat and other buckwheats. *Comments* Many blues and some hairstreaks share buckwheat hosts with this and the following species, but they are generally well marked dorsally, tended by ants, or feed on flowering parts.

## Gorgon Copper *Lycaena gorgon*

**Light green** or bluish-green, unmarked with **dense covering of fine white hair**. *Habitat* Dry open situations mainly in foothills, including canyons through chaparral, grassy hillsides, and rocky outcrops. *Found* March–July, depending on elevation. *Host* Nude and elongate buckwheats. *Comments* Eats top layers of leaves.

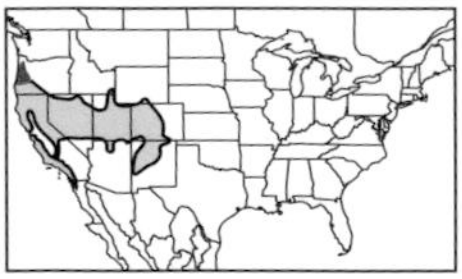

## Tailed Copper *Lycaena arota*

Green with double white dorsal stripe. No other copper is similar. Hostplant is unique for coppers. Soapberry Hairstreak is similar but uses different host. *Habitat* Frequently encountered along streamsides and other watercourses through foothill woodlands, but also in chaparral and oak openings, sagebrush steppes, and high mountain meadows. *Found* April–July depending on elevation. *Host* Gooseberries.

## Hermes Copper *Lycaena hermes*

Yellow-green with few or no markings. Host is unique for coppers. Brown Elfin mainly feeds on fruits, whereas Hermes feeds on leaves. *Habitat* Scrub and chaparral near the host only. *Found* April–June. *Host* Redberry. *Comments* Rare, local, and in need of help. An excellent candidate for an in-depth life history study.

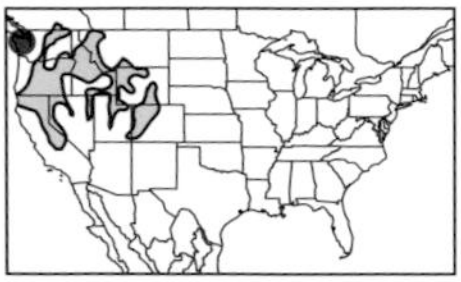

## Lilac-bordered Copper *Lycaena nivalis*

Green with maroon dorsal stripe. Great Copper occurs at lower elevations. Edith's Copper feeds on docks. *Habitat* Mountain meadows, openings in ponderosa pine and Douglas fir forest and sagebrush steppes. *Found* May–July. *Host* Knotweeds.

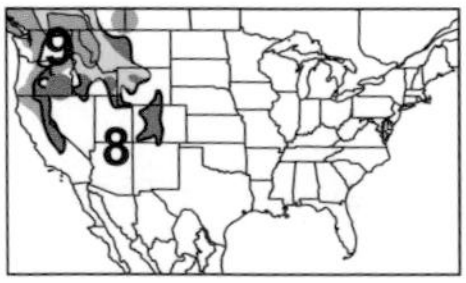


## 8. Lustrous Copper *Lycaena cupreus* Not shown.

## 9. Mariposa Copper *Lycaena mariposa* Not shown.

*Host* Docks and sorrels for Lustrous; blueberries for Mariposa. See pg. 156 for more information.

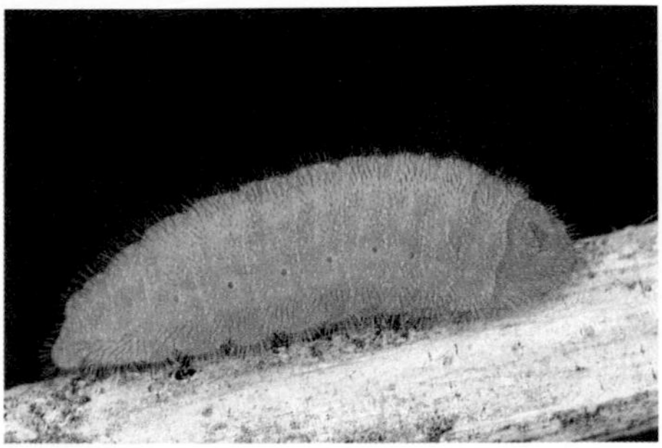

1. Ruddy Copper

2. Lilac-bordered Copper adult

3. Blue Copper

4. Gorgon Copper

5. Tailed Copper

6. Hermes Copper

7. Lilac-bordered Copper

10. American Copper chrysalis

## Hairstreaks See page 9 for general information.

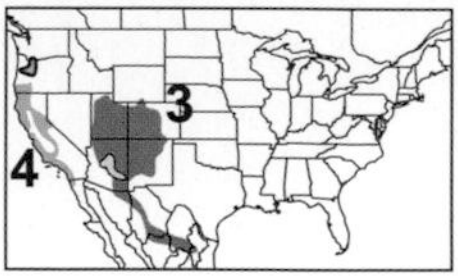


### 3. Colorado Hairstreak *Hypaurotis crysalus*
### 4. Golden Hairstreak *Habrodais grunus*

Both with long white hairs on undersides and somewhat flattened and elongated like coppers. Colorado is green with a pale sublateral line and wavy diagonal dashes on its sides. Golden is light bluish-green. *Habitat* Canyons and oak- covered hillsides. *Found* May–July, prior to flight in most of range. *Host* Colorado on Gambel oak, probably also silver leaf oak in Arizona; Golden on canyon oak, chinquapin (OR) and others. See pg. 156.

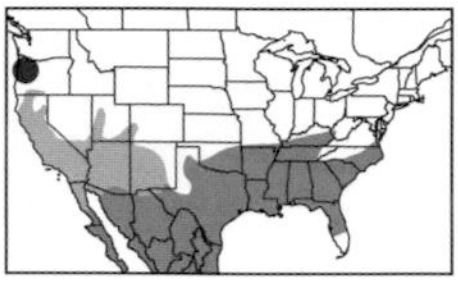

### Great Purple Hairstreak *Atlides halesus*

Green with distinctive diamond mark above head. Our only hairstreak on lowland mistletoes. *Habitat* Edges of moist woodlands, foothill canyons, streambanks, and desert areas with mistletoe-covered trees. *Found* March–Nov, in the South; July–Aug at more northern limits, on leaves. *Host* Mistletoes on ash, cottonwood, juniper, mesquite, and other hardwoods. *Comments* Chrysalids can be located under leaf litter or in bark crevices at the base of mistletoe-infested trees.

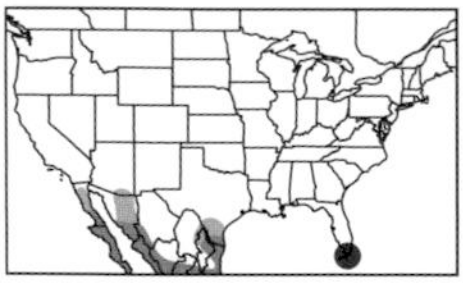

### *Silver-banded Hairstreak *Chlorostrymon simaethis*

Yellow-green to brownish with wavy dorsal band and two dark dorsal lines edged with reddish ovals. *Habitat* Thorn scrub and dry tropical woodland. *Found* March–Nov, inside balloons, eating unripened seeds. *Host* Balloon-vines. *Garden Tips* Balloon-vine. *Comments* Even caterpillars that live hidden lives inside the pods of balloon-vine are subject to danger. Note black spots toward rear end of the individual in photo 6. These spots are wounds left by a parasitic wasp—a dead caterpillar crawling!

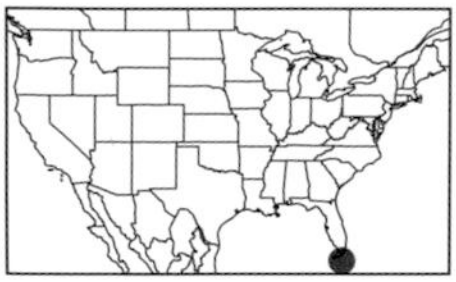

### Amethyst Hairstreak *Chlorostrymon maesites*

Not shown.

Reported to be green with dorsal chevrons and a few red lateral spots near the head. *Habitat* Tropical hammocks. *Found* Probably after flights during most of the year. *Host* Woman's-tongue. *Comments* Reported to eat flowers and buds, not leaves.

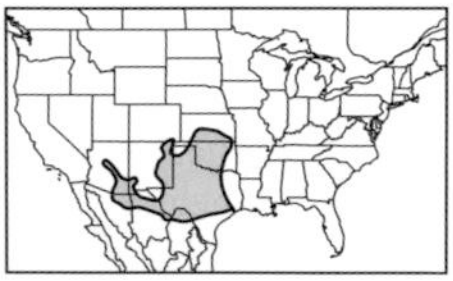

### Soapberry Hairstreak *Phaeostrymon alcestis*

Green with double white dorsal band. Not variable. Tailed Copper feeds on gooseberries. *Habitat* Sparse woods with host, including canyon watercourses and hedgerows. *Found* April–May, prior to single flight, on leaves. *Host* Western soapberry. *Comments* Prefers to eat new growth.

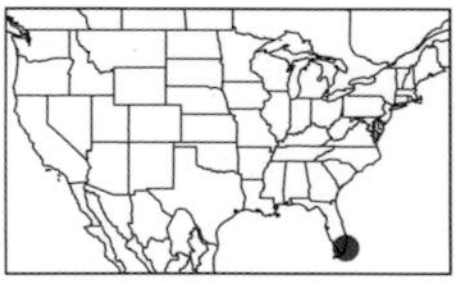

### *Atala *Eumaeus atala*

Distinctive. *Habitat* Anywhere in south Florida that its host occurs, including urban and suburban plantings. *Found* All year. *Host* Coontie and other ornamental cycads. *Garden Tips* Coontie. *Comments* Often feeds in groups. Chrysalids may hang suspended in rows under the stems of the host.

1. Atala chrysalids

2. Great Purple Hairstreak adult

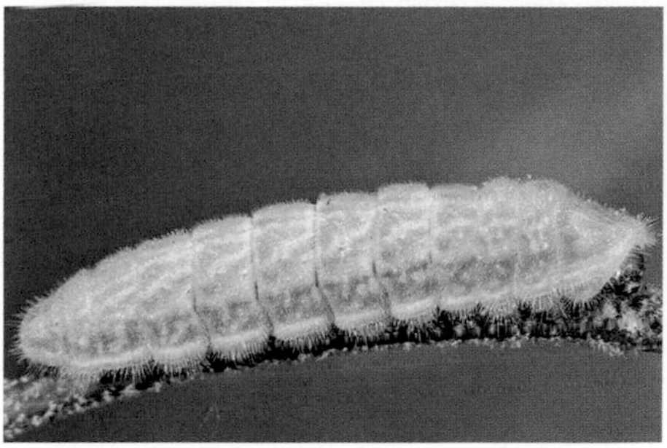

3. Colorado Hairstreak

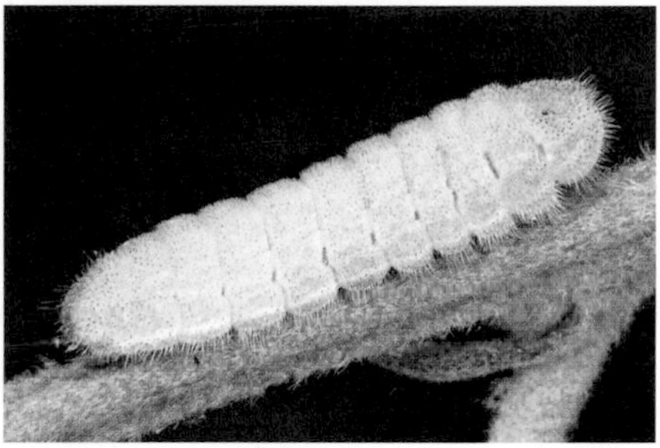

4. Golden Hairstreak

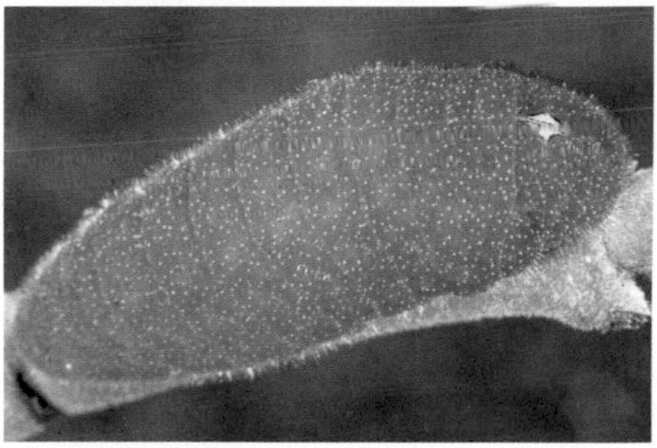

5. Great Purple Hairstreak

6. Silver-banded Hairstreak

7. Soapberry Hairstreak

8. Atala

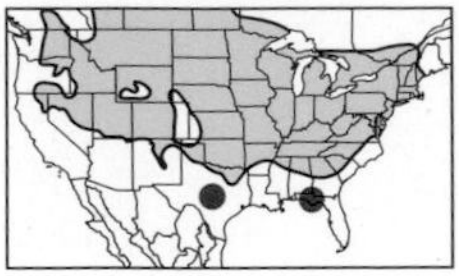

## Coral Hairstreak *Satyrium titus*

Pale green to yellow-green with **1–3 bright pink patches** on the back and at each end. *Habitat* Brushy fields, overgrown orchards, moist roadsides, etc. *Found* April–June, depending on region, on leaves prior to single flight. *Host* Wild cherries and wild plum. *Comments* Feeds at night, rests during the day at the base of small host trees.

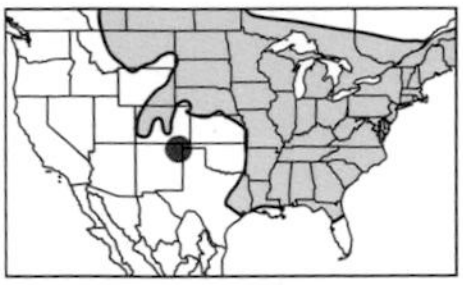

## Striped Hairstreak *Satyrium liparops*

Green, oblique lateral dashes may be faint or inconspicuous. Brown Elfin very similar. Spring Azure eats flowers and is usually tended by ants. *Habitat* Thickets, woodland openings and brushy edges. *Found* April–May, prior to flight, on leaves. *Host* Wild cherries, blueberries, and others.

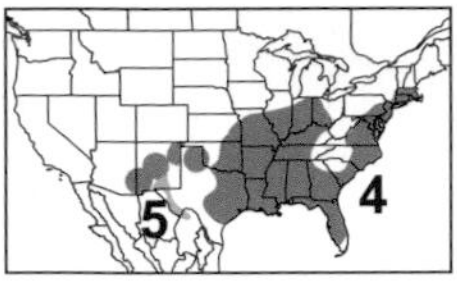


## 4. Oak Hairstreak *Satyrium favonius*

## 5. Poling's Hairstreak *Satyrium polingi* Not shown.

Oak Hairstreak is variable but **usually green with short reddish hairs**. White stripes may be present on back and sides. Poling's caterpillar is unknown. *Habitat* A variety of woodland edges including pine-oak woodlands and the edges of mixed deciduous woodlands in rich soils. *Found* 4. Springtime, as hostplant buds begin to open. 5. Unknown. *Host* Oaks.

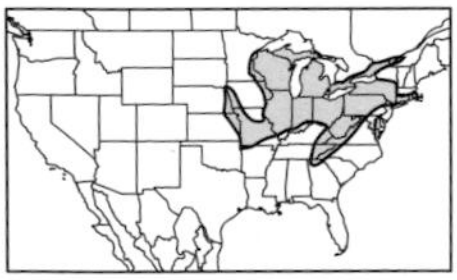

## Hickory Hairstreak *Satyrium caryaevorum*

Note the **dark, pale-edged, continuous, dorsal stripe** and a series of dark-green, pale-edged diagonal dashes. Banded Hairstreak, common and also found on hickories, has dark dorsal markings only at front and rear. *Habitat* Deciduous woodland edges and glades, often in richer soils than Banded Hairstreak. *Found* April–early June, prior to single flight. *Host* Hickories.

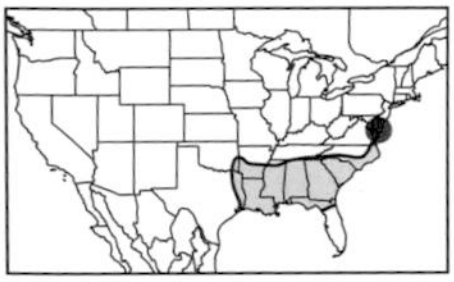

## Kings' Hairstreak *Satyrium kingi*

Green with lighter oblique white stripes. On common sweetleaf only. *Habitat* Moist woodlands where host occurs. *Found* March–June, varies with region, earlier on coastal plains, later in Appalachian Mountains. *Host* **Sweetleaf**.

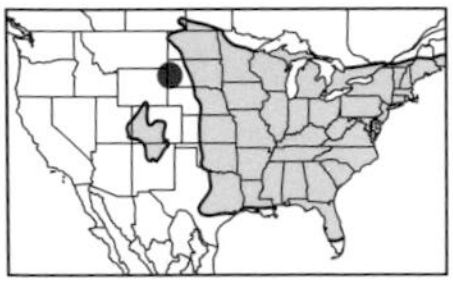

## Banded Hairstreak *Satyrium calanus*

Variable. Usually with **darker blotches in front and at rear**. Can be whitish or green or brown with or without yellow stripes and oblique dashes. *Habitat* Oak woodland edges, openings, and glades. *Found* March–May, prior to the single flight. *Host* Oaks and hickories.

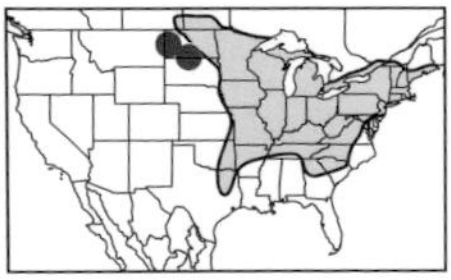

## Edwards' Hairstreak *Satyrium edwardsii*

Brown to reddish-brown with a series of light dashes along a lateral line. *Habitat* Woodlands with scrubby oaks and adjacent clearings. These are usually poor soil areas: pine barrens, rocky hilltops, prairie ridges, shale barrens, etc. *Found* April–May. *Host* Small oaks, especially scrub oak. *Comments* Feeds on leaves at night, tended by ants, spends day at base of tree.

1. Coral Hairstreak

2. Banded Hairstreak adult

3. Striped Hairstreak

4. Oak Hairstreak

6. Hickory Hairstreak

7. King's Hairstreak

8. Banded Hairstreak

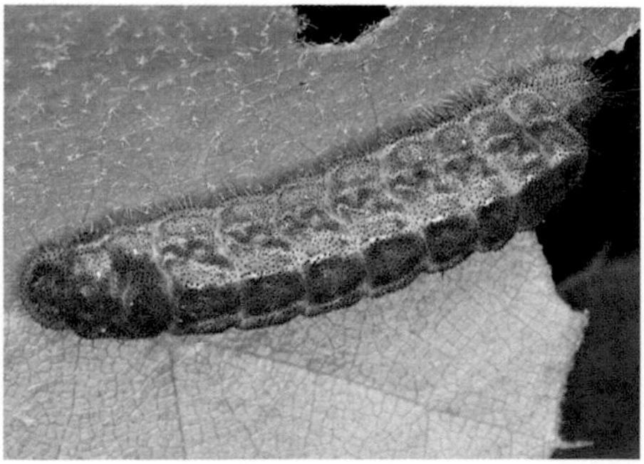

9. Edwards' Hairstreak

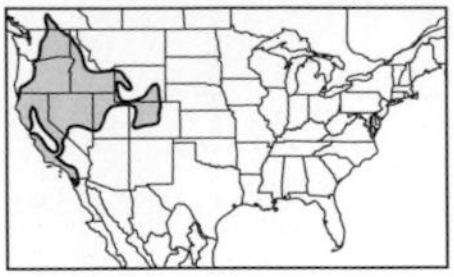

## **California Hairstreak** *Satyrium californica*

Tan or pink, with **gray-brown dorsal spots**. *Habitat* Chaparral, forest openings, foothill and lower mountain canyons, and sagebrush steppes. *Found* April–July, on leaves. *Host* Oaks, birch-leaf mountain mahogany. *Comments* Usually feeds at night. Sometimes tended by ants.

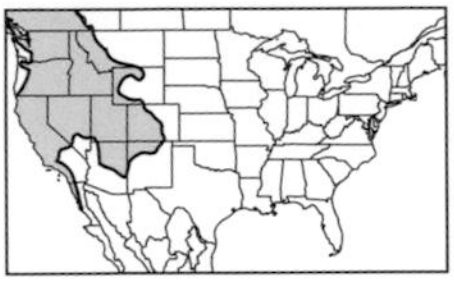

## **Sylvan Hairstreak** *Satyrium sylvinus*

Green with a pair of white dorsal stripes; flattened dorsally. Other similar caterpillars in range use different hosts. *Habitat* Willow-lined watercourses and springs. *Found* April–July, on leaves. *Host* Willows. *Comments* Eggs hatch with appearance of willow buds.

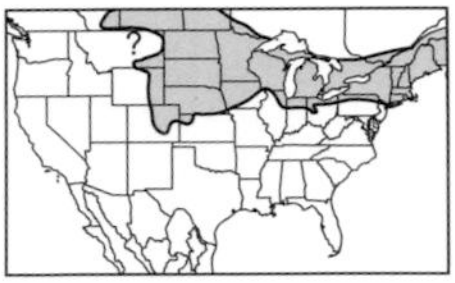


## **Acadian Hairstreak** *Satyrium acadica*

Green with pair of white dorsal stripes; flattened dorsally. Very similar to Sylvan Hairstreak. *Habitat* Open areas and thickets near streams and marshy places where willows grow. *Found* April–May, prior to single flight. *Host* Willows.

## **Mountain Mahogany Hairstreak** *Satyrium tetra*

Whitish-green with darker green dorsal stripes and **short orange hair** especially on thorax and behind head. *Habitat* Chaparral in open and slightly wooded areas. *Found* April–June, on leaves. *Host* Birch-leaf mountain mahogany.

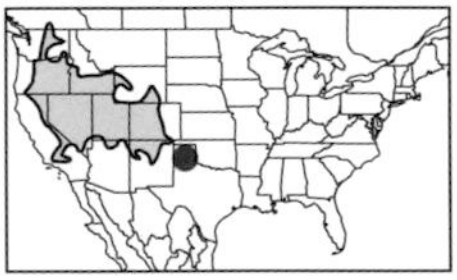

## **Behr's Hairstreak** *Satyrium behrii*

Numerous **white and yellow (or brown) diagonal lines**, little variation. Well-marked forms of Desert Elfin can be similar but use different host. *Habitat* Rocky arid areas along foothill ridges and low mountaintops, brushy openings in coniferous forests, and sagebrush steppes. *Found* March–July, on leaves and unopened flowers. *Host* Mountain mahogany in the Rockies, bitterbrush elsewhere.

## **7. Ilavia Hairstreak** *Satyrium ilavia*

## **8. Gold-hunter's Hairstreak** *Satyrium auretorum*

Not shown.

Green with cream or yellow lateral stripe, robust, but flattened dorsally. *Habitat* Oak-covered hills. *Found* May to early June, Ilavia on catkins. *Host* Ilavia on desert scrub oak; Gold-hunter's on California scrub oak and others.

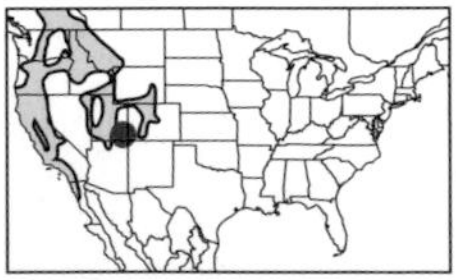

## **Hedgerow Hairstreak** *Satyrium saepium*

Green with pair of pale-yellow dorsal stripes, covered with **"frosting" of short white hairs**. Little variability. Brown Elfin lacks "frosting." *Habitat* Chaparral, open pine forest, oak woodlands, coastal sand dunes, etc. *Found* March–July, mainly on leaves but also buds. *Host Ceanothus*.

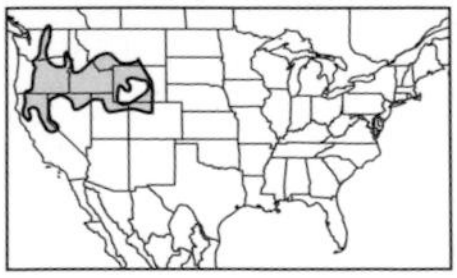

## **Sooty Hairstreak** *Satyrium fulginosum* Not shown.

Resembles blues. *Host* Lupines. See pg. 157.

1. California Hairstreak

2. Behr's Hairstreak adult

3. Sylvan Hairstreak

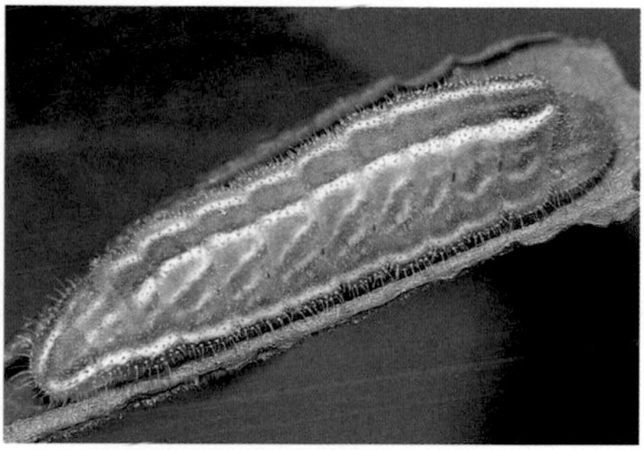

4. Acadian Hairstreak

5. Mountain Mahogany Hairstreak

6. Behr's Hairstreak

7. Ilavia Hairstreak

9. Hedgerow Hairstreak

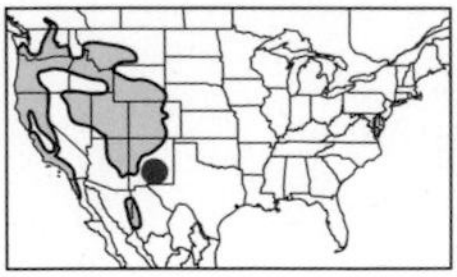

## Bramble Hairstreak *Callophrys dumetorum*

Variable. Usually green or red, sometimes with pair of white dorsal stripes and other faint cross markings. *Habitat* Chaparral, sand hills, rocky hills, sagebrush steppes, openings in foothill woods. *Found* April–July, after single flight, mostly on leaves but also on flowers. *Host* Buckwheats, deerweeds, and, in the Rockies, *Ceanothus*.

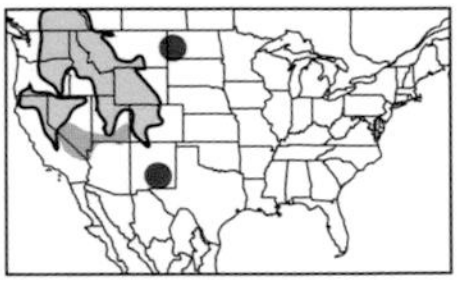

## Sheridan's Hairstreak *Callophrys sheridanii*

Green to pinkish-green or bluish-green with reddish stripes and dashes or chevrons. *Habitat* Woodlands, sagebrush and chaparral in Rockies; also rocky outcrops, open woodlands, scrub, and desert canyons. *Found* April–Aug after flights (up to 3 broods), primarily on leaves. *Host* Buckwheats

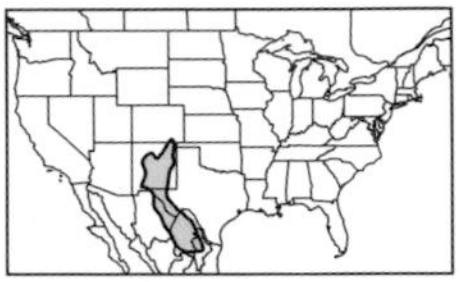

## Sandia Hairstreak *Callophrys mcfarlandi*

Pale yellow-green with faint white chevrons. *Habitat* Arid hillsides covered with Texas beargrass. *Found* April–June, after flights, on flowering stalks. *Host* Texas beargrass. *Comments* Tended by ants. Gray Hairstreak may also be found on beargrass blooms.

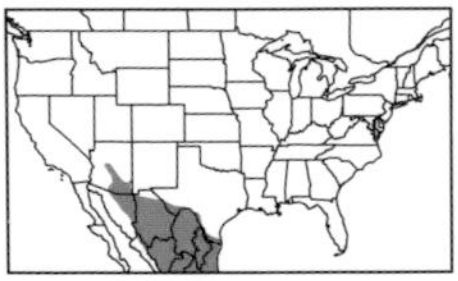

## Xami Hairstreak *Callophrys xami* Not shown.

Variable. Green, yellow-green, greenish-pink, or pale red, usually unmarked or with hints of oblique dorsal markings. No similar caterpillars occur on succulents within its range. *Habitat* Steep canyon cliffs with the host plant (Southwest) or scrubby sandy hills (southern Texas). *Found* April–Nov, after flights. *Host* Stonecrops and other related succulents.

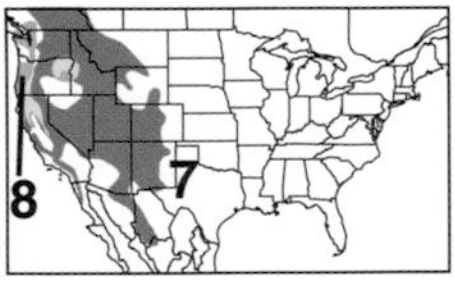


## 7. Thicket Hairstreak *Callophrys spinetorum*

## 8. Johnson's Hairstreak *Callophrys johnsoni*

Not shown.

Green or olive with red, green, and yellow markings, the back with **strongly raised chevrons creating a sawtooth appearance**. *Habitat* Openings in foothill and mountain coniferous forests. *Found* April–Oct, after flights, on leaves. *Host* **Dwarf mistletoes on conifers**.

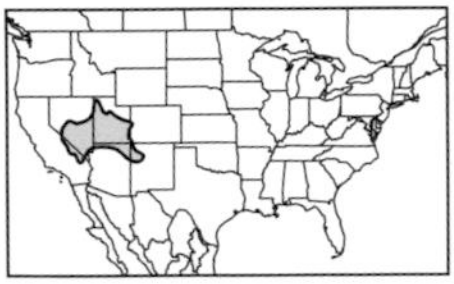

## Desert Elfin *Callophrys fotis*

Variable from unmarked green to green with white and yellow dorsal chevrons. Well-marked forms are similar to Behr's Hairstreak; but hosts are different. *Habitat* Canyons and rocky areas, especially in pinyon/sageland. *Found* April–June, after single flight, mainly on buds and flowers. *Host* **Cliff rose**.

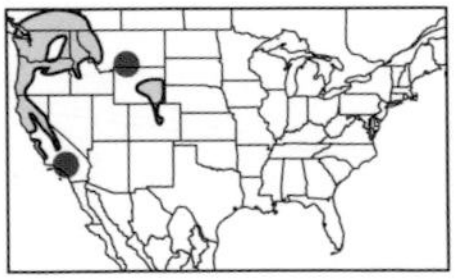

## Moss' Elfin *Callophrys mossii*

Variable. Can be orange, yellow, pink, or red. Most populations are red with white chevrons. *Habitat* Steep, rocky situations in coastal foothills and in the mountains near the hostplant. *Found* April–June, 1–4 weeks after single flight, longer for populations subject to coastal fog, on flowers or in bract centers of stonecrops. *Host* **Stonecrops**.

1. Bramble Hairstreak

2. Desert Elfin adult

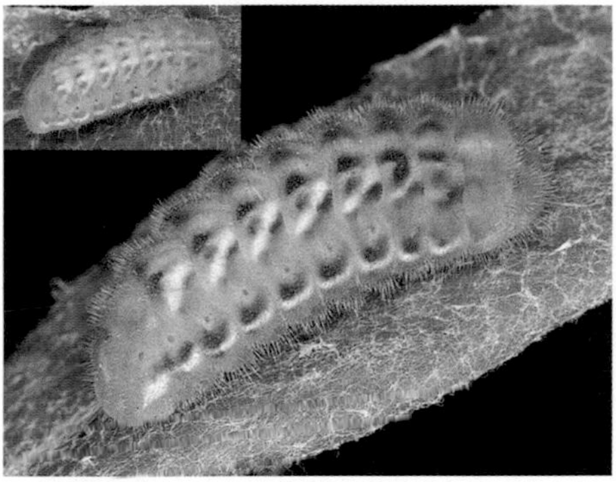
3. 'Coastal' Bramble Hairstreak

4. Sheridan's Hairstreak

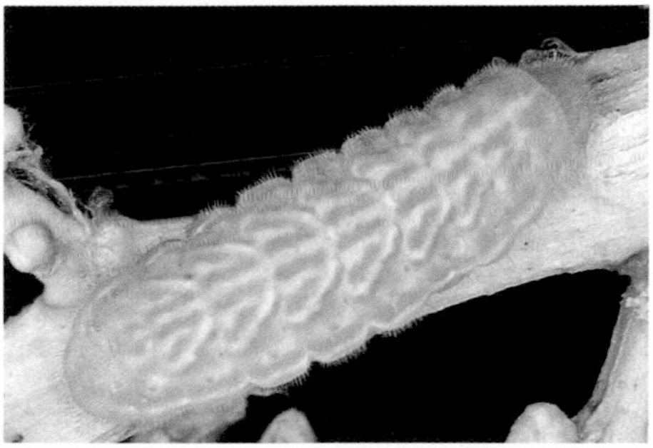
5. Sandia Hairstreak

7. Thicket Hairstreak

9. Desert Elfin

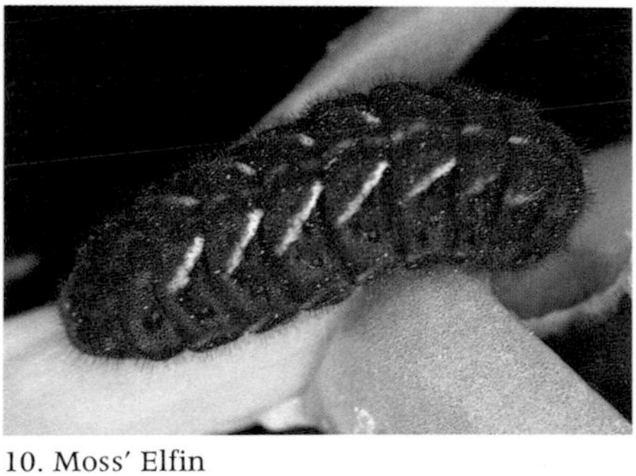
10. Moss' Elfin

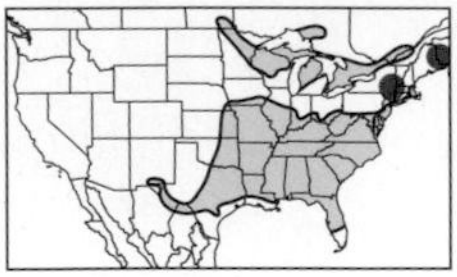

## Henry's Elfin *Callophrys henrici*

Variable. Pale to dark green or red; with pale, yellow-green or red-green oblique-ridged dashes dorsally, creating a **sawtooth appearance on the back,** and a yellow or reddish lateral stripe. *Habitat* A wide variety of woodlands with brushy understories; brushy barrens and bog edges. *Found* April–June, after single flight, mainly on flower buds. *Host* Redbud, American holly, European buckthorn, blueberries, Mexican buckeye and others. *Comments* Caterpillars develop rapidly.

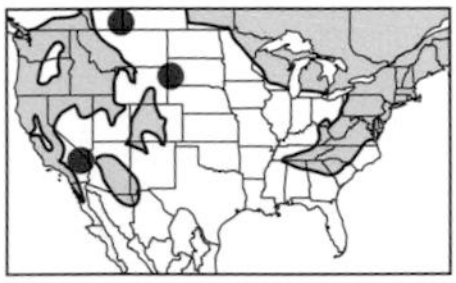

## Brown Elfin *Callophrys augustinus*

Yellowish-green, often unmarked but may sometimes have faint pale yellow and red chevrons dorsally. *Habitat* Extremely varied but generally found in acidic poor-soil woodlands, pine barrens, heath barrens, acid bogs, and extensive rocky outcroppings with blueberries (East), chaparral, brushy forest edges, and arid mixed woodlands (West). *Found* March–July, depending on region, on fruits, flowers, rarely leaves. *Host* Various heaths including blueberries, manzanitas; also buckwheats, dodder, *Ceanothus*, etc. *Comments* Propensity of this species (and Gray Hairstreak) to use a wide variety of plants used by other gossamer-wings simply means that you may not have the caterpillar you thought you had!

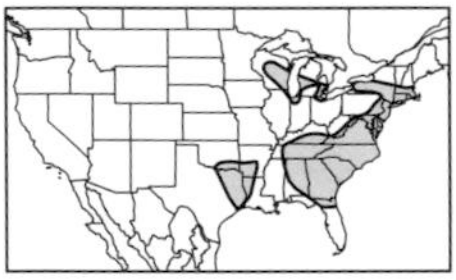

## Frosted Elfin *Callophrys irus*

Pale green with white lateral stripe and covered with short white hairs, giving it a whitish appearance. *Habitat* Sandy or rocky acidic areas cleared by fire or, much more often, by man; such as power-line cuts, railroad right-of-ways, and roadsides with good stands of its hostplant. *Found* May–June, after the single flight. *Host* Usually wild indigo, but also lupine. *Comments* Found in very small, local colonies.

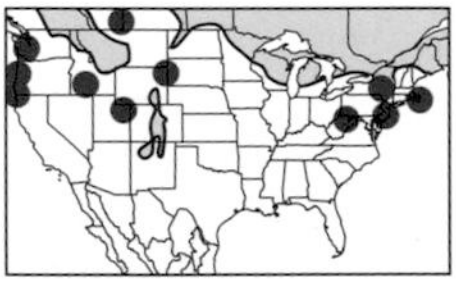

## Hoary Elfin *Callophrys polios*

Bright apple-green with pale lateral stripe. *Habitat* Dwarf pine barrens, ridges, and other barrens with an abundance of its hostplant. *Found* June–July, after single flight. *Host* **Bearberry** and trailing arbutus.

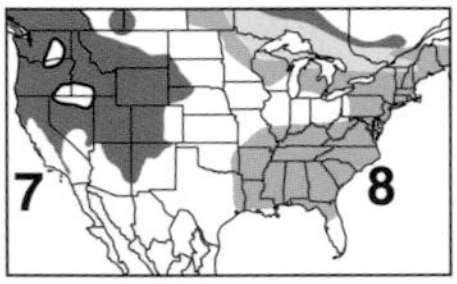


## 7. Western Pine Elfin *Callophrys eryphon*

## 8. Eastern Pine Elfin *Callophrys niphon* Not shown.

Green with broad white dorsal and lateral stripes. *Habitat* Pine forests, pine barrens, and deciduous woodlands mixed with pines. *Found* Western, June–July; Eastern, late April–June, after single flight. *Host* Pines. Western also reported on firs.

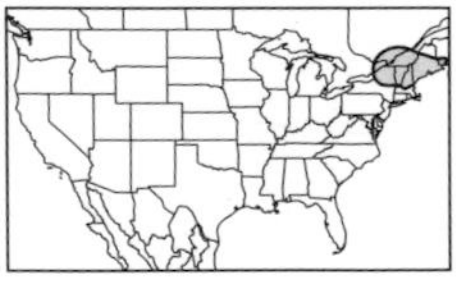

## Bog Elfin *Callophrys lanoraieensis*

Green with broad white dorsal and lateral stripes. Uses different host than the Western and Eastern Pine Elfins. *Habitat* Black spruce bogs. *Found* June–mid-August, after single flight. *Host* Black spruce.

1. Henry's Elfin

2. Western Pine Elfin adult

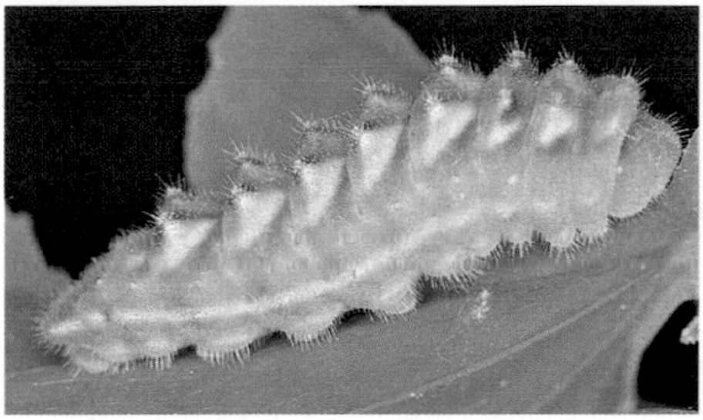

3. Henry's Elfin

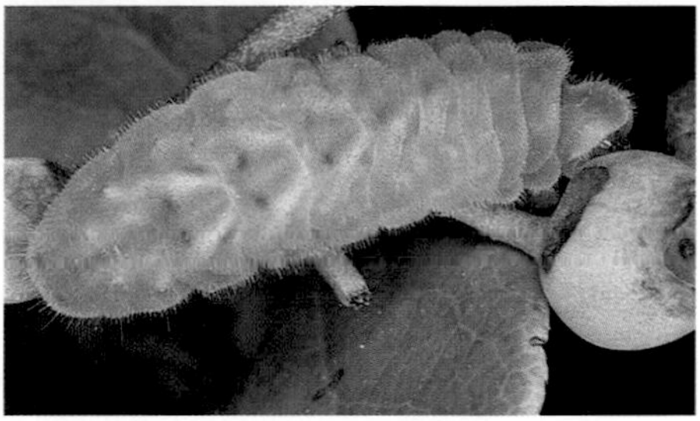

4. Brown Elfin

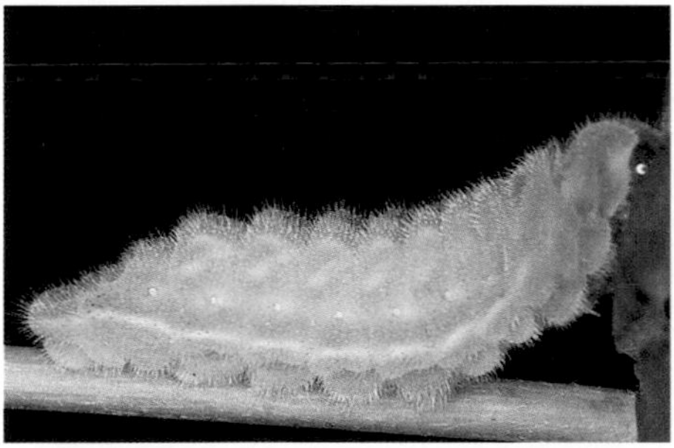

5. Frosted Elfin

6. Hoary Elfin

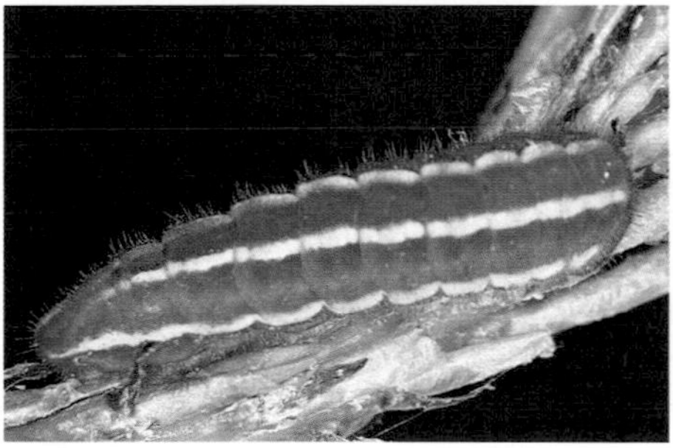

7. Western Pine Elfin

9. Bog Elfin

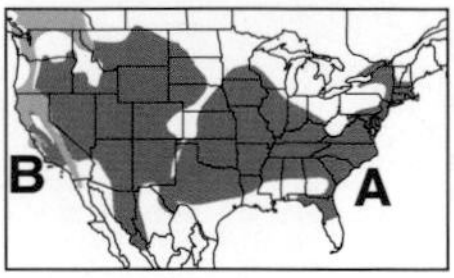


## Juniper Hairstreak *Callophrys gryneus*

Bright green or blue-green with white oblique bars on each segment and a broken white lateral line. *Habitat* Among stands of red cedar (East), junipers, cypresses, and cedars (West). *Found* April–Oct, after flights and, depending on region, on leaves. *Host* Red cedar in the East; junipers, cypresses, also western red, incense, and other cedars in the West. *Comments* Some treat the populations that feed on *Juniperus* (A on map) as different species from those that feed on non-*Juniperus* trees (B on map). The caterpillars are indistinguishable and rarely do two entities occur together, in which case each one has its own host. As you walk through and brush against junipers growing in thick stands you may find hitchhiking Juniper Hairstreak caterpillars.

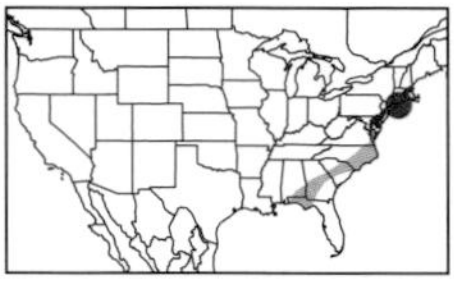

## Hessel's Hairstreak *Callophrys hesseli*

Bright green or blue-green with white oblique bars on each segment and a broken white lateral line. *Habitat* Atlantic white cedar swamps. *Found* May–June, and late July–Aug, after flights. *Host* **Atlantic white cedar.** *Comments* Second flight usually not as strong northward and caterpillars are few in number.

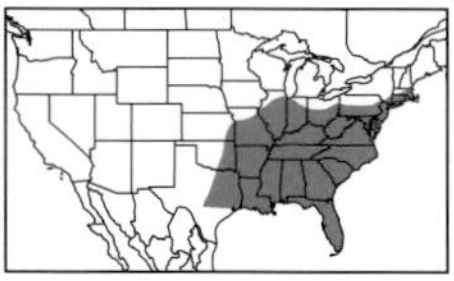

## White M Hairstreak *Parrhasius m-album*

Olive-green to yellow-green or red with dark, **rust-brown dorsal spots on front and rear**. Dark oblique dashes may also be present. *Habitat* Open, brushy areas adjacent to, or within, oak woodlands, especially on hilltops. *Found* April–Sept and Oct, after flights. *Host* Oaks.

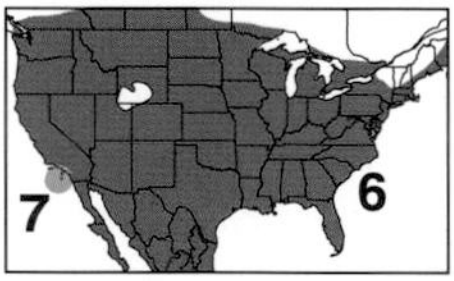


## 6. Gray Hairstreak *Strymon melinus*

## 7. Avalon Scrub-Hairstreak *Strymon avalona*

Not shown.

Extremely variable. Able to adapt to color of substrate upon which it is feeding. May or may not have chevron or oblique markings dorsally. Feeds primarily on flowering parts, which increases the chance of finding this species among those rarities with similar diets you might be otherwise seeking. *Habitat* Most common in disturbed open areas, but can be encountered in almost any habitat. *Found* March–Oct, usually on flowers. *Host* Gray on a wide variety of plants in many families, especially legumes and mallows. Avalon on silver-leaved lotus, deerweed, and giant buckwheat. *Comments* If you spend much time hunting for caterpillars, then you will probably discover a new Gray Hairstreak host.

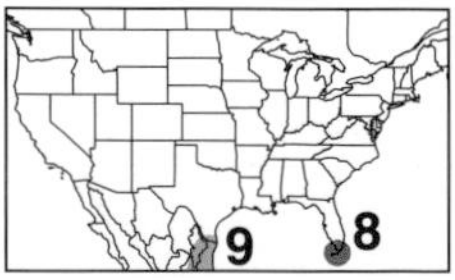


## *8. Martial Scrub-Hairstreak *Strymon martialis*

Waxy green with a faint lateral line. *Habitat* Coastal areas with hostplants. *Found* All year, on leaves; early instars, on flowers or fruits. *Host* **Bay cedar** and Florida trema.

## 9. Red-crescent Scrub-Hairstreak *Strymon rufofusca*

Not shown.

Caterpillar unknown. *Habitat* Thorn scrub. *Found* Most of year. *Host* Mallows.

1. Juniper Hairstreak

2. Juniper Hairstreak adult

3. Juniper Hairstreak

4. Hessel's Hairstreak

5. White M Hairstreak

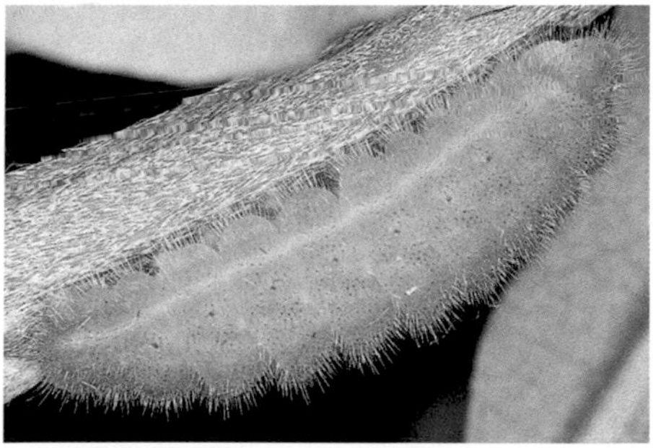

6. Gray Hairstreak

7. Gray Hairstreak

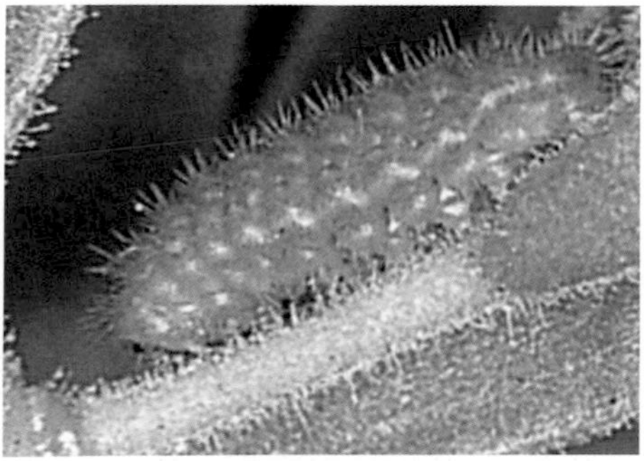

8. Martial Scrub-Hairstreak

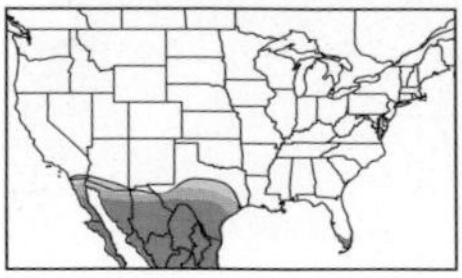

***Mallow Scrub-Hairstreak** *Strymon istapa*

Pale green or reddish with a dark dorsal line and pale lateral ridge. Probably not separable from Gray Hairstreak. Martial Scrub-Hairstreak is waxy green. *Habitat* Disturbed weedy fields and roadsides. *Found* Most of year in southernmost regions. *Host* Mallows, also bay cedar, pink purslane, and others.

**4. Lacey's Scrub-Hairstreak** *Strymon alea*

**5. Bartram's Scrub-Hairstreak** *Strymon acis*

Not shown.

See pg. 157 for more information. Lacey's is white with **green circles dorsally** and extra green marks laterally on the thorax and toward the rear. *Host* Lacey's on **southwestern bernardia**; **Bartram's on pineland croton**.

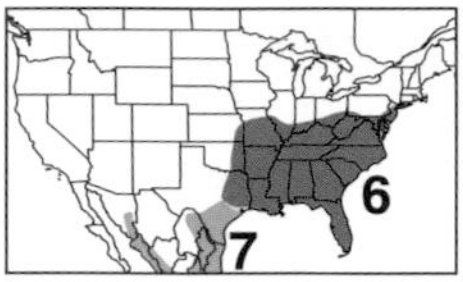


**6. Red-banded Hairstreak** *Calycopis cecrops*

**7. Dusky-blue Groundstreak** *Calycopis isobeon*

Not shown.

Gray to brown, covered with fine, short hairs. *Habitat* A wide variety of woodland openings and edges; In late summer found in an even broader range of habitats. *Found* March–Nov, after flights. *Host* Prefers rotting leaves on the ground, but also uses living leaves of sumacs, wax myrtle, and others. *Garden Tips* Wax myrtle for Red-banded.

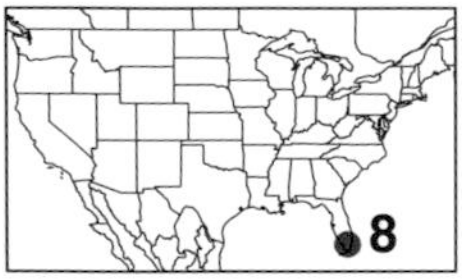


**8. Fulvous Hairstreak** *Electrostrymon angelia*

Not shown.

See pg. 157 for more information. *Host* Brazilian pepper and Jamaican dogwood.

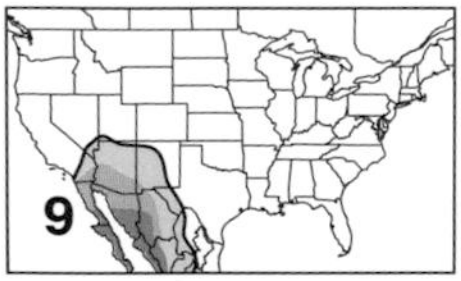


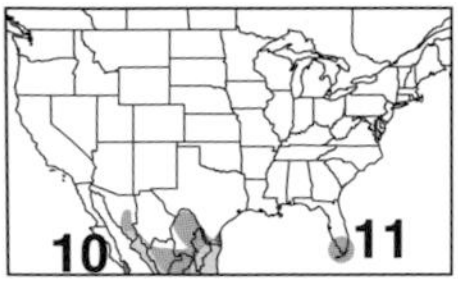


**9. Leda Ministreak** *Ministrymon leda*

**10. Clytie Ministreak** *Ministrymon clytie* Not shown.

**11. Gray Ministreak** *Ministrymon azia* Not shown.

See pg. 157 for more information. Leda Ministreak olive green, with white dorsal markings, the back deeply lobed or saw-toothed. Other species using same host lack the saw-toothed profile. *Habitat* Leda in arid region washes and canyons with mesquite. Clytie in thorn scrub. Gray in open scrub and disturbed area. *Host* Leda on mesquites, cat-claw acacia; Clytie on creeping mesquite; Gray on lead tree.

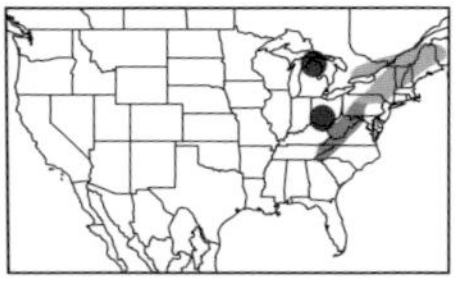

**Early Hairstreak** *Erora laeta*

Green to rust-brown with **dark-reddish blotches**. Very grainy appearance. *Habitat* Beech woodlands and edges of deciduous forest with beech. *Found* May–June, and Aug–Sept, after flights. *Host* Fruits of American beech and possibly galls, hazelnut, and yellow birch.

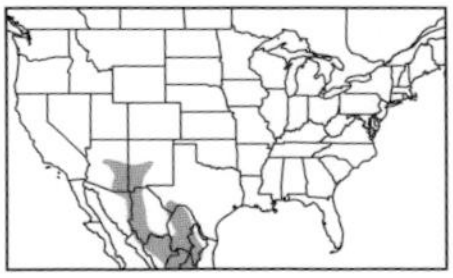

**Arizona Hairstreak** *Erora quaderna*

Yellow-green, mostly unmarked, with covering of short, brown hair, flattened dorsally. Very grainy appearance. Ilavia Hairstreak uses desert scrub oak in different habitat. *Habitat* Mid elevation oak-wooded canyons. *Found* April–Nov, after flights. *Host* Oaks, *Ceanothus*, perhaps manzanita.

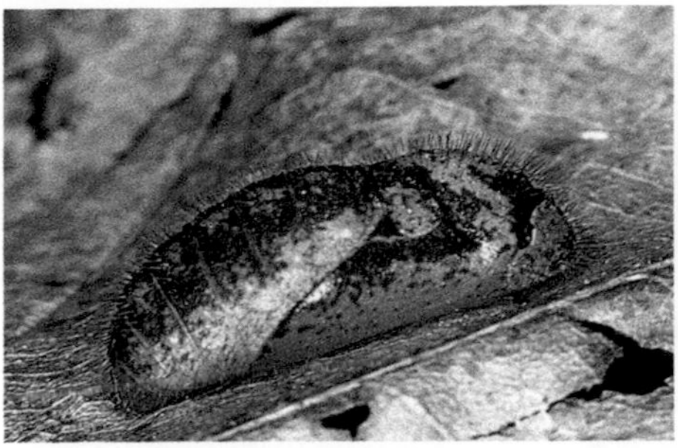

1. Red-banded Hairstreak chrysalis

2. Red-banded Hairstreak adult

3. Mallow Scrub-Hairstreak

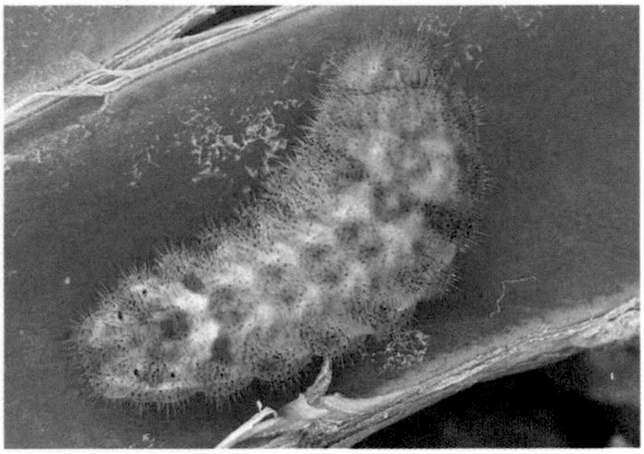

4. Lacey's Scrub-Hairstreak

6. Red-banded Hairstreak

9. Leda Ministreak

12. Arizona Hairstreak

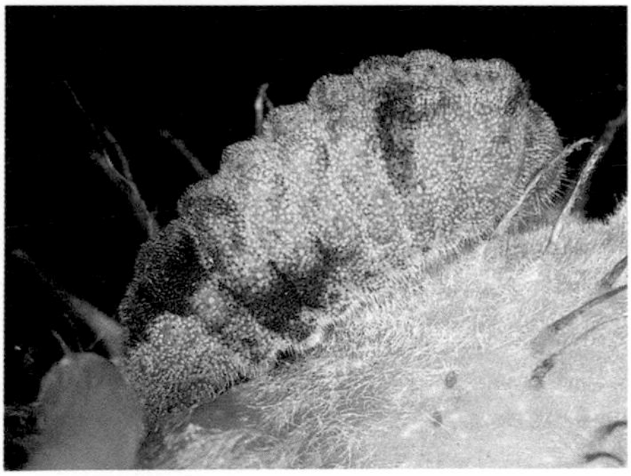

13. Early Hairstreak

## Blues (subfamily Polyommatinae)

See page 9 for more information about blues.

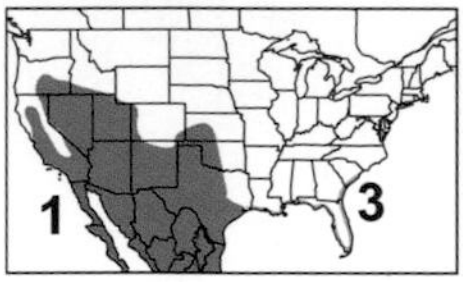


### 1. Western Pygmy-Blue *Brephidium exile*

### 3. Eastern Pygmy-Blue *Brephidium isophthalma*

Western Pygmy-Blue bluish-green to pinkish, very granular appearance, often with contrasting markings or chevrons. San Emigdio Blue lacks contrasting markings. Eastern Pygmy-Blue unmarked yellow-green. *Habitat* Western in deserts, disturbed areas in arid regions, roadsides, and coastal salt marshes; Eastern in coastal flats. *Found* Most of year. *Host* Western on saltbushes, lambsquarters, Russian thistle, sea purslane, and others; Eastern on glassworts. *Comments* The caterpillars are tiny, but attendant ants will make them easier to find if present.

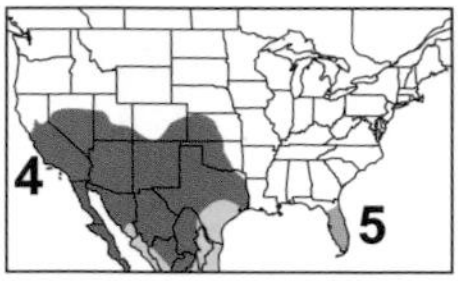


### 4. Marine Blue *Leptotes marina*

### 5. Cassius Blue *Leptotes cassius* Not shown.

Extremely variable in color and pattern, often matching the color of flowers it is feeding on. Usually has a dorsal stripe. Will not always be identifiable because it uses a wide variety of hosts. *Habitat* Generally distributed in open areas. *Found* April–Nov, all year in some areas, mostly on buds and flowers. *Host* Many different legumes, also leadwort (especially Cassius). *Garden Tips* Leadworts.

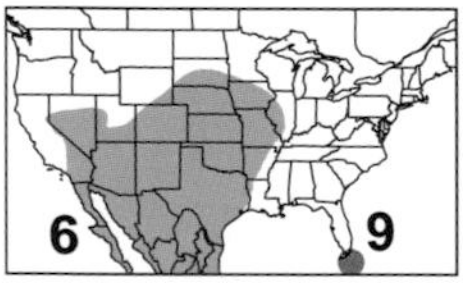


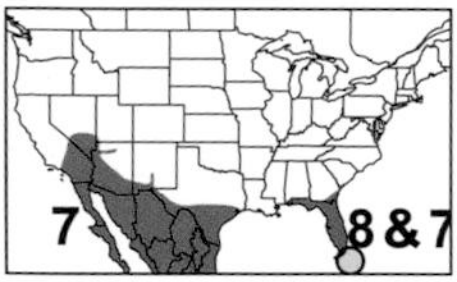


### 6. Reakirt's Blue *Hemiargus isola*

### 7. Ceraunus Blue *Hemiargus ceraunus*

### 8. Miami Blue *Hemiargus thomasi*

### 9. Nickerbean Blue *Hemiargus ammon* Not shown.

See pg. 158 for more information about these blues. Ceraunus and Reakirt's are variable—yellow, yellow-green, or red. Miami is green. **Nickerbean is dark green with lateral stripe.** *Host* Ceraunus and Reakirt's on many legumes; Miami on balloon-vine (inside the seed pod) and nickerbean (on new growth); Nickerbean on nickerbean and pineland acacia.

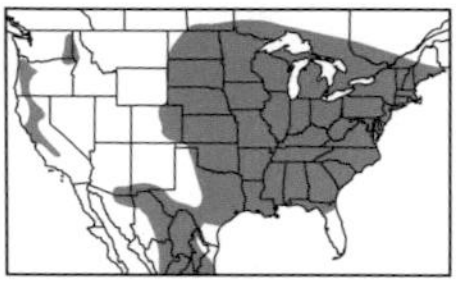

### 10. Eastern Tailed-Blue *Everes comyntas*

Variable, from pale green to bright green to yellow or red, with a series of pale lateral stripes and dashes. Probably not separable from other species using similar hosts. *Habitat* Open areas in general, especially disturbed areas. *Found* Feb–Nov, after flights. Extreme dates restricted to its southern range. *Host* Legumes. *Garden Tips* Clovers.

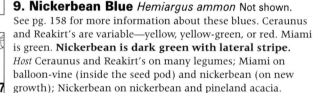

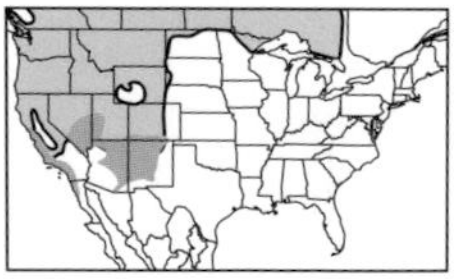

### 11. Western Tailed-Blue *Everes amyntula*

See Eastern Tailed-Blue. *Habitat* Many, including moist mountain meadows, chaparral, coastal dunes, sagebrush scrub, poplar woods, and redwood forest. *Found* March–Oct, often in inflated seedpods of milk vetches. *Host* Milk vetches, and other legumes. *Comments* Seals the entrance hole of the inflated seedpods with silk, denying entry by ants for unexplained reasons.

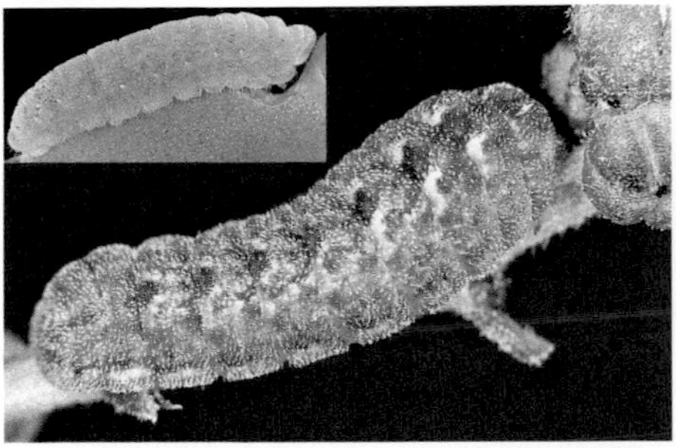

1. Western Pygmy-Blue (inset E. Pgymy-Blue)

2. Eastern Tailed-Blue adult

4. Marine Blue

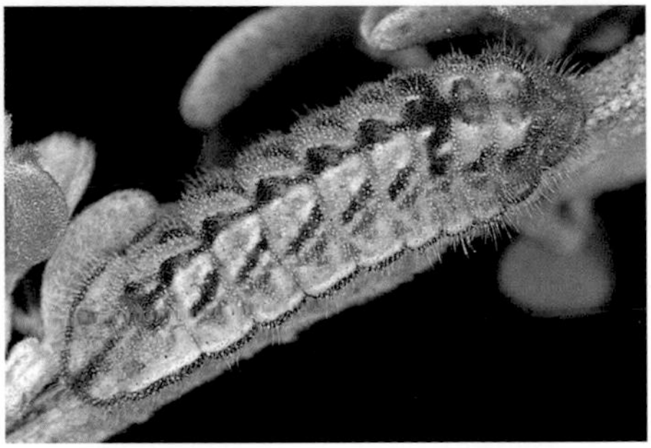

6. Reakirt's Blue

7. Ceraunus Blue

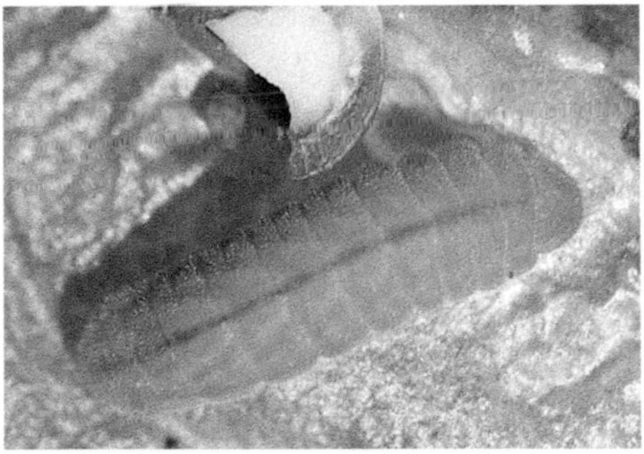

8. Miami Blue

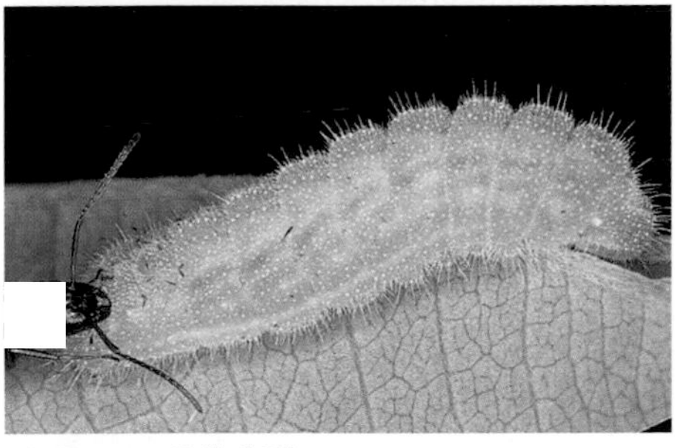

10. Eastern Tailed-Blue

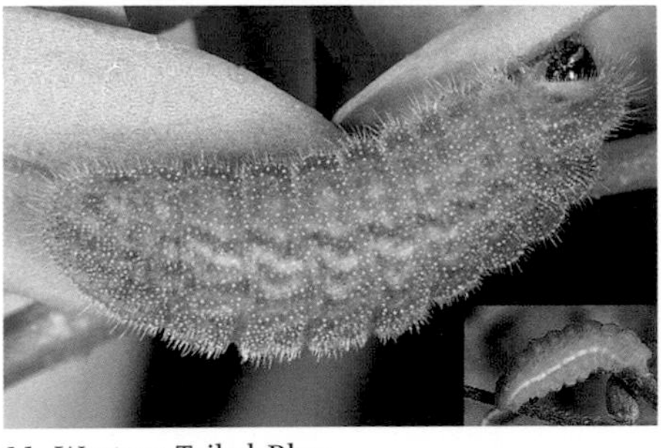

11. Western Tailed-Blue

**Azures** (genus *Celastrina*)

At least in the East, azures are primarily identifiable by host preference and flight time. Many populations have only a single flight each year, which often overlaps the flights of other populations during part of the spring to early summer period.

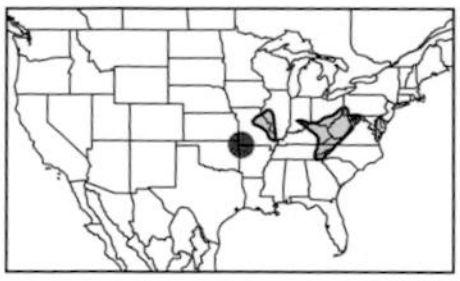

**Dusky Azure** *Celastrina nigra*

Green with few markings, and a faint sublateral line. Uses different host than other azures. *Habitat* Shaded woodland edges and clearings and roadsides, railroad right-of-ways, and power-line cuts. *Found* April–May, after single flight. *Host* Goatsbeard.

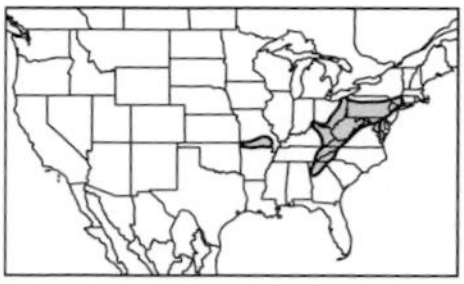

**Appalachian Azure** *Celastrina neglectamajor*

Variable. May be green to yellow-green to whitish or brown, with or without markings. Can easily be confused with "Summer" Spring Azure, which shares the same host, especially toward late spring. Other blues on the same host usually have stronger markings. *Habitat* Rich transition zone woodlands and their borders. *Found* May–June, after single flight. *Host* Black cohosh.

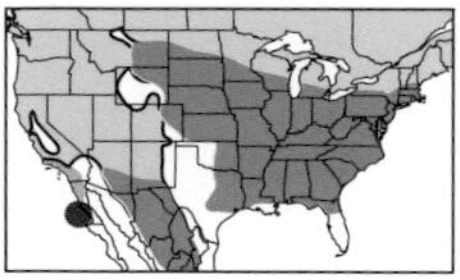

**Spring Azure** *Celastrina ladon*

Variable. Whitish, green or reddish-brown, with or without markings. Atlantic population is usually whitish. *Habitat* Deciduous woodlands and woodland edges and openings; also barrens, roadsides, and large parks (East), desert and mountain woodlands, chaparral, desert scrub (West). *Found* April–Oct (East), Jan–Nov (West), after flights. Northern populations April–May; Atlantic populations May–June; Cherry Gall populations late May–June; Summer populations late-May–Oct. *Host* Extremely varied. In the West *Ceanothus*, buckthorns and many others. In the East, often on dogwood flowers, occasionally viburnums and cherry, and on many others. Northern populations often on blueberries and cherry. Atlantic populations primarily use hollies. Cherry Gall populations primarily use cherry galls formed by eriophyid mites, and also cherry flowers:—usually on black cherry or choke cherry. *Comments* Our understanding of the population genetics of this species is still rudimentary. The named populations of Spring Azures, adapted to particular hosts or groups of hosts, may eventually be considered to be anything from full species to less than subspecies. In the West the relationship between populations is also complex.

1. Dusky Azure

2. Appalachian Azure adult

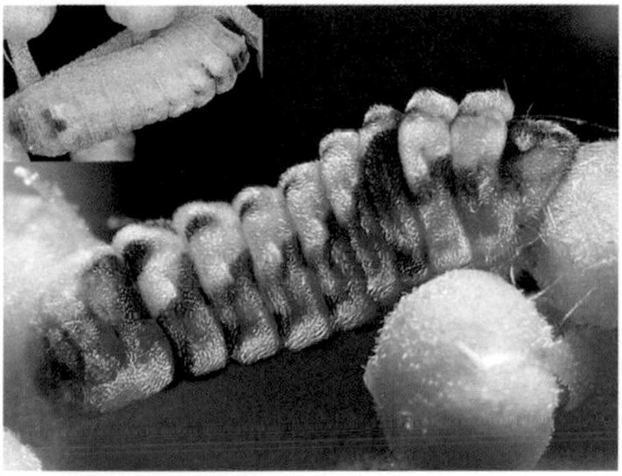

3. Appalachian Azure

4. Spring Azure, 'Northern' population

5. Spring Azure, 'Summer' population

6. Spring Azure, 'Summer' population

7. Spring Azure, 'Atlantic' population

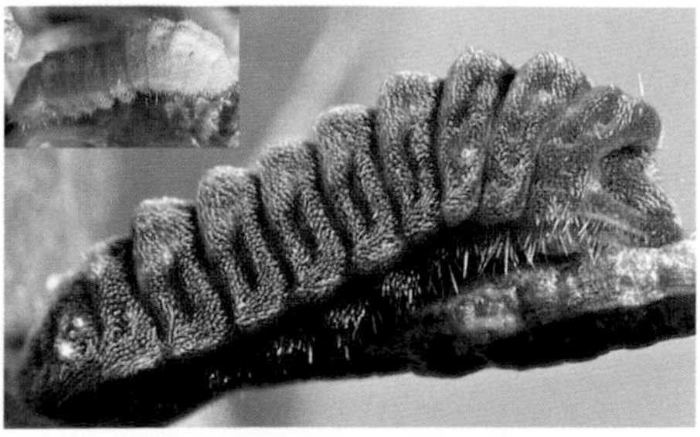

8. Spring Azure, 'Cherry Gall' population

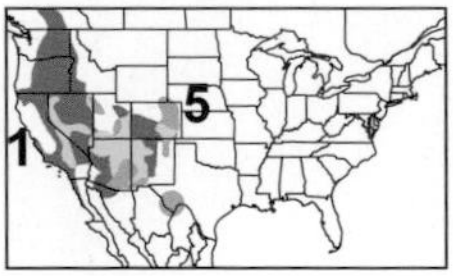


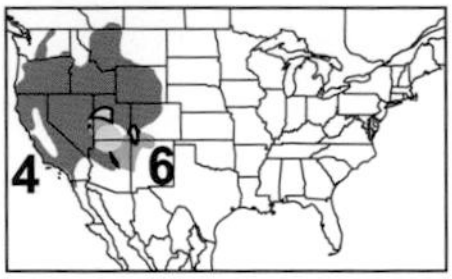


**1. Square-spotted Blue** *Euphilotes battoides*

**4. Dotted Blue** *Euphilotes enoptes* Not shown.

**5. Rita Blue** *Euphilotes rita* Not shown.

**6. Spalding's Blue** *Euphilotes spaldingi* Not shown.

See pg. 159 for more information about buckwheat blues. Variable. Can be white, pink, red, yellow or brownish, unmarked or marked with dorsal chevrons. *Habitat* Dry situations with buckwheats. *Found* May–Oct, after the single flight for each population. *Host* Buckwheats, with most local populations using only one species. *Comments* Individuals in the two photos of Square-spotted Blue came from adjacent plants.

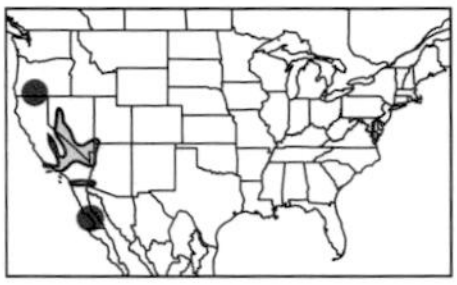

**Small Blue** *Philotiella speciosa*

Green to greenish-red or yellowish-red, marked or unmarked. May not be separable from Ceraunus, Dotted or Acmon Blues at western edge of Colorado Desert where all use reniform buckwheat. In Mojave Desert no other blues use punctured bract. *Habitat* Mainly desert gulches and hillsides but also known from mixed chaparral woodland. *Found* April–July, after single flight, on flowers and seeds. *Host* Punctured bract, reniforme buckwheat and related plants. *Comments* Hibernates as chrysalis and is able to survive years of drought in desert areas, yet rests close to the surface where summertime temperatures are brutally hot! To our knowledge, not tended by ants.

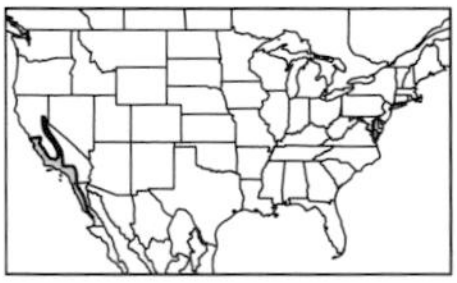

**Sonoran Blue** *Philotes sonorensis*

Variable. Green to yellow to red, marked or unmarked. The only blue in its range to completely bore into the leaves of succulents. Moss' Elfin uses similar hosts in some cases but feeds on exterior parts. *Habitat* Canyons and cliffs with the host plant. *Found* March–June, after single flight. *Host* Dudleyas, especially *Dudleya cymosa*. *Comments* Tended by ants.

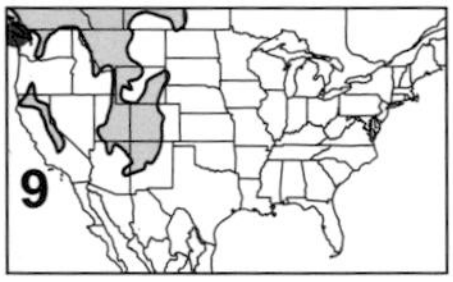

**9. Arctic Blue** *Agriades glandon*

**10. Heather Blue** *Agriades cassiope*

Green with red dorsal stripe edged with white. The extent of variation if any is unknown to us. *Habitat* 9. in wet mountain meadows (CA/OR), also rocky areas above treeline for Rocky Mountains, and prairie gulches, hillsides elsewhere. 10. on rocky slopes with seeps, at or above treeline. *Found* May–June (9) or June–July (10), prior to single flight, early instars follow flight then overwinter. *Host* Rock-primroses, shooting stars (CA/OR), and saxafrage for 9. Mountain heather for 10.

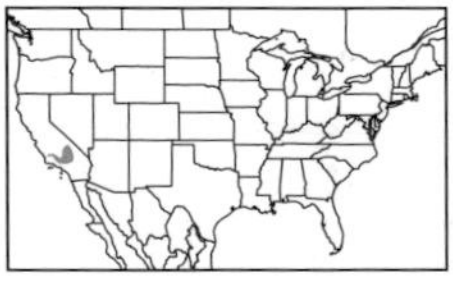

**San Emigdio Blue** *Plebejus emigdionis*

Pale bluish-green to grayish-pink, granular appearance, unmarked. Western Pygmy-Blue is comparable, but often with pink blotches. *Habitat* Arid washes and hillsides, but absent from vast areas containing the hostplant. *Found* March–Sept, especially June, mostly after flights, on leaves. *Host* **Fourwing saltbush**. *Comments* Found during the day at the base of host in the company of ants.

1. Square-spotted Blue

2. Sonoran Blue adult

3. Square-spotted Blue

7. Small Blue

8. Sonoran Blue

9. Arctic Blue

10. Heather Blue

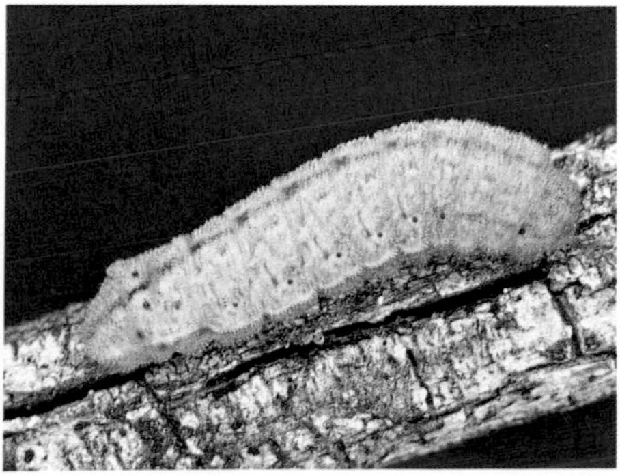

11. San Emigdio Blue

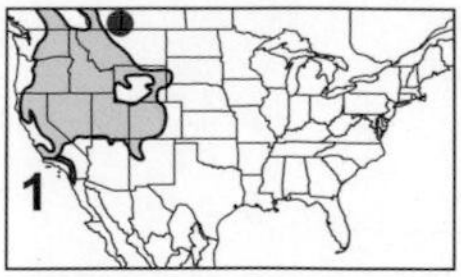


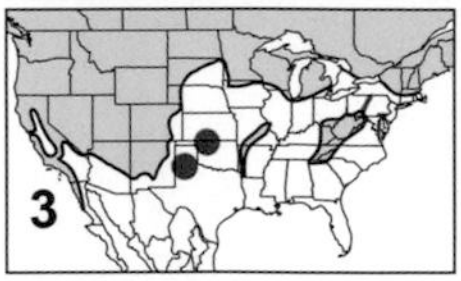


## 1. Arrowhead Blue *Glaucopsyche piasus*

## 3. Silvery Blue *Glaucopsyche lygdamus*

Variable, green to yellow to pink to purplish-gray to reddish-brown with variable amounts of dorsal chevron markings. Probably not separable from other lupine feeders where ranges overlap. *Habitat* A variety of open areas. *Found* Varies with region, March–July, after single flight, on flowers and seeds. *Host* Wood vetch, tufted vetch, deer vetches, milk vetches, and lupines. Only lupines for Arrowhead Blue.

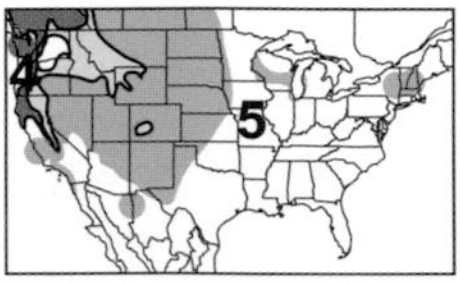


## 4. Northern Blue *Lycaeides idas*

## 5. Melissa Blue *Lycaeides melissa*

Green, markings faint if present. *Habitat* Northern Blue mainly in moist meadows and bogs in coniferous forest. Melissa in a wide variety of drier habitats. *Found* Northern, June–July, prior to single flight. Melissa, March–Oct, primarily after flights. *Host* Many legumes, including deer vetches, milk vetches, and lupines. *Comments* Overwinter as eggs.

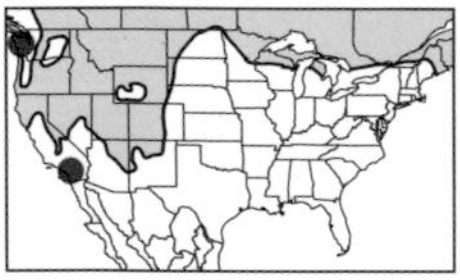

## Greenish Blue *Plebejus saepiolus*

Green or pale green with white lateral line edged with red or pink. *Habitat* Moist meadows, mainly in the mountains up to and above treeline; boggy areas in sagebrush. *Found* April–June, prior to flight, on leaves or flowers. *Early Instars* After flight, in flower heads of host where they overwinter. *Host* Clovers.

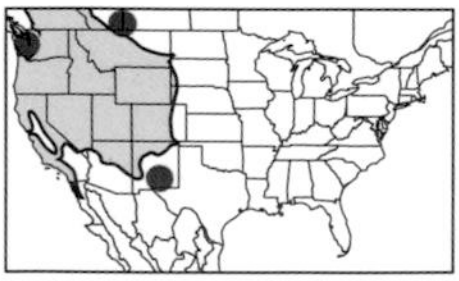

## Boisduval's Blue *Plebejus icarioides*

Variable, green to yellow to pink to reddish-brown with variable amounts of dorsal chevron markings. Very similar to other lupine feeders. *Habitat* Woodland openings, prairies, and sagebrush scrub. *Found* March–June, depending on region, prior to flight, on leaves, then flowers or seeds. *Early Instars* After flight, July–Sept, overwintering in debris at base of host. *Host* Lupines. *Comments* Often found in ant burrows being tended by ants at the base of host lupine.

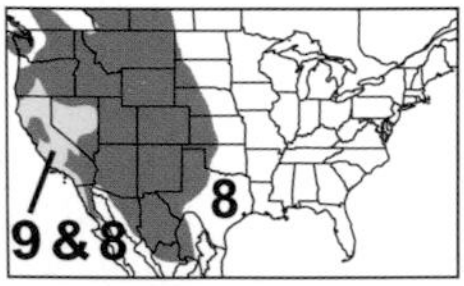


## 8. Acmon Blue *Plebejus acmon*

## 9. Lupine Blue *Plebejus lupinus* Not shown.

Variable, green to yellow to pink to reddish-brown with variable amounts of dorsal chevron markings. Lupine Blue more often with white or yellow lateral line. Probably not separable from similar species (listed above) that share hosts. *Habitat* A wide variety of open habitats. *Found* Feb–Nov, both prior to and after the several flights, on leaves, flowers, or seeds. *Host* Only buckwheats for Lupine Blue. Acmon also on deer vetches and milk vetches.

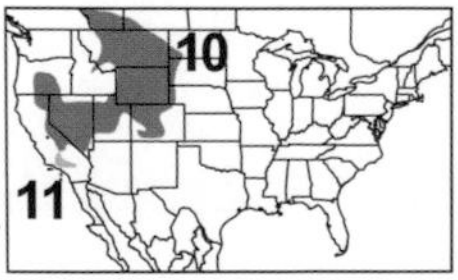


## 10. Shasta Blue *Plebejus shasta* Not shown.

## 11. Veined Blue *Plebejus neurona* Not shown.

*Host* Shasta Blue on vetches, clovers, and lupines. Veined Blue on Kennedy's and Wright's buckwheats. See pg. 159.

1. Arrowhead Blue

2. Melissa Blue, mated adults

3. Silvery bluc

4. Northern Blue

5. Melissa Blue

6. Greenish Blue

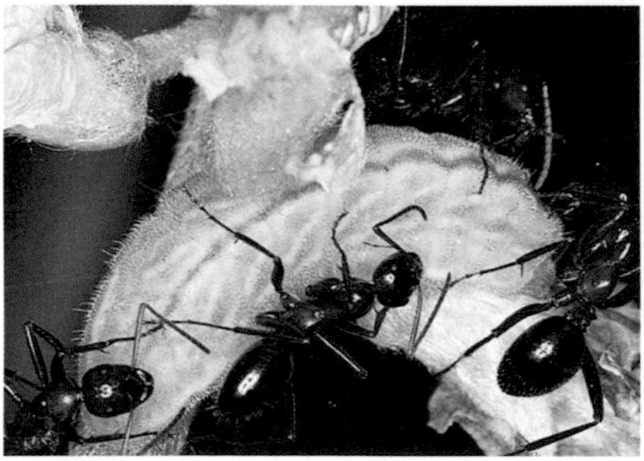

7. Boisduval's Blue

8. Acmon Blue

## **Metalmarks** (family Riodinidae)

See page 10 for general information about metalmarks.

### **Calephelis Metalmarks**

All are whitish-green with long dorsal and lateral hair tufts.
See page 159 for more information.

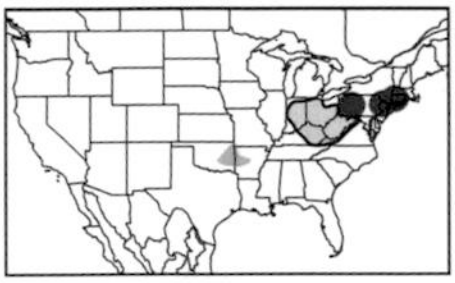

**Northern Metalmark** *Calephelis borealis*
*Habitat* Open glades or ridges (power-line cuts) within limestone soil woodlands. *Found* April–May, prior to single flight. *Early Instars* July–Sept, after flight. *Host* **Round-leaved ragwort**. *Comments* Overwinters as 5th or 6th instar, and may require up to 8 or 9 instars to complete growth.

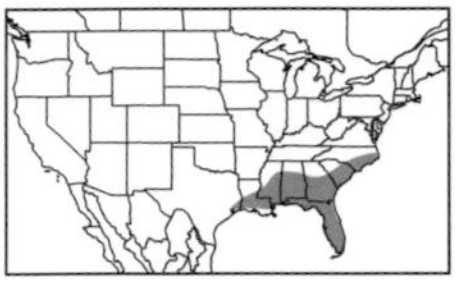

**Little Metalmark** *Calephelis virginiensis*
Whitish-green with long dorsal and lateral hair tufts. Separate from other metalmarks by range. *Habitat* Open pine flats. *Found* March–Nov between flights. *Host* **Yellow thistle**.

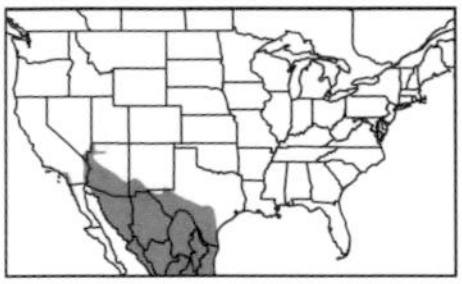

**Fatal Metalmark** *Calephelis nemesis*
The tips of the dorsal hair tufts are swept back. *Habitat* Thorn scrub, chaparral, arid region riparian areas. *Found* Probably most of the year. *Host* Seepwillow, virgin's bower, and California encelia. *Garden Tips* Virgin's bower. *Comments* Does not hibernate but feeds continuously and slowly during winter months ensconced in dead, twirled leaves of the host.

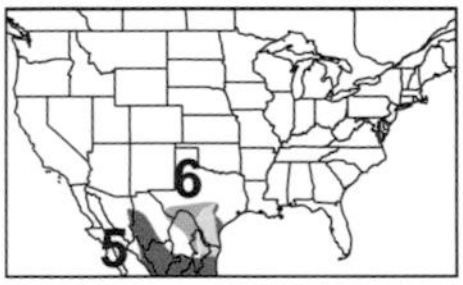


**5. Rounded Metalmark** *Calephelis perditalis*
**6. Rawson's Metalmark** *Calephelis rawsoni*
Not shown.
Rounded caterpillar whitish-green with long, white dorsal and lateral hair tufts. Rawson's caterpillar unknown. *Habitat* Rounded in tropical woodland and thorn scrub. Rawson's in moist situations in arid regions. *Found* Probably most of year. *Host* Mistflowers. *Comments* And you thought that the adults were difficult to identify!

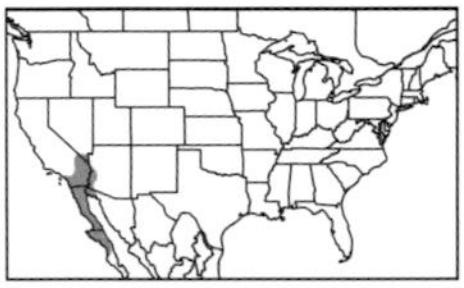

**Wright's Metalmark** *Calephelis wrighti*
*Habitat* Desert canyons and washes. *Found* Probably most of the year, although feeding may cease during extended dry periods. *Host* **Sweetbush**.

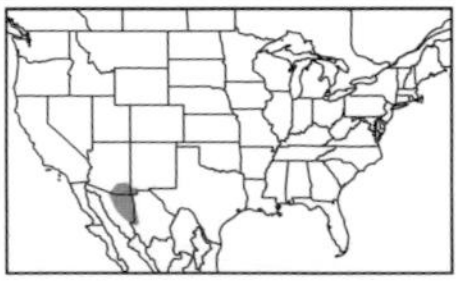

**Arizona Metalmark** *Calephelis arizonensis*
Uses different host than Fatal Metalmark, but the two can be found together as adults with both hosts nearby. *Habitat* Moist situations in arid regions. *Found* May feed most of year in some habitats. *Host* Beggar ticks associated with riparian bottoms.

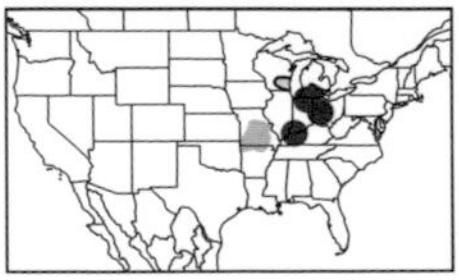

**Swamp Metalmark** *Calephelis muticum*
Prefers different host and entirely different habitat than Northern Metalmark. *Habitat* Moist to wet meadows in peatland. *Found* May–Aug, prior to flights. *Host* **Swamp thistle**.

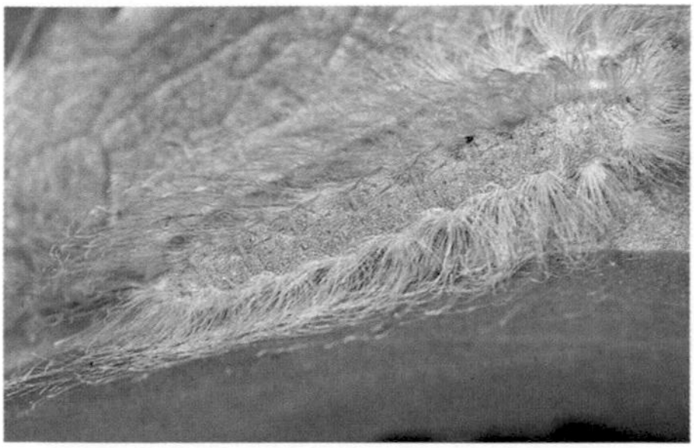

1. Northern Metalmark

2. Northern Metalmark adult

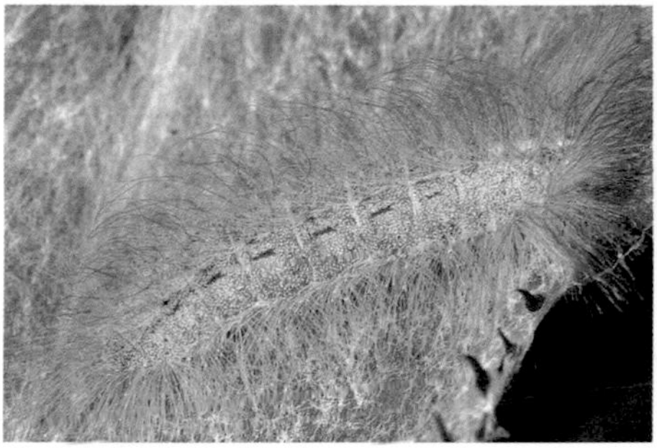

3. Little Metalmark

4. Fatal Metalmark

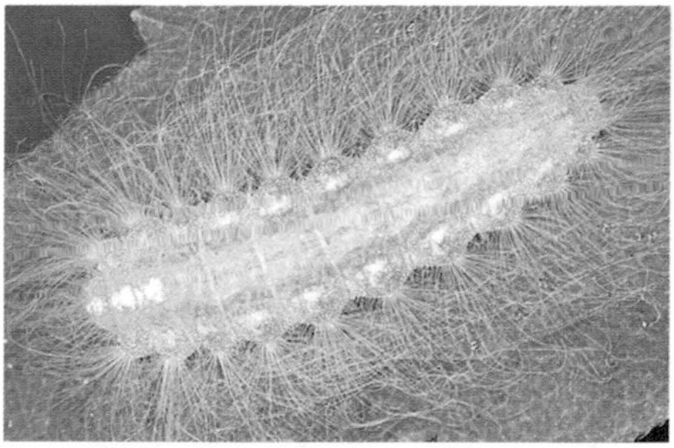

5. Rounded Metalmark

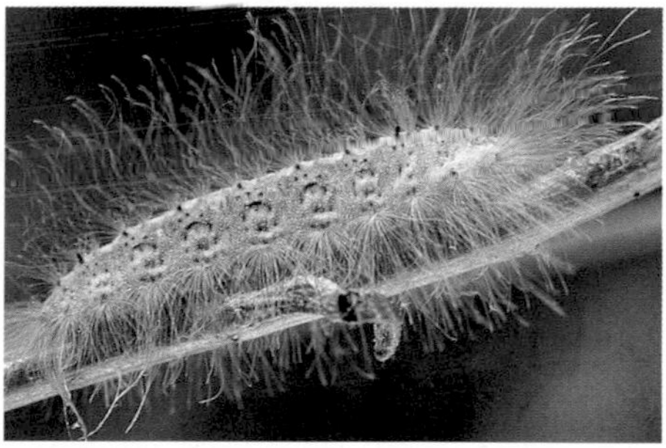

7. Wright's Metalmark

8. Arizona Metalmark

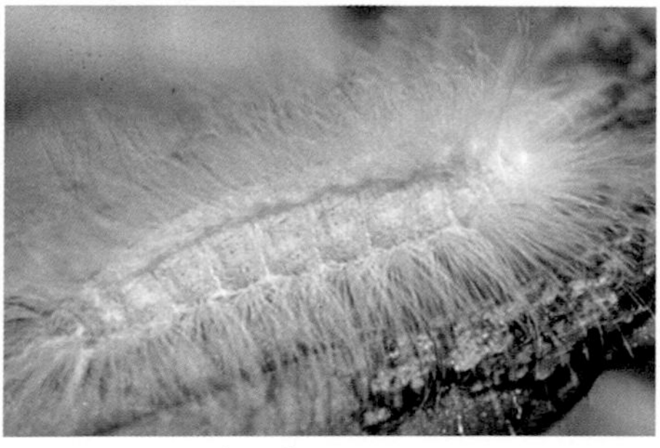

9. Swamp Metalmark

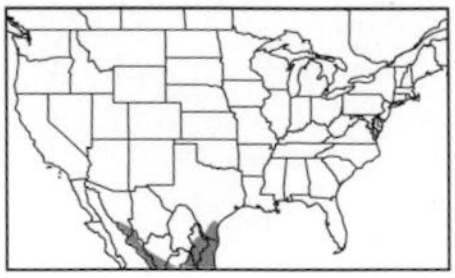

**Red-bordered Metalmark** *Caria ino*
Greenish-white, covered with long whitish hairs with **orange tuft of hairs above the head**. *Habitat* Thornscrub and tropical thorn forest. *Found* April–Dec, after flights. *Host* Spiny hackberry. *Comments* Lives inside a nest of tied leaves of the host, where pupation eventually takes place.

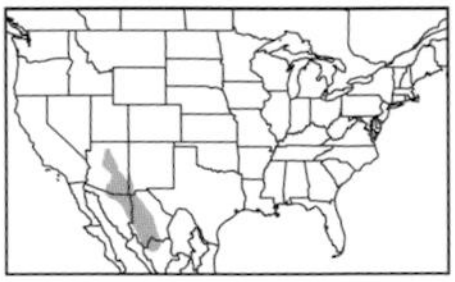

**Zela Metalmark** *Emesis zela*
Next to last instar is tan with long hair tufts. Otherwise, unreported. *Habitat* Sycamore-lined canyons through mid-elevation oak woodlands. *Found* Probably after both flights. *Host* Still unknown in nature; oak is eaten in lab. *Comments* A life history waiting to be discovered.

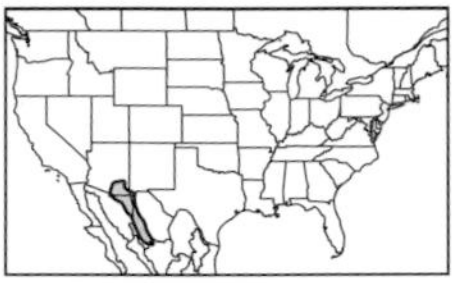

**Ares Metalmark** *Emesis ares*
Light green covered with a bloom of short white hair. *Habitat* Sycamore-lined canyons through mid-elevation oak woodlands. *Found* June–July, prior to single flight. *Early Instars* Sept–Oct, after flight. *Host* Emory oak and possibly other oaks. *Comments* Development is slow. Lives in nest on the host.

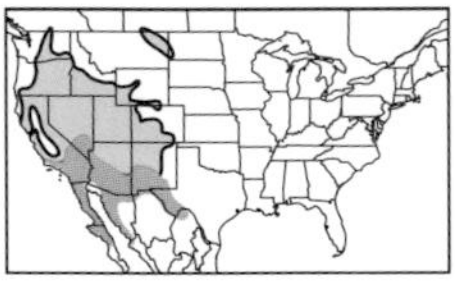

**Mormon Metalmark** *Apodemia mormo*
Distinctive. Brownish-purple with black and yellow. *Habitat* Arid regions, including sand dunes, rocky canyons and hillsides, sagebrush steppes. *Found* Prior to flights, varies with region. *Host* Mostly buckwheats, but ratany is preferred in parts of the southwest and Texas. *Comments* Older caterpillars feed at night.

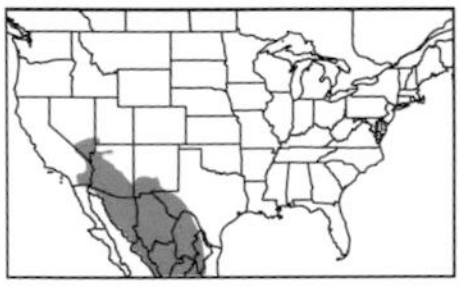

**Palmer's Metalmark** *Apodemia palmeri*
Green, sometimes pinkish, with creamy subdorsal stripes, covered in very short, fine, white hair. *Habitat* Arid regions, especially low deserts. *Found* All year. *Host* Mesquites, including velvet, honey, and screwbean. *Comments* Ties leaflets together at tips creating a "rib cage"–like shelter for concealment.

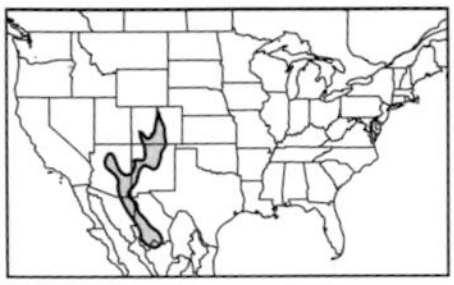

**Nais Metalmark** *Apodemia nais*
Pink to brown with two dorsal rows of black hair tufts. *Habitat* Open areas with the host in mountain pine forest, 6000–9000′; also mid-elevation woodland 3500–5000′ (TX). *Found* April–July, prior to single flight. *Early Instars* Probably after flight. *Host Ceanothus fendleri*; Harvard's plum in Texas.

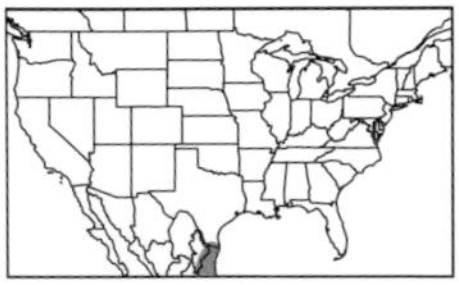

***Red-bordered Pixie** *Melanis pixe*
White or whitish-gray, with thin yellow and black stripes and long lateral tufts of white hairs. *Habitat* Urban or suburban areas where the host (non-native) has been planted or established. *Found* Most of the year. *Host* Guamuchil. *Garden Tips* Guamachil. *Comments* Will colonize isolated groups of host trees.

**Blue Metalmark** *Lasaia sula* Not shown.
Caterpillar unknown. *Habitat* Coastal scrub *Host* Creeping mesquite.

1. Red-bordered Metalmark

2. Red-bordered Pixie adult

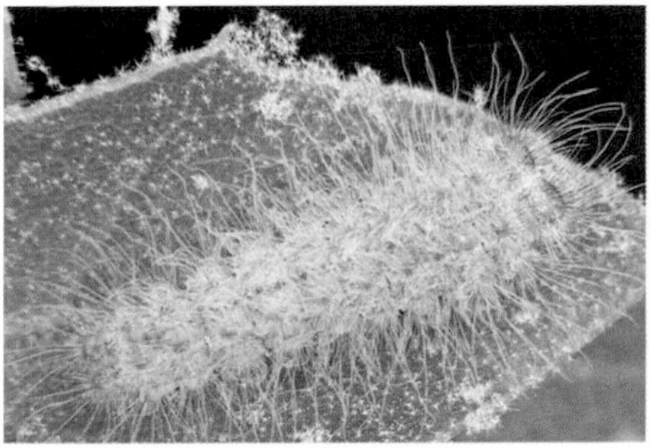

3. Zela Metalmark (penultimate instar)

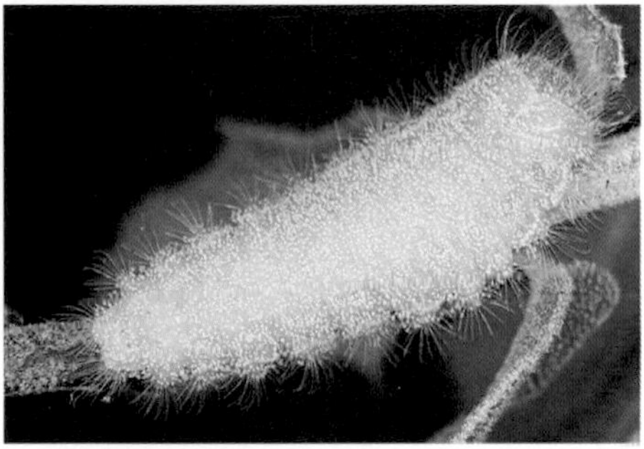

4. Ares Metalmark

5. Mormon Metalmark

6. Palmer's Metalmark

7. Nais Metalmark

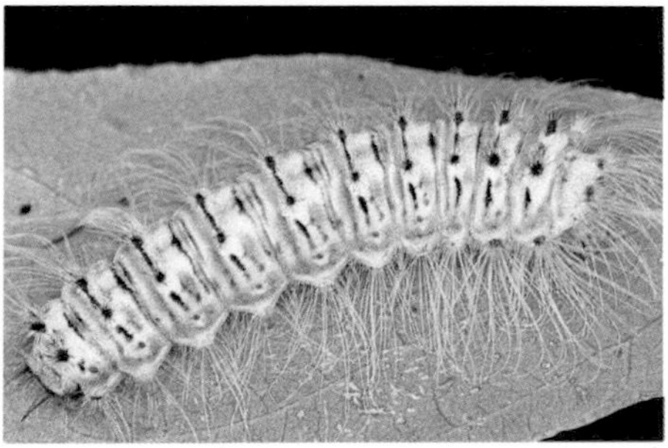

8. Red-bordered Pixie

## **Brushfoots** (family Nymphalidae)

See page 10 for general information about brushfoots.

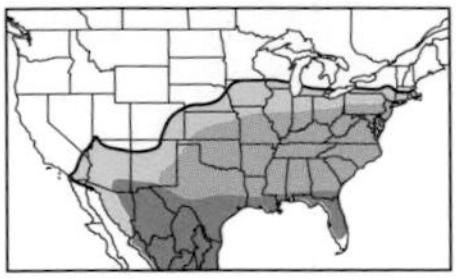

### ***American Snout** *Libytheana carinenta*

Yellow-green with fine longitudinal stripe. Rests with front of the body hunched and often with only the back half of the body and the tip of the head touching the host. Resembles *Colias* sulphurs rather than other brushfoots but uses different host. *Habitat* Thickets, open woodlands and desert canyons with hackberries. *Found* March–Oct southward; June–Sept northward. *Host* Hackberries. *Comments* Often responds to summer rains with dramatic population explosions, during which caterpillars can be found feeding and pupating on branches and leaves of the host.

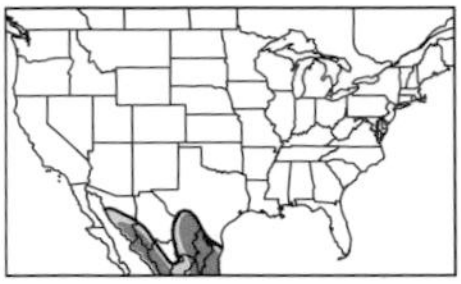

### **Mexican Silverspot** *Dione moneta*

Black with yellow dorsal and lateral bands. Rare and known to breed in south Texas only on rare occasions. *Habitat* Subtropical oak woodland and lowland scrub. *Found* May–Dec in southern Texas. *Host* Passion-vines.

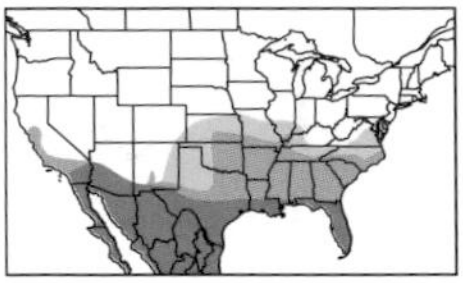

### ***Gulf Fritillary** *Agraulis vanillae*

Variable. Generally orange/purple, sometimes with subdorsal white stripe. Pair of **head spines not clubbed**. *Habitat* Coastal areas, thorn scrub and gardens. *Found* Most of year in warmer regions, Feb–Dec elsewhere. *Host* Passion-vines. *Garden Tips* Passion-vines, especially maypop. *Comments* Despite the "nasty" looking spines and obvious warning colors, these caterpillars are reported to be eaten by birds.

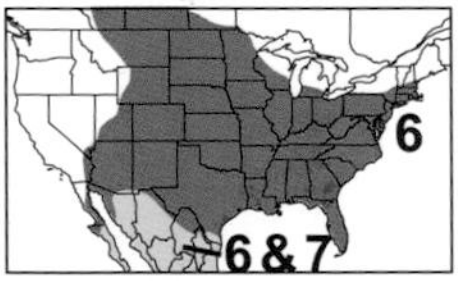


### **6. Variegated Fritillary** *Euptoieta claudia*

### **7. Mexican Fritillary** *Euptoieta hegesia* Not shown.

Slightly variable. Red with white longitudinal stripes edged with black. Pair of **head spines clubbed at tip**. *Habitat* Many open situations. *Found* April–Oct, late summer to fall northward. *Host* Passion-vines, violets, flaxes, green violet, et al. *Garden Tips* Passion-vines. *Comments* Variegated caterpillars on flax in the Southwest are grazers, feeding on one flax plant then moving on to another; hence they are often found on the ground.

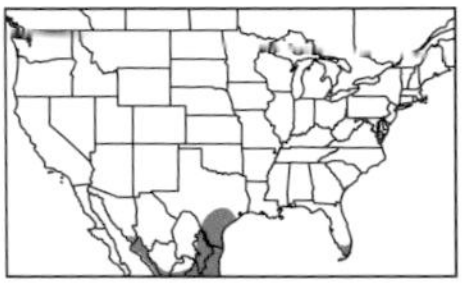

### ***Julia Heliconian** *Dryas iulia*

Variable. Usually blackish with varying amounts of white or cream striping dorsally and usually with a row of white spots along lower body. **Spines are very long**, longer than those of Gulf Fritillary. *Habitat* Open disturbed situations near tropical and subtropical woodlands. *Found* Most of year. *Host* Passion-vines. *Garden Tips* Corky-stemmed passion-vine and maypop.

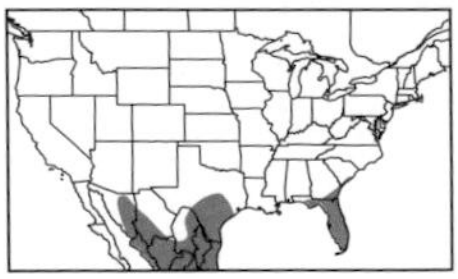

### ***Zebra Heliconian** *Heliconius charithonia*

Distinctive. Crimson Patch is similar but uses different host. *Early Instars* Somewhat resemble Gulf Fritillary. *Habitat* Woodland and hammock edges. *Found* Most of year in southern FL and TX; late summer to fall northward. *Host* Passion-vines. *Garden Tips* Corky-stemmed passion-vine and maypop.

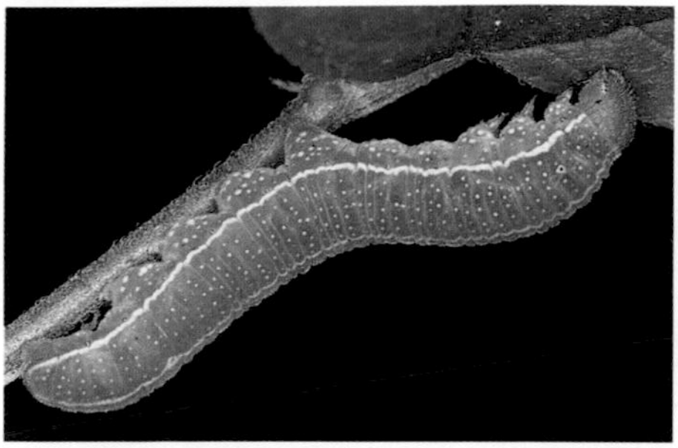

1. American Snout

2. Gulf Fritillary adult

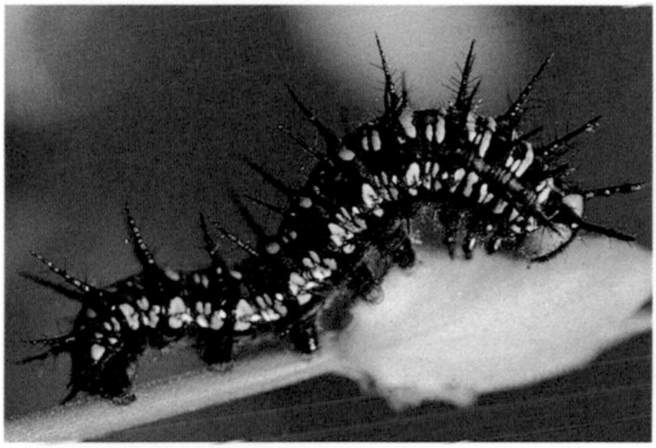

3. Mexican Silverspot

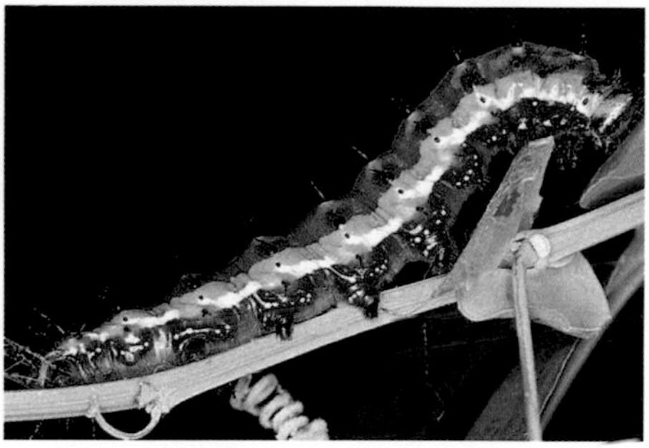

4. Gulf Fritillary

5. Gulf Fritillary

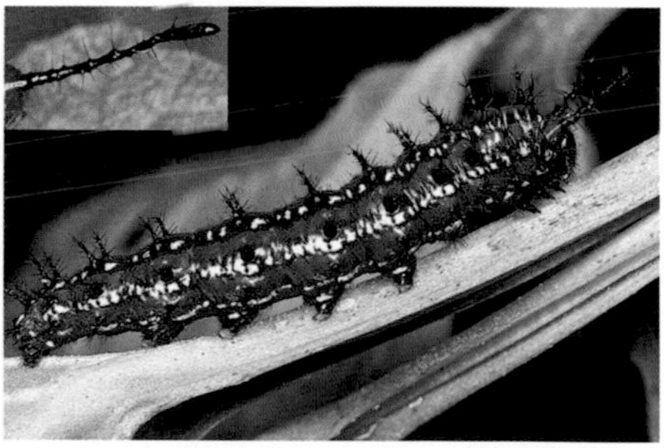

6. Variegated Fritillary

8. Julia Heliconian

9. Zebra Heliconian

## Greater Fritillaries

A popular group. Eight are treated on this and the following page and the other 6 on pages 160–61. Most are black with lighter markings and 3 rows of branching spines on either side of the body. All feed on violets, but the specific violets used are poorly known. Winter is spent as a first instar, unfed caterpillar. The best time to search for these caterpillars is a month or so before adults take flight, which is mainly late spring or early summer. Pupation is inside a simple tent made with strands of silk stretched between surrounding surfaces. Eggs are not laid on the host but on debris near where violets will emerge the following spring. These caterpillars are very secretive, feeding for brief periods, then returning to their hiding place under the host leaves or nearby vegetation. Few people have found them in the wild.

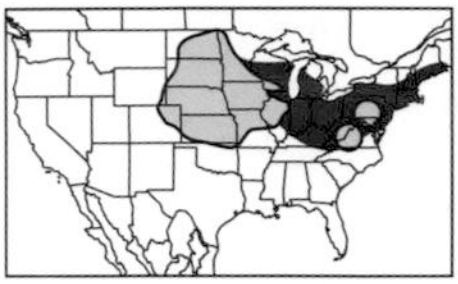

### Regal Fritillary *Speyeria idalia*

Broad yellow to orange dorsal stripe with yellow to orange transverse stripes. *Habitat* Tall-grass prairie, wet fields and meadows. *Found* April–June. *Host* Violets.

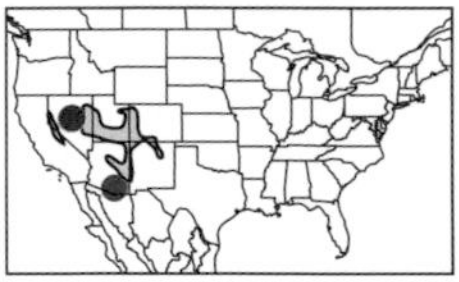

### Nokomis Fritillary *Speyeria nokomis*

Broad yellow to orange dorsal stripe with yellow to orange transverse stripes. *Habitat* Moist meadows near streams and wet fields. *Found* April–June. *Host* Violets. *Comments* Female caterpillars feed ten days longer than do males.

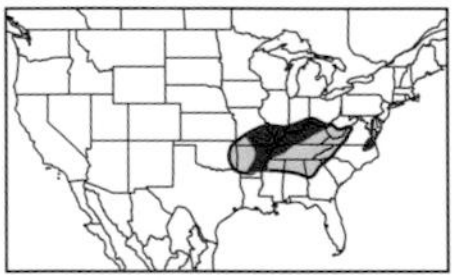

### Diana Fritillary *Speyeria diana*

Uncommon. Blackish-purple. Branching **black spines with basal ⅓ red** to deep orange. *Habitat* Rich, moist mountain forests. *Found* March–June, prior to flight. *Host* Violets. *Comments* Usually found on woodland violets.

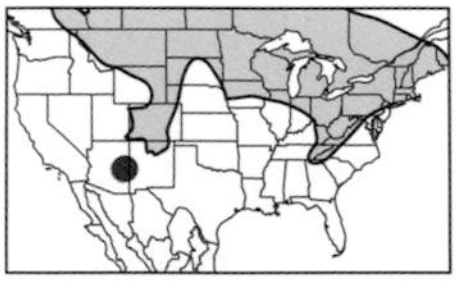

### Aphrodite Fritillary *Speyeria aphrodite*

Black, **base of spines brown** to orange-brown. *Habitat* Prefers more wooded, cooler areas than Great Spangled Fritillary, but often found with it; in the Midwest, moist prairies are also used. *Found* April–June, prior to flight. *Host* Violets.

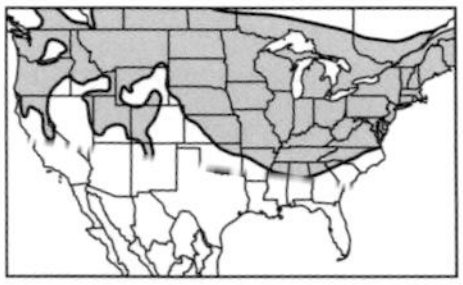

### Great Spangled Fritillary *Speyeria cybele*

Common. Branching **black spines with basal ⅓ red** to deep orange. *Habitat* Open fields and meadows, roadsides, etc., wet meadows; prairies, in the West. *Found* March–June. *Host* Violets.

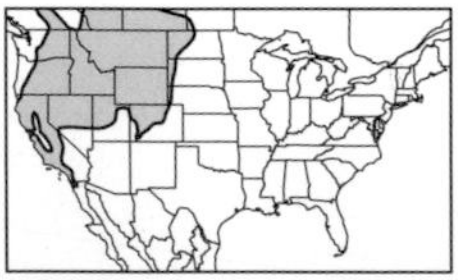

### Callippe Fritillary *Speyeria callippe*

Black with paler twinned dorsal stripes and orange or yellow spines. *Habitat* Prairie hills, ridges, and chaparral. *Found* March–May; early June at higher elevations. *Host* Violets.

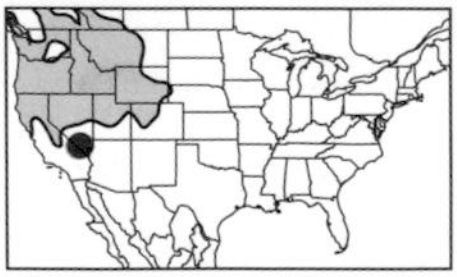

### Zerene Fritillary *Speyeria zerene*

Black with pale yellow twinned dorsal stripes, the top two rows of spines black, middle row may be black or yellow, bottom row yellow. *Habitat* Woodland openings, grasslands, coastal meadows, and dunes. *Found* May–July. *Host* Violets.

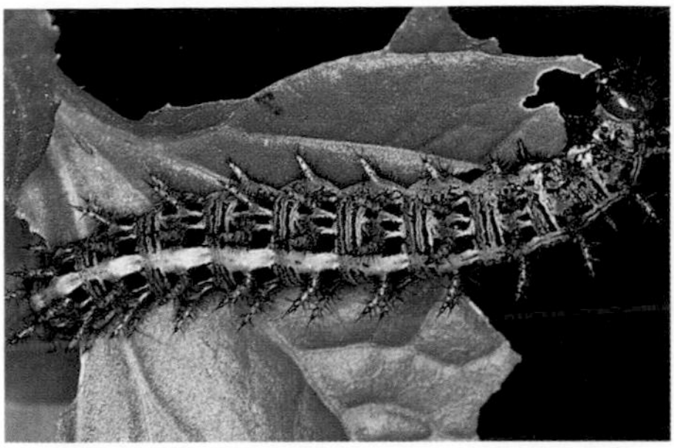
1. Regal Fritillary

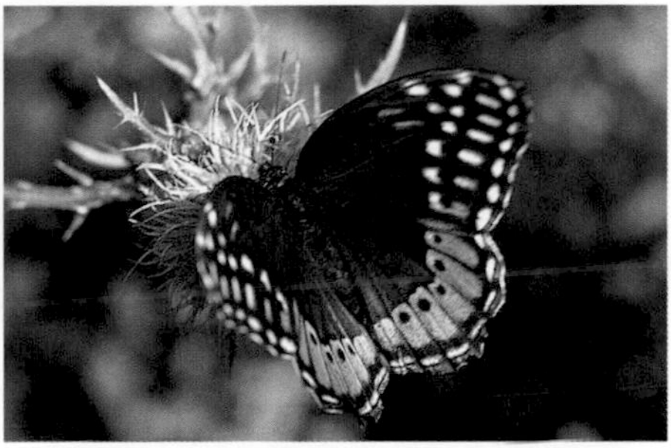
2. Diana Fritillary adult

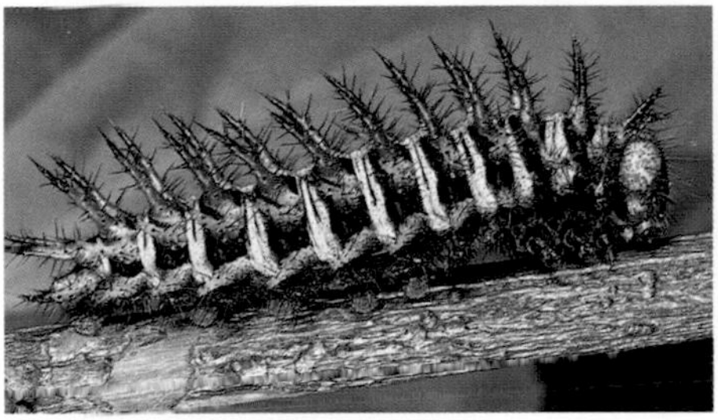
3. Nokomis Fritillary

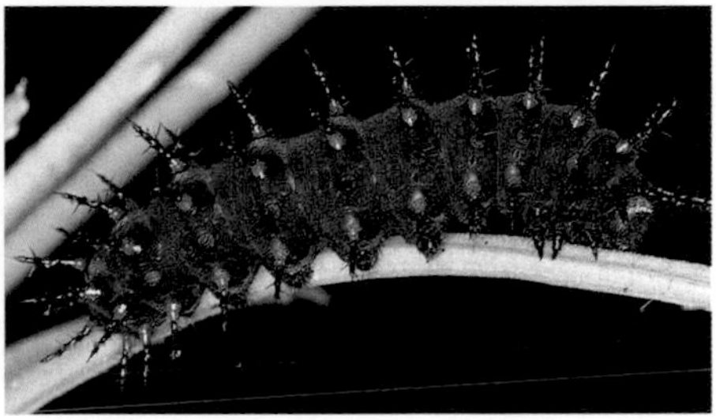
4. Diana Fritillary

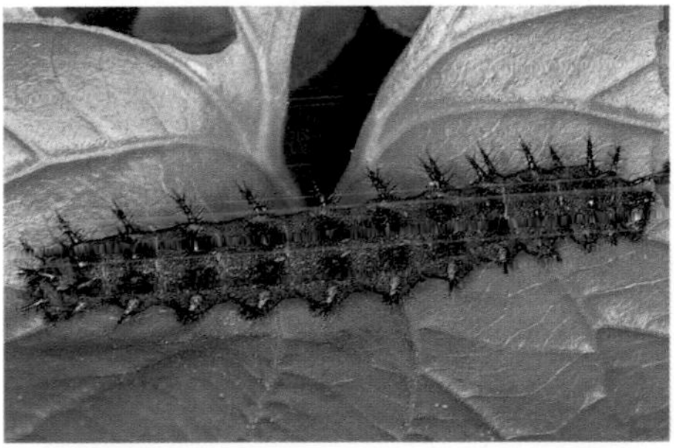
5. Aphrodite Fritillary

6. Great Spangled Fritillary

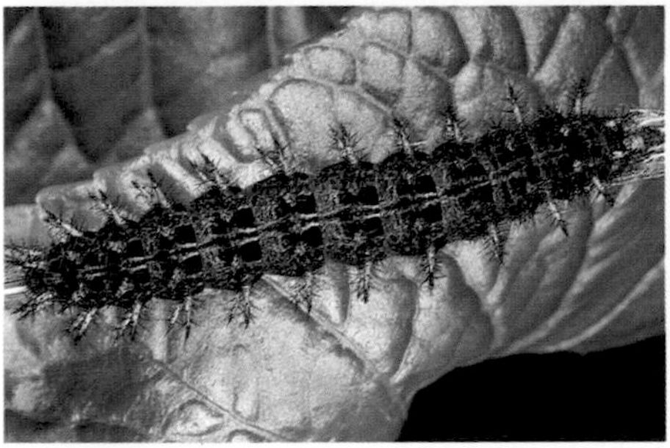
7. Callippe Fritillary

8. Zerene Fritillary

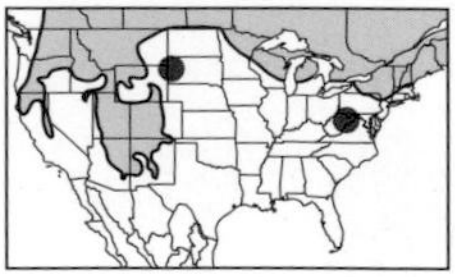

**Atlantis Fritillary** *Speyeria atlantis*
Brownish-black, finely striated, marked with black blotches; spine bases black. *Habitat* Open, mixed woodlands, especially in glades and boggy areas. *Found* April–June. *Host* Violets.

**Lesser Fritillaries.** See page 161 for more information.

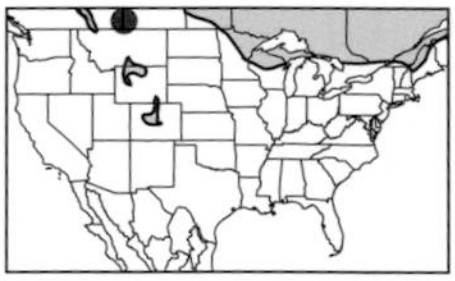

**Bog Fritillary** *Boloria eunomia*
Brown to reddish-brown with paler spines. *Habitat* Acid bogs, bogs, and moist areas above timberline. *Found* April–June. *Host* Shrubby willows, violets, and small cranberry.

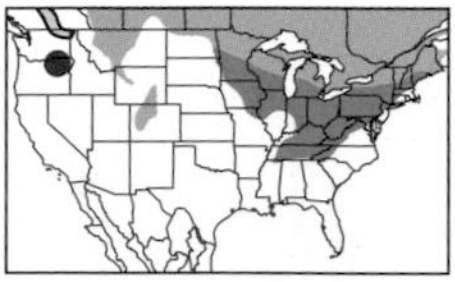

**Meadow Fritillary** *Boloria bellona*
Mottled gray to dark brown. Spines yellow-brown to brown, cream and black at their base (see inset to photo 5). *Habitat* Grassy fields and meadows, especially moist ones. *Found* March–April, June–Aug. *Host* Violets.

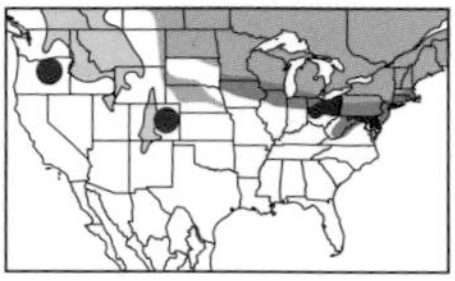

**Silver-bordered Fritillary** *Boloria selene*
Gray, brown, or yellowish with black blotches. **Long spines** yellow to orange, extend over head. *Habitat* Bogs, wet meadows, wet prairies, marshes. *Found* April, June–Sept. *Host* Violets.

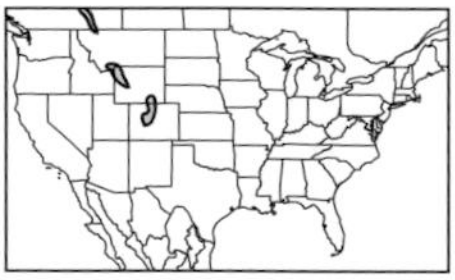

**Frigga Fritillary** *Boloria frigga*
Black with bold white or purple stripes. Spines black. *Habitat* Black spruce bogs, willow thickets, and swamps. *Found* May–July. *Host* Shrubby willows and dwarf birch.

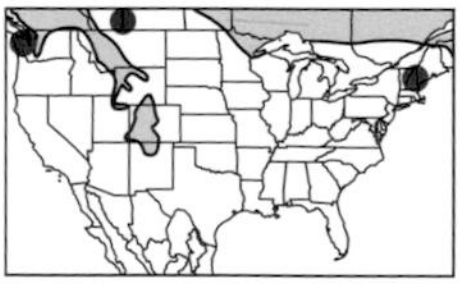

**Purplish Fritillary** *Boloria montinus*
Gray with fine black dorsal stripes, gray spines, and black head. *Habitat* Moist meadows and bogs, and openings in coniferous forest. *Found* April–June. *Host* Willows, violets, and western bistort.

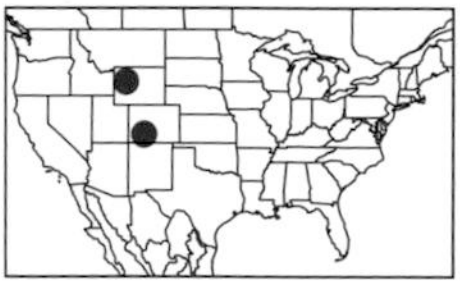

**9. Dingy Fritillary** *Boloria improba* Not shown.
Mottled dark brown with creamy subdorsal line edged with black. *Habitat* Moist areas above treeline with the host. *Found* May–June. *Host* Prostrate alpine willows including snow willow.

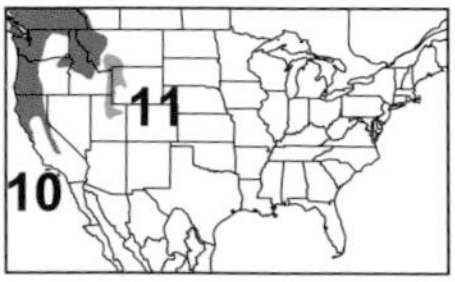


**10. Pacific Fritillary** *Boloria epithore* Not shown.
**11. Relict Fritillary** *Boloria kriemhild* Not shown.
Relict, mottled black, white subdorsal line. Pacific is blackish-gray, streaked with white subdorsally, and with a reddish-brown sublateral stripe; spines, orange or reddish. *Habitat* Moist forest openings and mountain meadows. *Found* April–June. *Host* Violets.

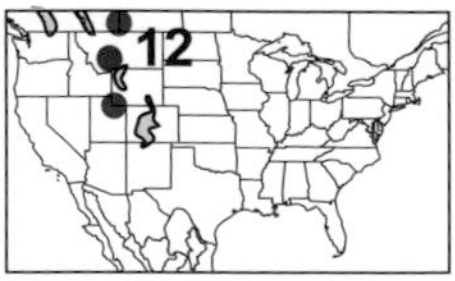


**12. Freija Fritillary** *Boloria freija* Not shown.
Mottled dark and light brown. *Habitat* Sub-alpine willow thickets. *Found* April–June. *Host* Bearberry and blueberry

**Alberta Fritillary** *Boloria alberta* Not shown.
**Astarte Fritillary** *Boloria astarte* Not shown.
Caterpillars are unknown. See page 161.

1. Atlantis Fritillary

2. Silver-bordered Fritillary adult

3. Silver-bordered Fritillary chrysalis

4. Bog Fritillary

5. Meadow Fritillary

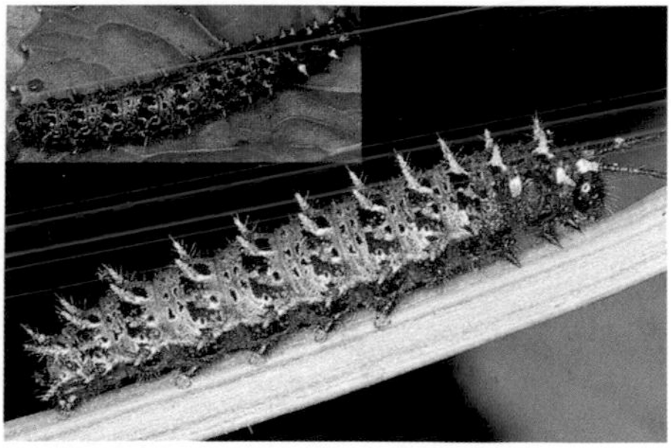

6. Silver-bordered Fritillary

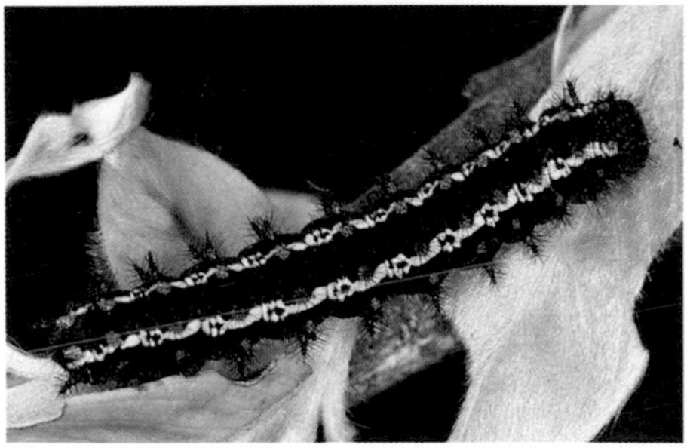

7. Frigga Fritillary

8. Purplish Fritillary

## Checkerspots. See page 161 for general information.

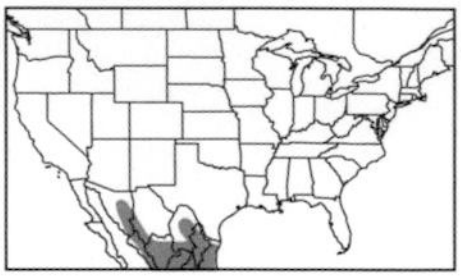

### Elf *Microtia elva*

Mottled white and black; basal area of spines on first three segments black; **head orange**. *Habitat* Thorn scrub. *Found* Most of year in Mexico, after summer rains in southern Texas and southern Arizona. *Host* Tetramerium, and probably other plants in the Acanthus family. *Comments* A rarity, with only a slight possibility for caterpillars to be found north of Mexico.

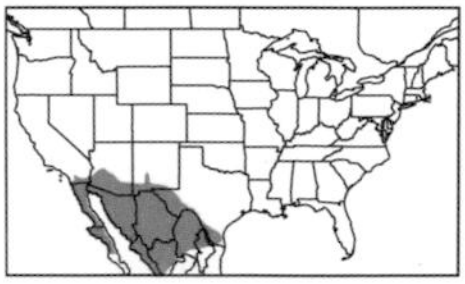

### Tiny Checkerspot *Dymasia dymas*

Mottled black and white, **black spines on first three segments**, all others grayish-black, all with orange at base; head black with white markings. Elada Checkerspot has all spines grayish-black. *Habitat* Low-elevation arid regions, including desert canyons and foothills. *Found* Most of year in southern regions; Feb–Nov elsewhere. *Host* Chuparosa, tubetongue, and *Tetramerium*.

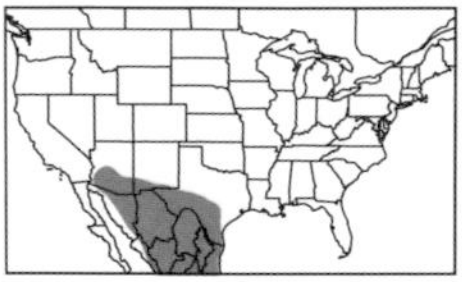

### Elada Checkerspot *Texola elada*

Mottled black and white; **grayish-black spines on all segments**; head black with white markings. *Habitat* Thorn scrub and oak-juniper woodland. *Found* March–Oct. *Host* Desert-honeysuckle in Arizona, other Acanthus family plants in Texas. *Comments* Little or no feeding occurs during dry spells in Arizona.

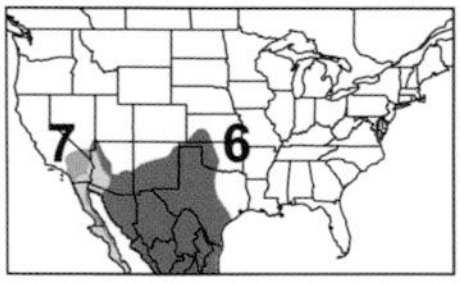


### 6. *Bordered Patch *Chlosyne lacinia*

Three forms (plus intermediates); mostly orange with black (photo 5 inset), black with orange dorsal band (photo 5), or mostly black (photo 6); head red-orange. *Habitat* A wide variety of open situations. *Found* All year. *Host* Aster family, especially sunflowers. *Comments* Early instars feed together.

### 7. California Patch *Chlosyne californica* Not shown.

Similar but has black head and mainly uses desert sunflower.

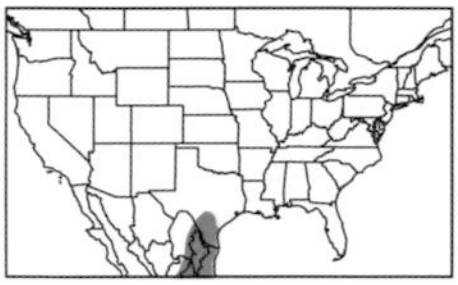

### Crimson Patch *Chlosyne janais*

Grayish-green to white with black crossbands between segments; head orange. Zebra Heliconian lacks black crossbands. *Habitat* Tropical woodland and thorn scrub. *Found* June–Oct. *Host* Acanthus family. *Comments* Older caterpillars hide by day, feed at night. When disturbed young caterpillars jerk their bodies in unison or drop on silk lines.

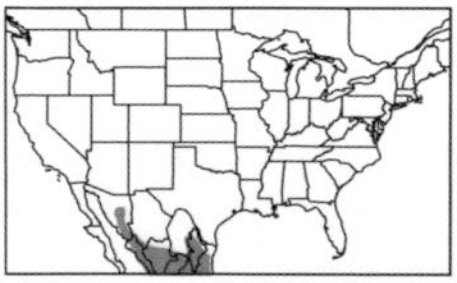

### Rosita Patch *Chlosyne rosita*

Bluish or whitish-gray with creamy sublateral stripe; head orange. *Habitat* Thorn scrub. *Found* Mainly July–Oct. *Host* Twinseeds. *Comments* A rare stray. Caterpillars only recorded from the United States once or twice.

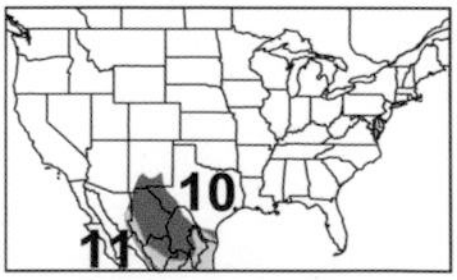


### 10. Definite Patch *Chlosyne definita* Not shown.

### 11. Banded Patch *Chlosyne endeis* Not shown.

Caterpillars unreported. *Host* Definite on *Stenandrium*. Banded on *Carlowrightia parviflora*.

1. Elf

2. Bordered Patch adult

3. Tiny Checkerspot

4. Elada Checkerspot

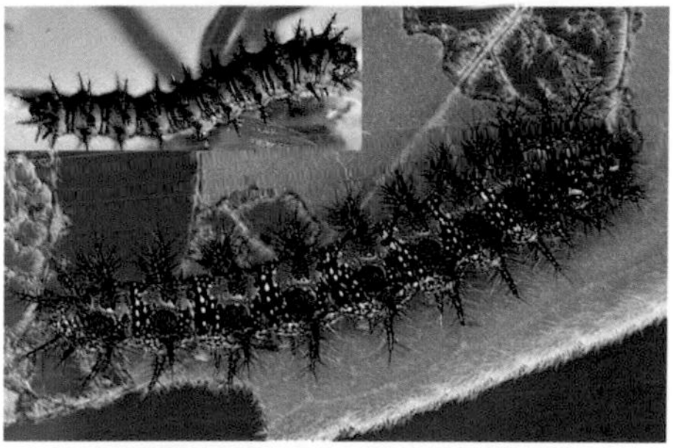

5. Bordered Patch

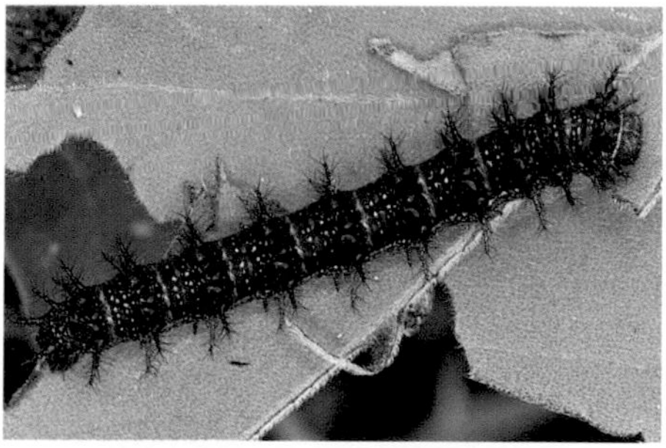

6. Bordered Patch

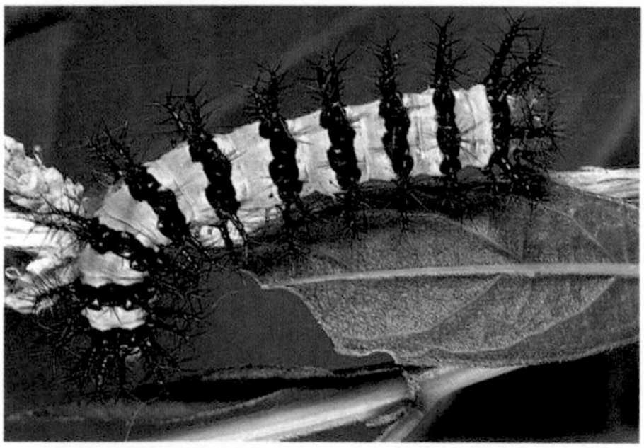

8. Crimson Patch

9. Rosita Patch

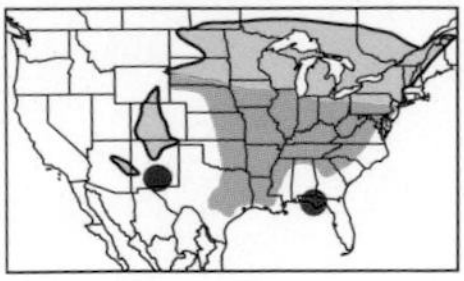

## Silvery Checkerspot *Chlosyne nycteis*

Black with orange sublateral band; **spines in orange band are brown**. *Habitat* Open deciduous woodlands and edges; stream edges in open country. *Found* March–June. *Host* Aster family, including sunflowers, brown-eyed susan, and others. *Comments* When disturbed caterpillars readily curl up and drop off of host to the ground.

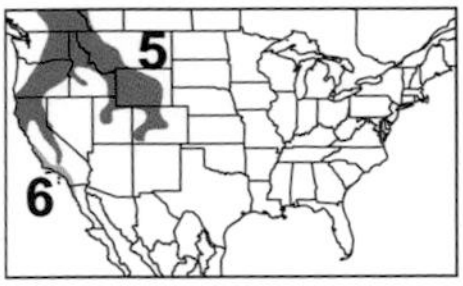


## 5. Northern Checkerspot *Chlosyne palla*

## 6. Gabb's Checkerspot *Chlosyne gabbii* Not shown.

Black with white spots and brown dashes at the base of the mid-dorsal spines. Hoffmann's Checkerspot lacks brown dashes along mid-dorsal spines. Rockslide Checkerspot occurs at higher elevations. *Habitat* A wide variety from woodland openings to sagebrush. *Found* Feb–July, depending on elevation. *Host* Mainly asters and rabbitbrushes. Gabb's especially on beach aster.

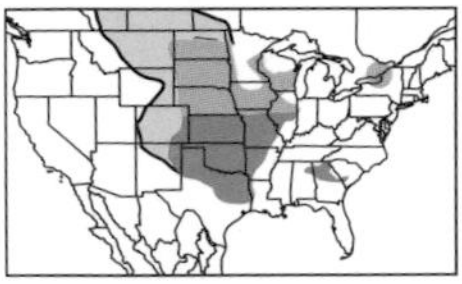

## Gorgone Checkerspot *Chlosyne gorgone*

Similar to Bordered Patch with three forms, ranging from mostly black to mostly orange; spines and head black. Bordered Patch has red-orange head. *Habitat* A great variety of open situations including prairies. *Found* April–July. *Host* Sunflowers and related plants.

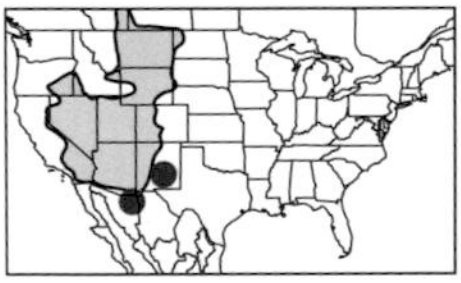

## Sagebrush Checkerspot *Chlosyne acastus*

Black with white spots and brown dashes at the base of the mid-dorsal spines. Northern Checkerspot occurs in more moist habitats and in rare instances where the two co-occur Northern Checkerspot flies later. *Habitat* Open arid areas, including desert flats and hills, sagebrush steppes, and juniper, pinyon, oak, or mixed conifer woodland. *Found* Feb–July; rarely a second flight, Aug. *Host* Desert aster, rabbitbrushes, and others.

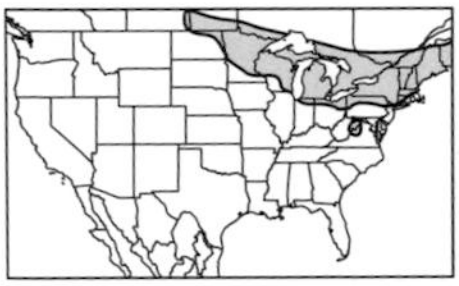

## Harris' Checkerspot *Chlosyne harrisii*

Orange with black striations; head and spines black. Very similar to Baltimore Checkerspot (pg. 90), which overlaps this species in range and habitat. *Habitat* Wet shrubby meadows and marsh borders with its host. *Found* April–May and July–Aug. *Host* Flat-topped white aster. *Comments* Eggs are laid in a cluster and young caterpillars are gregarious.

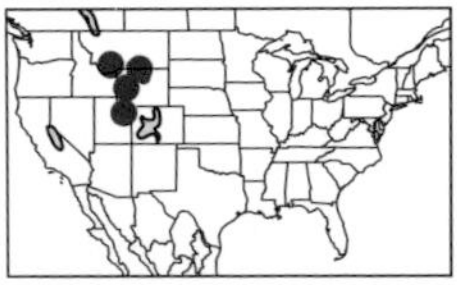

## Rockslide Checkerspot *Chlosyne whitneyi*

Not shown.

Rockslde Checkerspot is very similar to Northern Checkerspot but is found at high altitudes on rockslides. *Habitat* High elevation rockslides and scree slopes above timberline. *Found* July, probably biennial in some areas. *Host* Fleabanes.

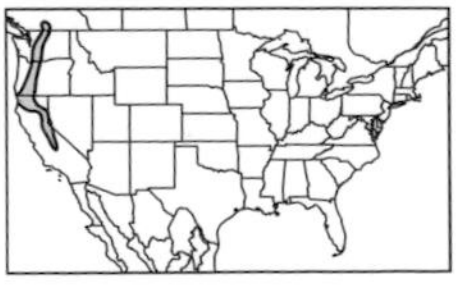

## Hoffmann's Checkerspot *Chlosyne hoffmanni*

Not shown.

Similar to Northern Checkerspot, but **lacks the brown dashes along mid-dorsal spines**. *Habitat* Openings in mountain coniferous forest. *Found* May–July. *Host* Asters.

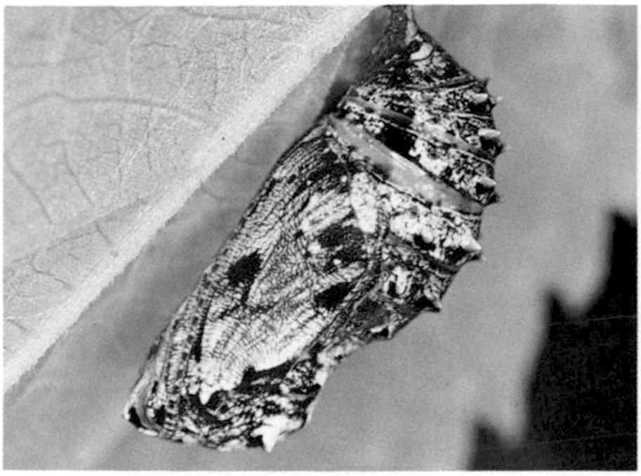

1. Silvery Checkerspot chrysalis

2. Sagebrush Checkerspot adult

3. Silvery Checkerspot

4. Silvery Checkerspot

5. Northern Checkerspot

7. Gorgone Checkerspot (head facing left)

8. Sagebrush Checkerspot

9. Harris' Checkerspot

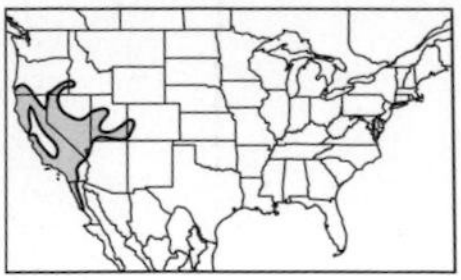

**Leanira Checkerspot** *Chlosyne leanira*
Black, with orange dorsal and lateral bands or spots, and white spots; **head black**. Desert populations lack white spots. Fulvia Checkerspot has orange head. *Habitat* Varied, including open pine and aspen woodlands, sagebrush steppes, chaparral, desert hills, and sand dunes. *Found* March–July, rarely later, 2–3 weeks prior to flight(s). *Host* Paintbrushes. *Comments* Appearance of caterpillars following winter unpredictable and related to climatic conditions from previous months or, in some cases, years.

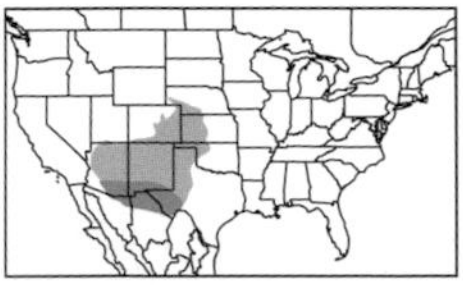

**Fulvia Checkerspot** *Chlosyne fulvia*
Black with **orange dorsal band; head orange**. Black Checkerspot has yellow not orange markings. Leanira Checkerspot in deserts similar but has black head, and ranges do not overlap. *Habitat* Open pinyon-juniper hillsides, prairie, or grassland, often in limestone, volcanic, or sandstone areas. *Found* Jan–Nov, before and after flights. *Host* Paintbrushes, and bird's beak. *Comments* Some Southwestern populations use paintbrush in the spring, then switch to bird's beak after summer rains.

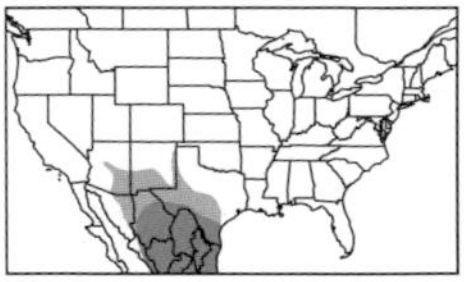

**Theona Checkerspot** *Chlosyne theona*
Black, mottled with gray; head orange. Variable Checkerspots (next page) have black heads. *Habitat* Thorn scrub, arid grasslands, open juniper woodland, and brushy canyons. *Found* Feb–Nov, most of year in south Texas. *Host* Paintbrushes, silverleafs, and others.

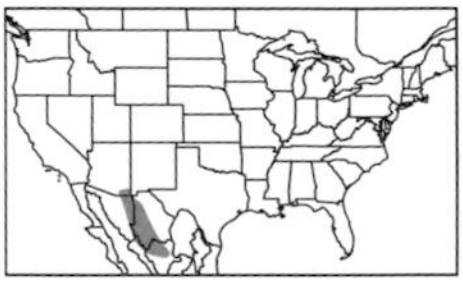

**Black Checkerspot** *Chlosyne cyneas*
Black with **yellow dorsal and lateral bands** and spots; head orange. *Habitat* Moist oak-pine woodland. *Found* March–Nov, best prior to several flights. *Host* Paintbrushes. *Comments* Restricted to four mountain ranges in the Southwest and usually found at higher elevations than Fulvia Checkerspot.

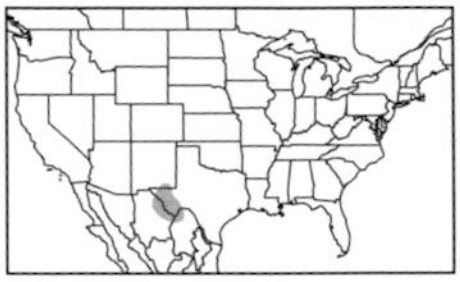

**Chinati Checkerspot** *Chlosyne chinatiensis*
Black, speckled with lighter spots; spines and head black. *Habitat* Arid limestone hills with stands of the host. *Found* March–April, prior to first brood; subsequent broods dependent upon rainfall. *Host* Silverleafs.

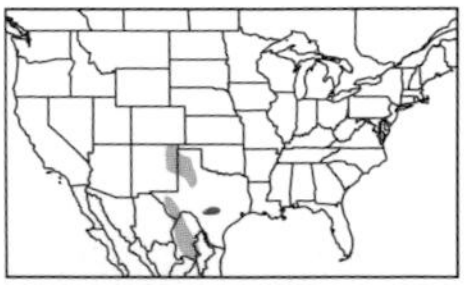

**Dotted Checkerspot** *Poladryas minuta*
Orange with short black spines. *Habitat* Limestone ridges with sparse woods. *Found* March–Aug, on leaves. *Host* Beardtongues.

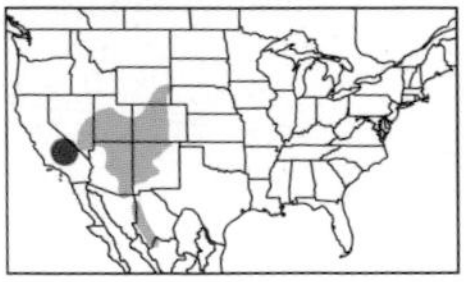

**Arachne Checkerspot** *Poladryas arachne*
Striped white and black with the top row of spines black. Variable and Edith's Checkerspots (pg. 90) have most of the top row of spines orange or at least orange at the base. *Habitat* Mountain meadows, pinyon-juniper woodland and arid grasslands. *Found* March–Aug, especially 2–3 weeks before flights. *Host* Beardtongues.

1. Leanira Checkerspot

2. Theona Checkerspot adult

3. Fulvia Checkerspot

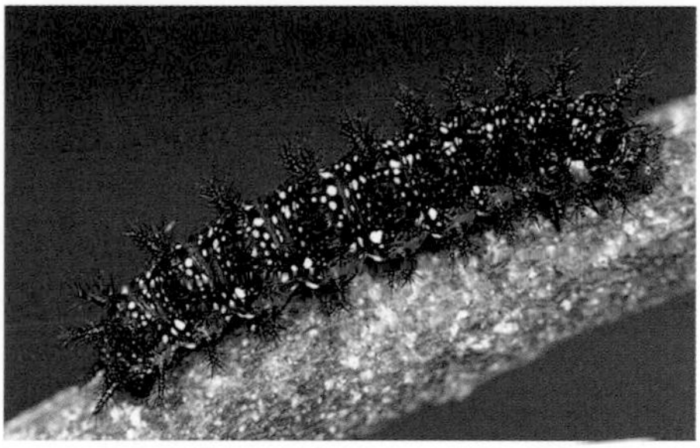

4. Theona Checkerspot

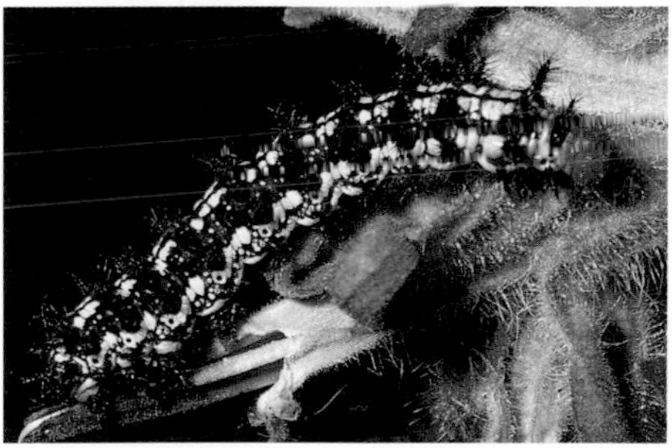

5. Black Checkerspot

6. Chinati Checkerspot

7. Dotted Checkerspot

8. Arachne Checkerspot

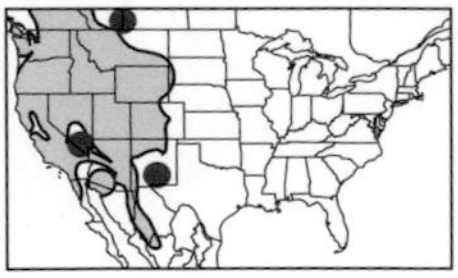

## Variable Checkerspot *Euphydryas chalcedona*

Variable. Body black, gray or mottled with varying numbers of white blotches or bands; head black. Top row of spines orange or at least with some orange at base. May resemble Edith's Checkerspot in some populations, although Edith's Checkerspot caterpillars usually precede Variable Checkerspot caterpillars by 3 weeks in habitats where they occur together. Arachne Checkerspot has top row of spines black, black at base. *Habitat* many open situations, including mountain meadows, desert canyons, chaparral, and high-elevation barrens. *Found* Jan–July, depending on elevation, mainly prior to flight. *Host* Beardtongues, paintbrushes, monkeyflowers, snowberries, and others. *Comments* Included here are "Anicia" Variable Checkerspot and other populations considered by some to be distinct species.

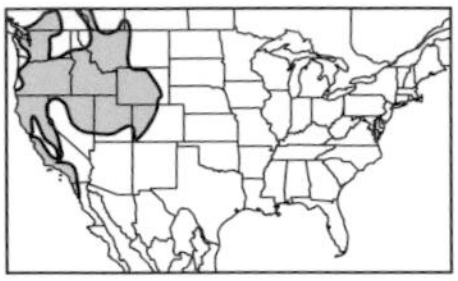

## Edith's Checkerspot *Euphydryas editha*

Variable. Body black, gray or mottled with varying numbers of white blotches or bands; head black. Top row of spines orange or at least with some orange at base. Some populations may resemble Variable Checkerspot, although Variable Checkerspot caterpillars usually appear later in most habitats where they occur together. *Habitat* Mainly in localized stressed habitats, including ocean bluffs, serpentine chaparral, desert edges, ridgetops in sagebrush scrub, and rocky outcrops above treeline. *Found* Feb–July, depending on elevation, best prior to single flight. *Host* Paintbrushes, Chinese house, lousewort, plantain, and others. *Comments* One of the few butterflies that uses annuals as hosts, primarily post-hibernation.

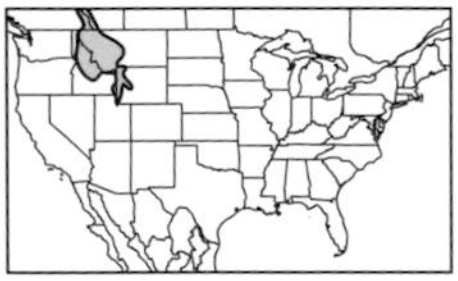

## Gillett's Checkerspot *Euphydryas gillettii*

Black with broad yellow dorsal stripe and white lateral stripe; head and spines black. *Habitat* Moist openings and meadows near streams and conifers, sometimes sagebrush edges. *Found* April–May, prior to flight. *Early Instars* July–Aug, after flight. *Host* Mainly twinberry honeysuckle; other plants are used after hibernation. *Comments* Young caterpillars live gregariously in a communal silk feeding-web on the leaf on which they emerged and then overwinter in the web.

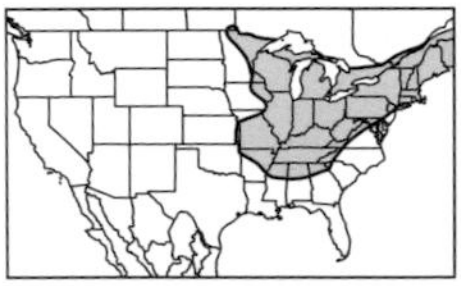

## Baltimore Checkerspot *Euphydryas phaeton*

Orange with black striations, **head and rear black**. Harris' Checkerspot (pg. 86) is very similar but uses a different host. *Habitat* Usually marshes/wet meadows with turtlehead, but also found in dry fields. *Found* March–April. *Early Instars* June–Aug, after single flight. *Host* Turtlehead before overwintering, English plantain, viburnum, and others prior to single flight.

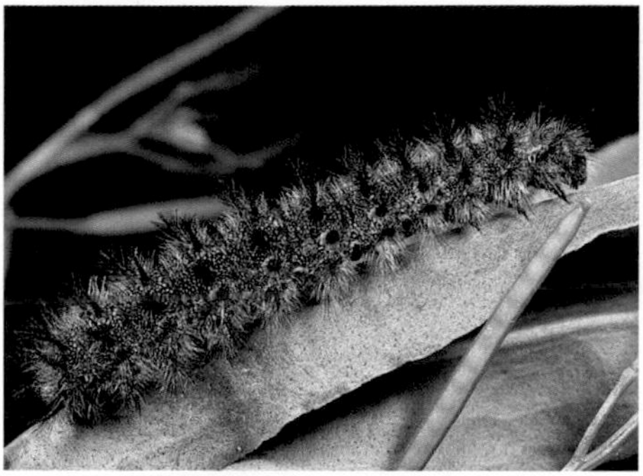

1. 'Anicia' Variable Checkerspot

2. Baltimore Checkerspot adult

3. 'Anicia' Variable Checkerspot

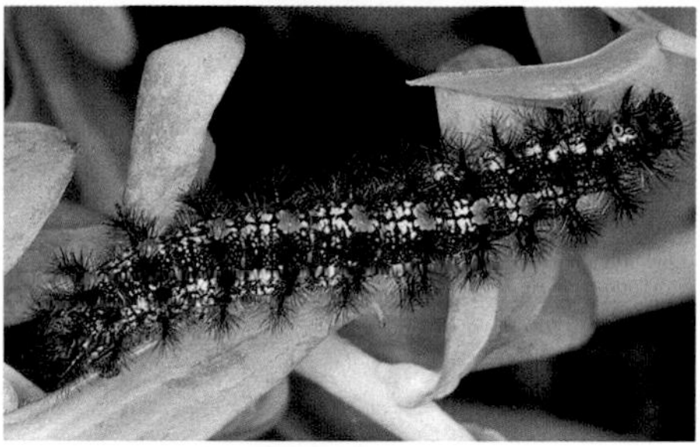

4. 'Chalcedon' Variable Checkerspot

5. 'Chalcedon' Variable Checkerspot

6. Edith's Checkerspot

7. Gillett's Checkerspot

8. Baltimore Checkerspot

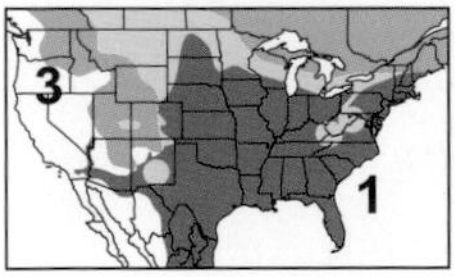


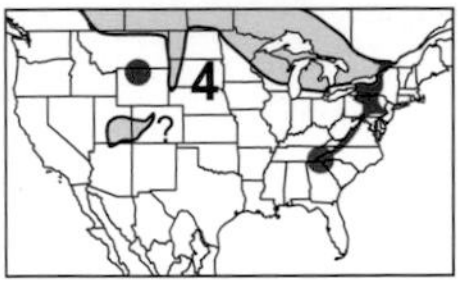


***1. Pearl Crescent** *Phyciodes tharos*

**3. Northern Crescent** *Phyciodes selenis* Not shown.

**4. Tawny Crescent** *Phyciodes batesii*

Pearl and Northern Crescents are identical. Brown with brown spines and **white lateral line;** head black with white markings. Tawny Crescent averages darker. *Habitat* Pearl Crescent is widespread in open situations, fields, meadows, power-line cuts, suburbia, etc. Northern more closely tied to openings in and near woodlands, Tawny in moist woodland or dry pastures. *Found* All year in the South, April–Oct northward. *Host* Asters. *Comments* Pearl and Northern are common; Tawny is rare.

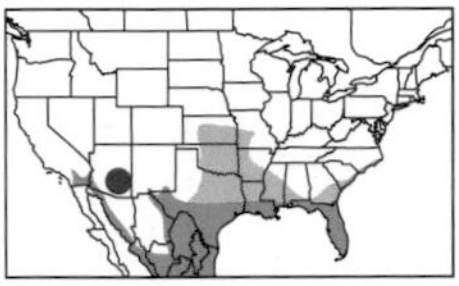

***Phaon Crescent** *Phyciodes phaon*

Brown with **black dorsal and lateral stripes**. *Habitat* Moist open situations with its low, mat-forming host. Often along roadsides, trails or lake beds. *Found* All year in Florida and south Texas; March–Oct elsewhere. *Host* **Frogfruits**.

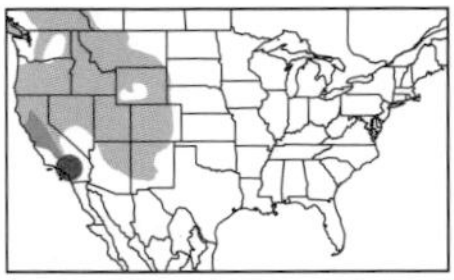

**Field Crescent** *Phyciodes campestris*

Blackish-brown with fine black striations and two cream to gray stripes along back; **head black with light crescent around eyes**. *Habitat* Plains to mountain fields, meadows and brushy coastal habitats. *Found* March–Oct. *Host* Asters.

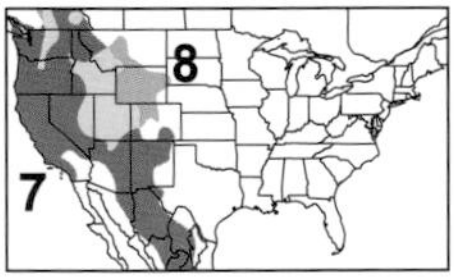


**7. Mylitta Crescent** *Phyciodes mylitta*

**8. Pale Crescent** *Phyciodes pallida* Not shown.

Mylitta Crescent is black, mottled with white dots, with fine whitish dorsal stripes and a wide lateral stripe speckled with black, containing orange spines. Pale Crescent caterpillar is unknown, probably similar. *Habitat* Open situations, from mountain meadows to agricultural fields. *Found* Mylitta most of year southward, spring to fall northward. Pale Crescent April–July. *Host* Thistles. *Comments* Mylitta prefers wetter habitats than does Pale Crescent, although they may occur together.

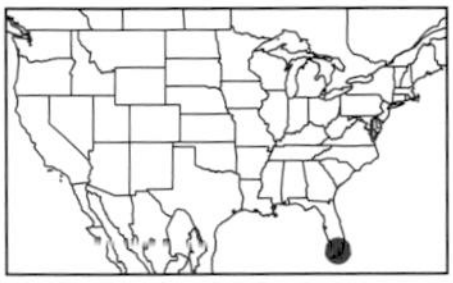

**Cuban Crescent** *Phyciodes frisia*

Mottled brown and black with **orange spines above head;** lateral band more brown than orange. *Habitat* Edges of subtropical hammocks and other open situations. *Found* Most of year in the South. *Host* Shrimpflower and crimson dicliptera.

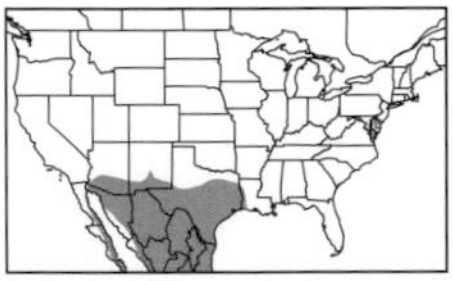

***Texan Crescent** *Phyciodes texana*

Gray, mottled black with a **broad buff lateral band**. *Habitat* Open woodland, thorn scrub, roadsides and parks. *Found* Feb–Oct, before and after flights. *Host* Acanthus family, including *Dicliptera*, *Jacobinia*, and *Ruellia*. *Garden Tips* Twinseeds.

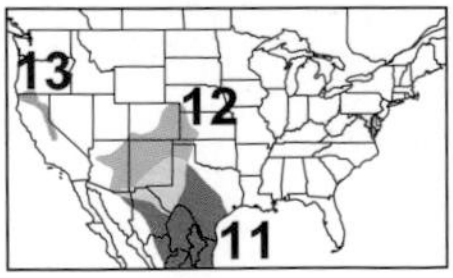


**11. Vesta Crescent** *Phyciodes vesta* Not shown.

**12. Painted Crescent** *Phyciodes picta* Not shown.

**13. California Crescent** *Phyciodes orseis* Not shown.

See p. 161–62 for more information about crescents.

1. Pearl Crescent

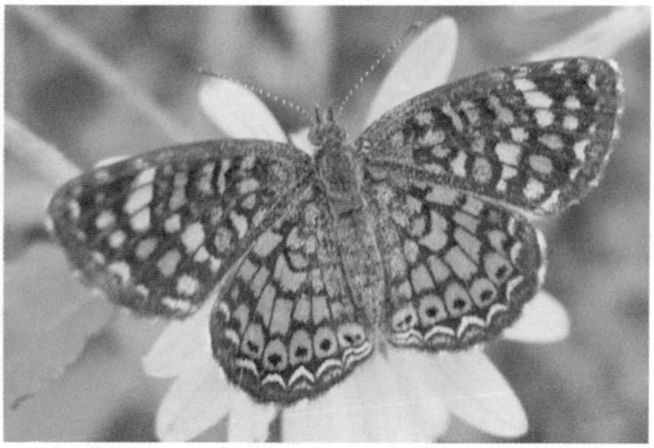

2. Vesta Crescent adult

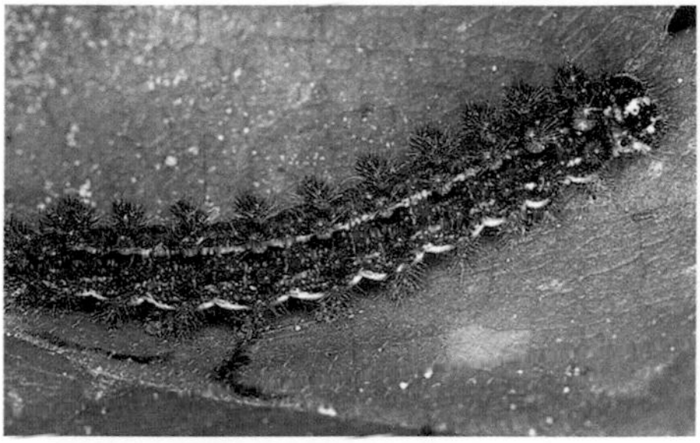

4. Tawny Crescent

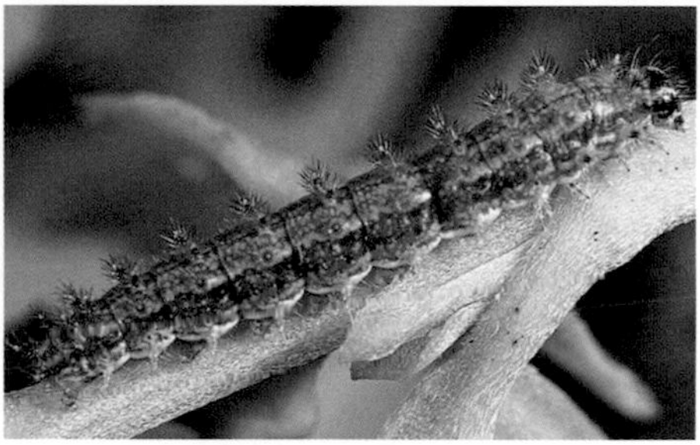

5. Phaon Crescent

6. Field Crescent

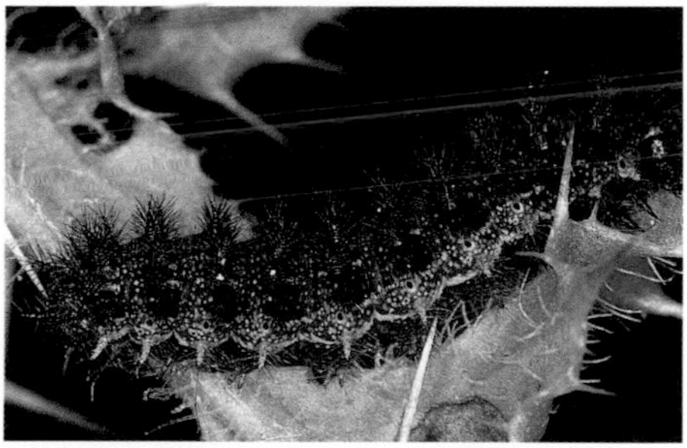

7. Mylitta Crescent

9. Cuban Crescent

10. Texan Crescent

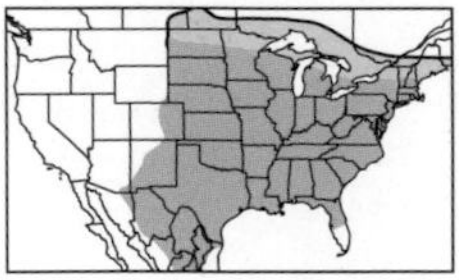

***Question Mark** *Polygonia interrogationis*

Variable. Gray, black, or yellow with cream to orange stripes. May have four rows of orange spines behind head. *Habitat* Woodlands and adjacent openings. *Found* May–June, July–Sept, between flights. *Host* Elms, hackberries, nettles, and others.

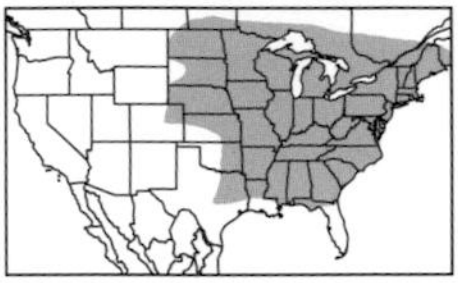

**Eastern Comma** *Polygonia comma*

Variable. Black, brown, yellow, or cream with white or yellow spines and a white lateral band. *Habitat* Woodlands and adjacent openings. *Found* May–June, July–Aug, between flights. *Host* Elms, nettles, and hops. *Comments* Lives in leaf shelter formed by fastening leaf edges together.

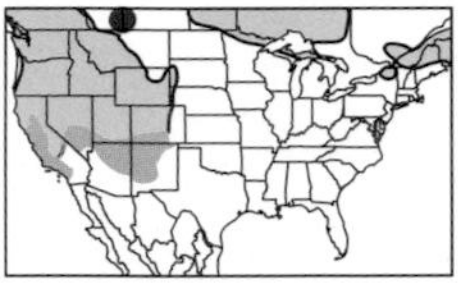

**Hoary Comma** *Polygonia gracilis*

Variable. Gray with white dashes dorsally and **four rows of orange spines behind head;** rear spines white or black. Green Comma and Question Mark use different hosts. *Habitat* Northern hardwood forests. *Found* May–July. *Host* Gooseberries.

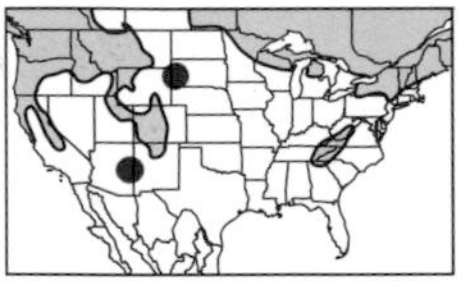

**Green Comma** *Polygonia faunus* Not shown.

Very similar to Hoary Comma. *Habitat* Streamsides and openings in coniferous or mixed mountain woodlands. *Found* Mainly May–July depending on region. *Host* Willows, birches, alders, and others.

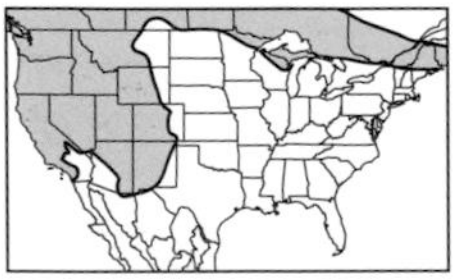

**Satyr Comma** *Polygonia satyrus*

Black and white (proportions variable) with white spines and black lateral band with orange. See Red Admiral pg. 99. *Habitat* Riparian areas in woodlands and in semi-desert regions. *Found* March–Nov, after flights. *Host* Nettles and willows. *Comments* Lives in drooping leaf shelter formed by eating through leaf base and fastening edges of the leaf together.

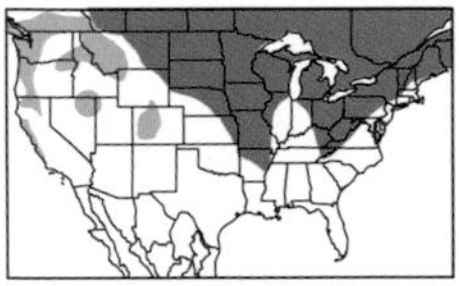

**6. Gray Comma** *Polygonia progne*

**7. Oreas Comma** *Polygonia oreas* Not shown.

Variable. Yellow to brown, with black dashes on sides; spines are yellow to brown. *Habitat* Mainly rich, deciduous woodlands with northern elements. *Found* May–June, July–Sept, between flights. *Host* Gooseberries. *Comments* Do not make nests.

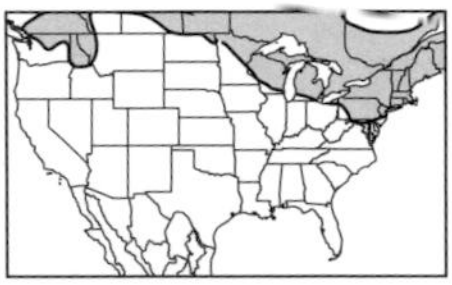

**Compton Tortoiseshell** *Nymphalis vau-album*

Variable. Greenish to black with several rows of pale yellow to white dashes along the body. Similar to commas. *Habitat* Woodlands. *Found* May–June. *Host* Birches and willows. *Comments* Gregarious.

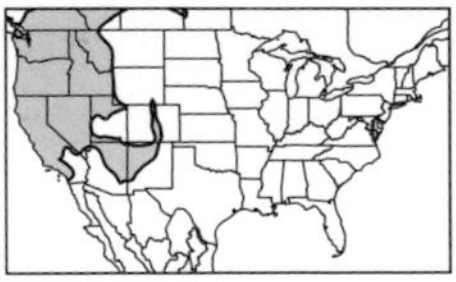

**California Tortoiseshell** *Nymphalis californica*

Variable. Dorsal yellow stripe divided by black mid-dorsal line; head black; spines black or orange. Inset shows early instar. *Habitat* Coniferous and mixed woodland. *Found* March–Aug, prior to or after flights. *Host Ceanothus*. *Comments* Gregarious until late in development.

1. Question Mark

2. Hoary Comma adult

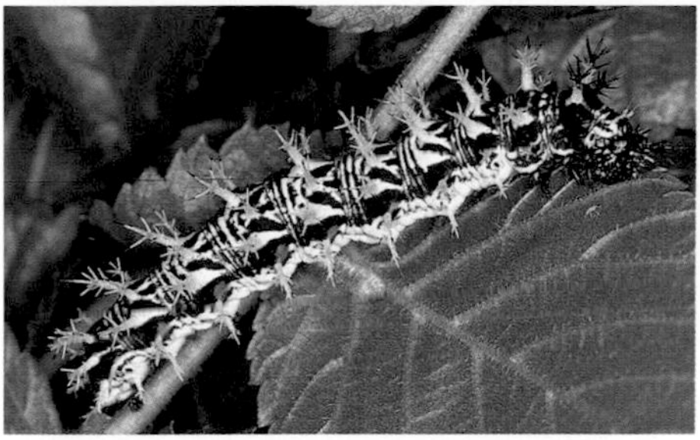

3. Eastern Comma

4. Hoary Comma

5. Satyr Comma

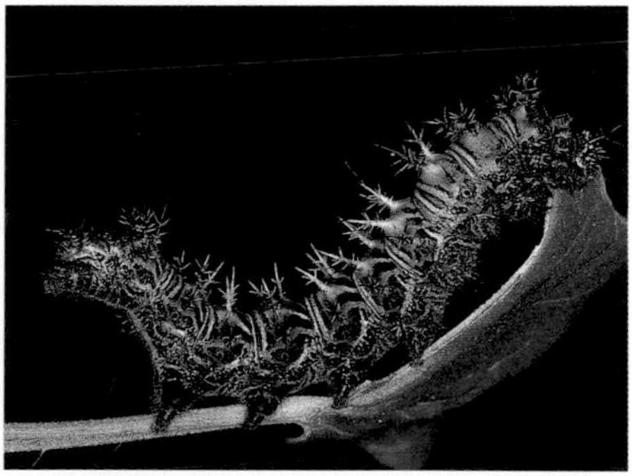

6. Gray Comma

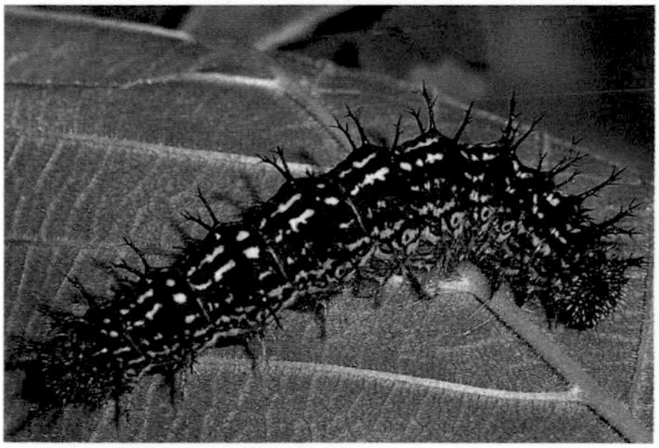

9. Compton Tortoiseshell

10. California Tortoiseshell

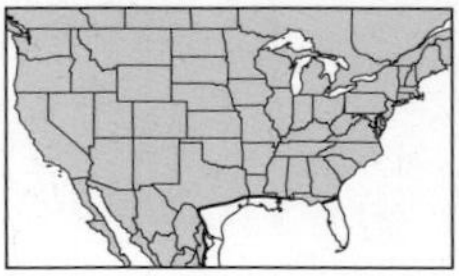

**Mourning Cloak** *Nymphalis antiopa*
Black, speckled with small white dots and **red-orange dorsal spot band;** head black and without spines; **prolegs orange/red**. *Habitat* Characteristic of hardwood forests, but may be found many habitats including fields and suburbs. *Found* Feb–July, prior to single flight. *Host* Willows, poplars, birches, and others. *Comments* Caterpillars gregarious throughout development.

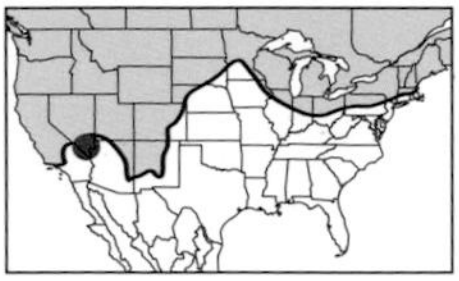

**Milbert's Tortoiseshell** *Nymphalis milberti*
Black, speckled with small white dots, with **greenish-yellow lateral stripes**, grayish-green ventrally; head black and lacks spines. *Habitat* Mountains, open fields, usually moist, near woodlands or streams. *Found* April to early June, and July–Aug, between flights. *Host* Nettles. *Comments* Early instars are gregarious, often found in silken web; older caterpillars are more solitary.

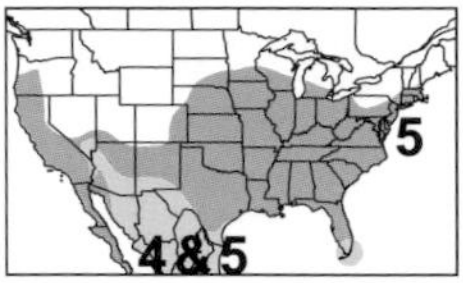


**4. Tropical Buckeye** *Junonia genoveva*
***5. Common Buckeye** *Junonia coenia*
Black with **blue dashes at base of dorsal spines**, white lateral stripes or dashes with lateral spines orange at base; head with two short black spines on top. Tropical often darker than Common Buckeye. *Habitat* Open and/or disturbed areas. *Found* Most of year. *Host* Plantains, monkeyflowers, speedwells, and others. *Comments* Common Buckeye much more common. *Garden tips* For Common Buckeye try frogfruits or snapdragons.

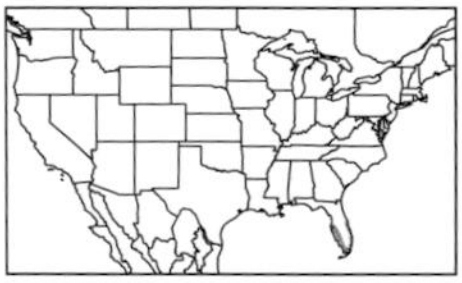

**Mangrove Buckeye** *Junonia evarete* Not shown.
Variable and not really separable from Common Buckeye, but sometimes darker. *Habitat* Edges of black mangrove swamps and adjacent areas. *Found* Most of year. *Host* **Black mangrove**.

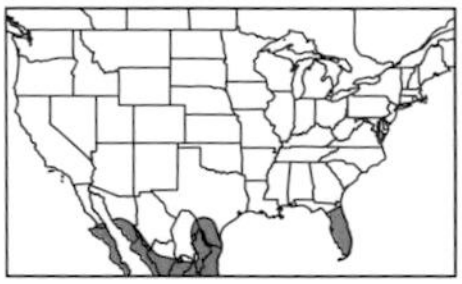

***White Peacock** *Anartia jatrophae*
Black, often with broken bands of small white or silver dots; **spines orange at base; head with long clubbed horns**. *Habitat* Open and/or disturbed areas in the tropics and subtropics. *Found* All year in south Florida and Texas. *Host* Water hyssop, frogfuits, various verbenas, and acanthus family plants. *Garden tips* Smooth water hyssop and green shrimp plant.

**Banded Peacock** *Anartia fatima*
Black, **speckled with small white/silver dots**, black lateral band prominent; spines reddish-brown to brown at base; **head with long clubbed horns**. *Habitat* A variety of subtropical open areas and disturbed places. *Found* A rare but regular stray. *Host* Ruellias.

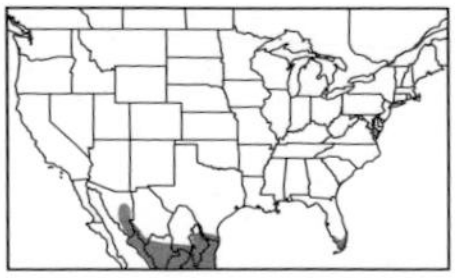

**Malachite** *Siproeta stelenes*
Black with bright pink, orange, or red dorsal spines, often black-tipped and swollen at the base; head with two red or black, long clubbed spines. *Habitat* Tropical hardwood hammocks; overgrown avocado and citrus groves. *Found* All year. *Host* Green shrimp plant and relatives. *Garden tips* Green shrimp plant.

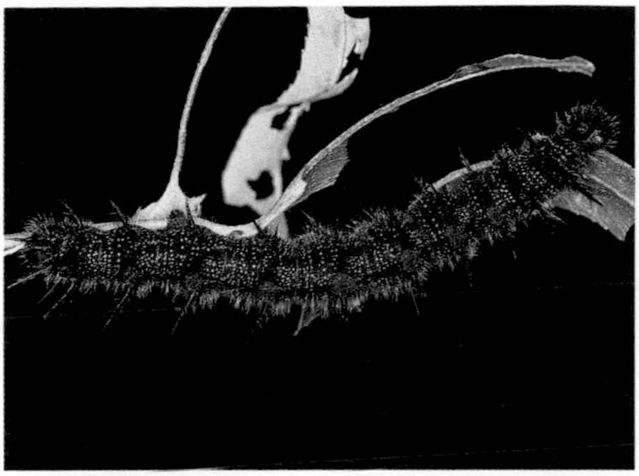

1. Mourning Cloak

2. Common Buckeye adult

3. Milbert's Tortoiseshell

4. Tropical Buckeye

5. Common Buckeye

7. White Peacock

8. Banded Peacock

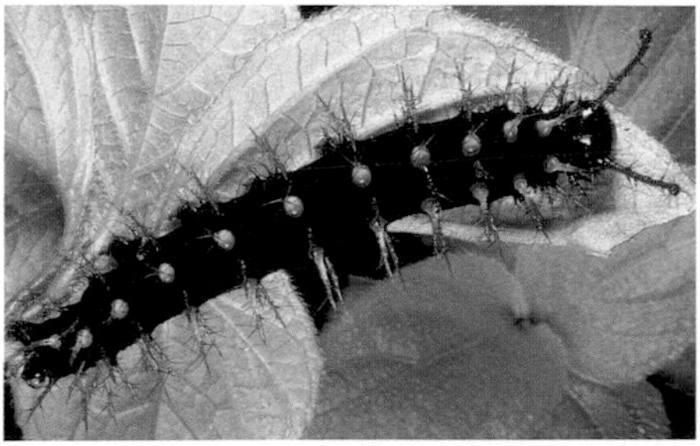

9. Malachite

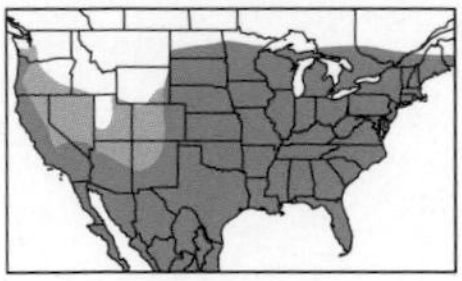

***American Lady** *Vanessa virginiensis*
Black to reddish with cream to yellow bands and white spots on seven abdominal segments. *Habitat* Open spaces, including fields, meadows, roadsides, and coastal dunes. *Found* Most of year in South, farther north mainly May–June, late July–Aug, in a silken nest on host. *Host* Pearly everlastings, cudweeds, pussytoes and others. *Garden Tips* Pearly everlastings. *Comments* A widespread and common garden butterfly.

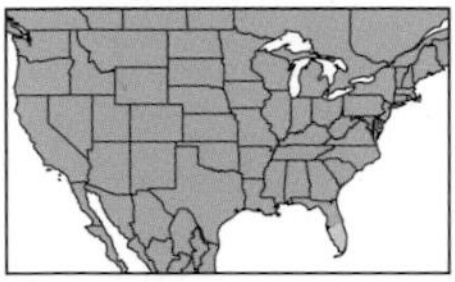

***Painted Lady** *Vanessa cardui*
Variable. Brown with yellow sublateral line and fine yellow dashes on lateral surface and forming a broken double stripe dorsally; spines and hair gray to white. *Habitat* Can be found in any type of open habitat. *Found* Feb–Sept, between flights, in a silken nest on the host. *Host* Prefers thistles and related plants. Feeds on a variety of plant genera during migrations and at other times when adults are in abundance. *Garden Tips* Hollyhocks and thistles. *Comments* Along with the Gray Hairstreak, Painted Ladies have the most varied diet of North American butterflies. Each year, Painted Ladies stream out of northern Mexico during March and April in impressive migratory swarms. Both the numbers of butterflies and the extent of the territory they reach, varies widely from year to year.

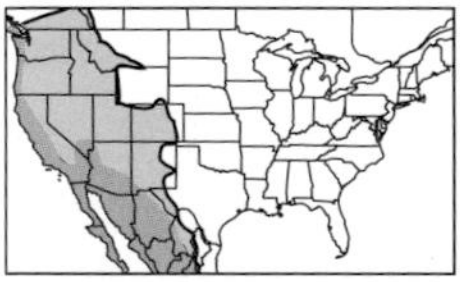

**West Coast Lady** *Vanessa annabella*
Variable. Black to tan or reddish, heavily speckled with yellow or orange dots and dashes. Could be confused with Painted Lady. *Habitat* Open situations, often disturbed, and gardens. *Found* April–Oct, between flights, in a silken nest on the host. *Host* Mallows and nettles. *Garden Tips* Hollyhocks, globe mallows.

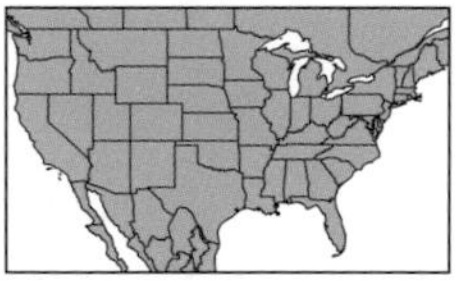

***Red Admiral** *Vanessa atalanta*
Variable. Black speckled with white dashes; spines black; head black without spines on top. Satyr Comma has pair of spines on head. *Habitat* Open situations including fields, beaches, suburbia, woodland roads, and moist meadows near woodlands. *Found* Most of year in South, between flights northward, May–Sept, in nest of tied leaves of the host. Mainly common to abundant. *Host* Nettles. *Garden Tips* Pellitory and false nettle. *Comments* Another species that is strongly migratory, repopulating northern regions each spring.

1. American Lady pupa

2. Painted Lady adult

3. American Lady

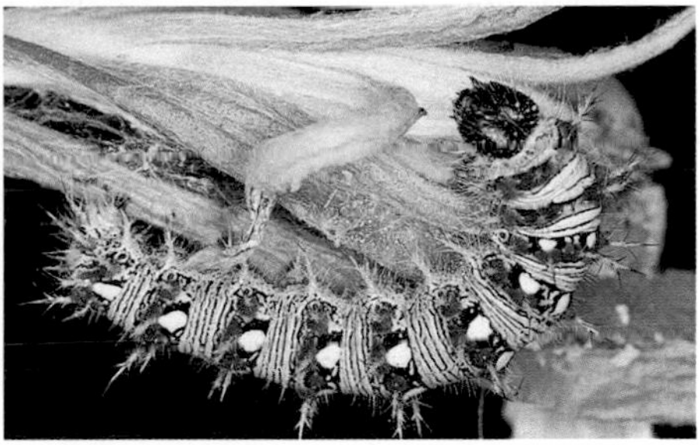
4. American Lady

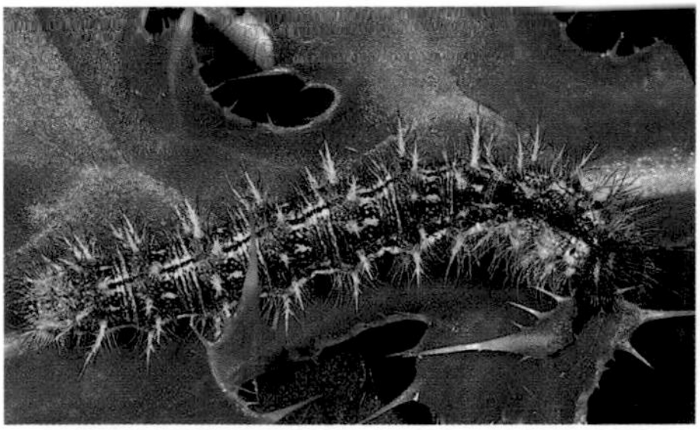
5. Painted Lady

6. Painted Lady

7. West Coast Lady

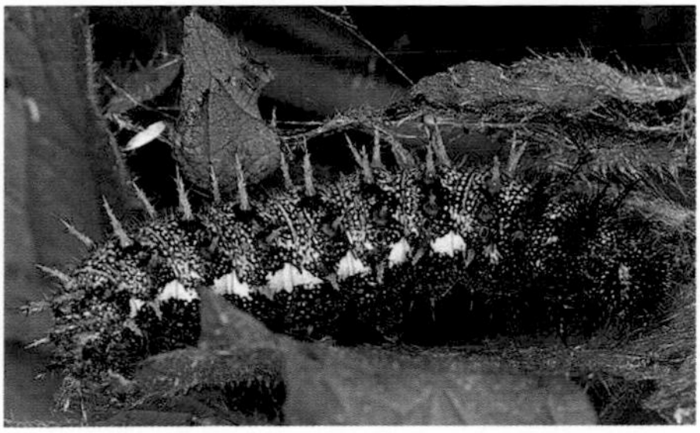
8. Red Admiral

## Admirals & Relatives

Beautifully mottled and ornate caterpillars of the admiral group resemble bird droppings. This distinctive group has bristled tubercles rather than branched spines. The tubercles behind the head and at the rear are the most developed and have a horn-like appearance. Eggs are laid singly, usually at the outer tip of the host leaves. Young caterpillars also resemble small bird droppings and overwinter in a hibernaculum, a rolled leaf partially chewed and attached to the stem of the host with silk. The hosts are trees, but seedlings as well as short shrubby species are used so locating caterpillars is not difficult. The four admirals themselves are variable. Brown, reddish-brown to olive green with a cream or pink saddle-like patch on top of the back and often behind the head, a long pair of tubercles behind the head, shorter ones at the rear, and raised humps elsewhere along the back.

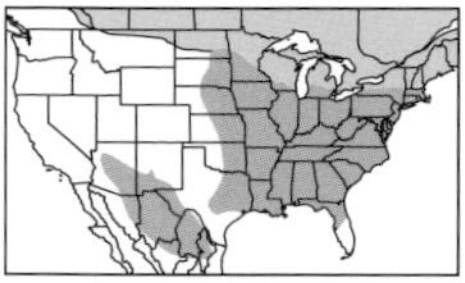

### Red-spotted Admiral *Limenitis arthemis*

Weidemeyer's Admiral usually found at higher elevations where ranges overlap. *Habitat* Rich moist woodlands are preferred, but the species is widely distributed and often found in suburban areas. *Found* Feb–Sept, before, between, and after flights. *Host* Cherries, poplars, willows, and other trees. *Comments* This species, and other admirals, can be found during winter months on the hosts, inside a hibernaculum.

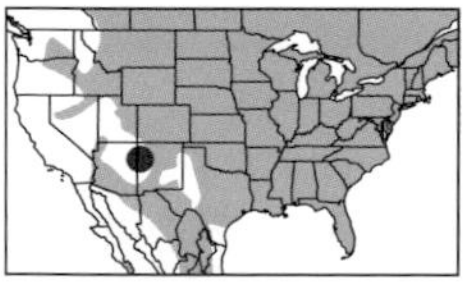

### Viceroy *Limenitis archippus*

Variable. Similar to Red-spotted Admiral, **usually green** (photo 4) but also brown to olive-brown (early instar, photo 5), **often with shorter head tubercles**. Weidemeyer's Admiral occur at higher elevations where ranges overlap. *Habitat* Open areas adjacent to watercourses or wet areas with willows including canals. *Found* Feb–Sept, before, between, and after flights. *Host* Willows.

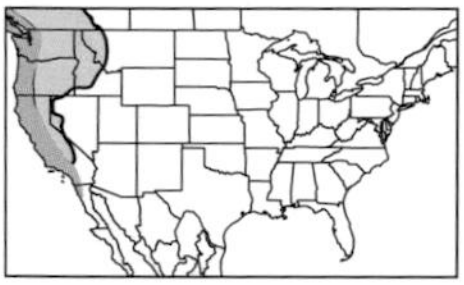

### Lorquin's Admiral *Limenitis lorquini*

Probably not reliably distinguishable from Weidemeyer's Admiral. *Habitat* Openings and edges of moist forest and riparian areas. *Found* Mar–Sept, before, between, and after flights. *Host* Mainly willows, but also aspens and others.

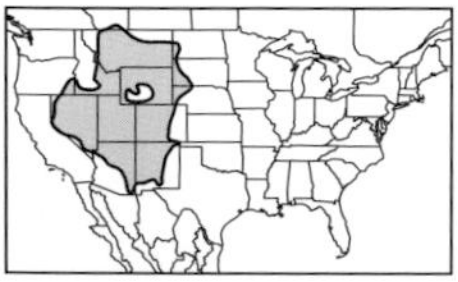

### Weidemeyer's Admiral *Limenitis weidemeyerii*

*Habitat* Moist mountain forest, washes, and riparian areas in arid country. *Found* May–June, before and after single flight, later in rare cases of second brood. *Host* Willows, aspens, serviceberries, and others.

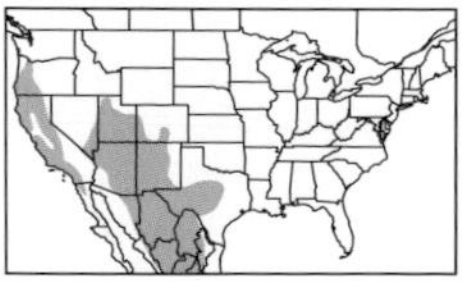

### California Sister *Adelpha bredowii*

Green to yellow with 4 to 6 pairs of sparsely branched green, yellow-orange, or tan spines. *Habitat* Oak woodland, often in canyons or washes. *Found* Mar–Nov, before, between, and after flights, often on upper side of host leaves. *Host* Oaks. *Comments* Overwinters as a partially grown caterpillar but not in a hibernaculum.

1. Red-spotted Admiral

2. California Sister adult

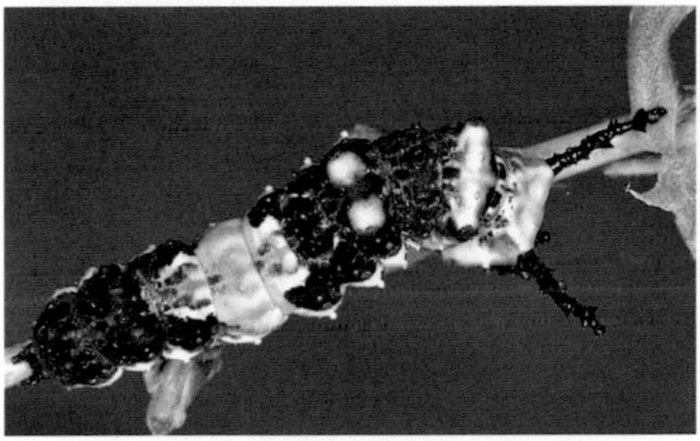

3. Red-spotted Admiral

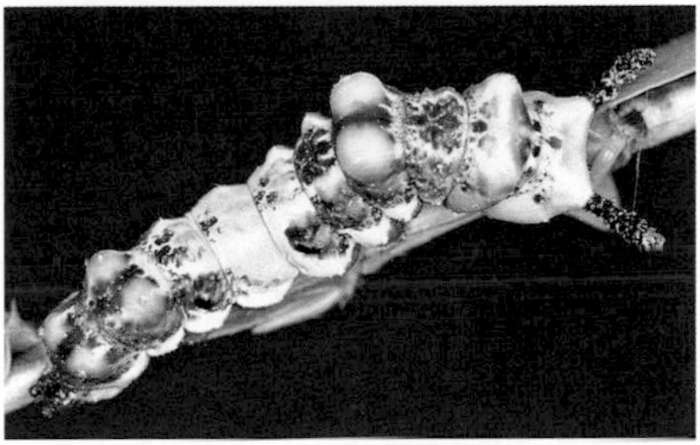

4. Viceroy

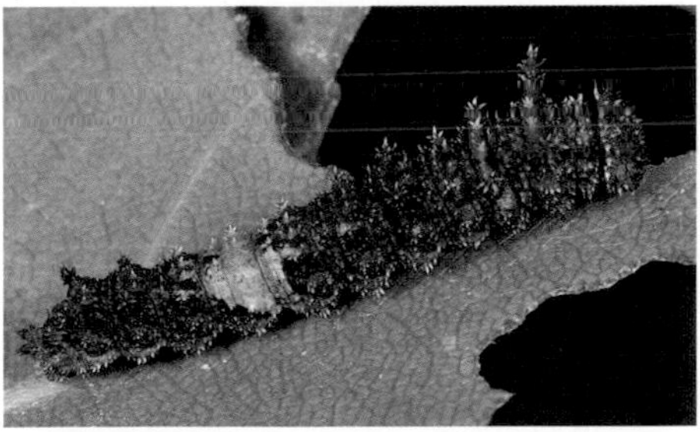

5. Viceroy (earlier instar)

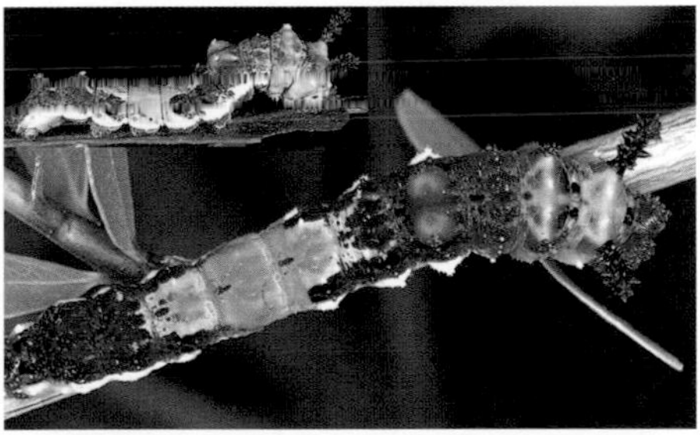

6. Lorquin's Admiral

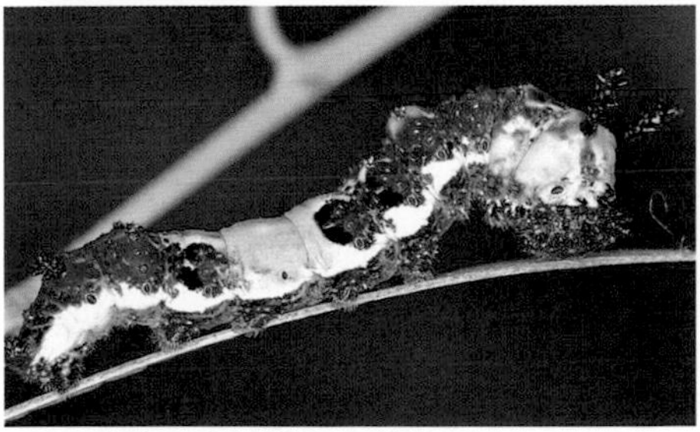

7. Weidemeyer's Admiral

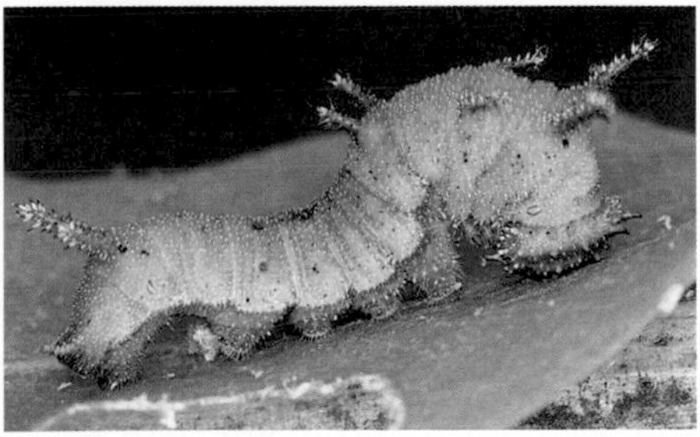

8. California Sister

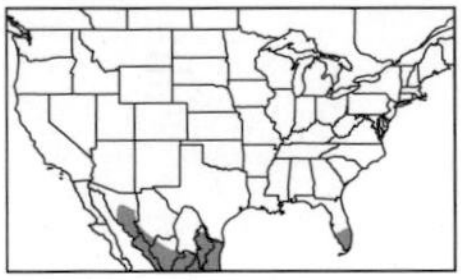

***Ruddy Daggerwing** *Marpesia petreus*
**Distinctive color pattern**. White or yellow with green, orange, and tan blotches; four long unbranched dorsal horns, head with long horns not clubbed at tips. *Habitat* Edges and openings of tropical woodlands and gardens. *Found* All year in Florida; strays to southern Texas and southern Arizona. *Host* **Figs**. *Garden Tips* Figs. *Comments* Caterpillars cut the midrib, near the base of the leaf, to stop the flow of gummy latex.

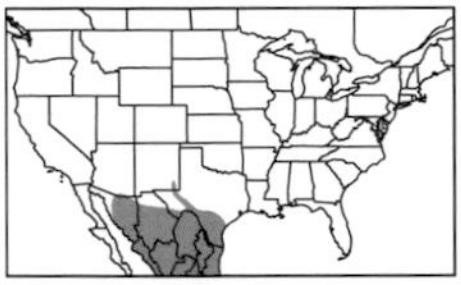

**Common Mestra** *Mestra amymone*
**Brown with green dorsal oblong patches**; head with long brown horns, black clubs at tips. *Habitat* Openings and edges of tropical woodlands and thorn scrub. *Found* July–Oct, after flights in southern Texas; mainly Aug–Oct in southern Arizona. *Host* Noseburns.

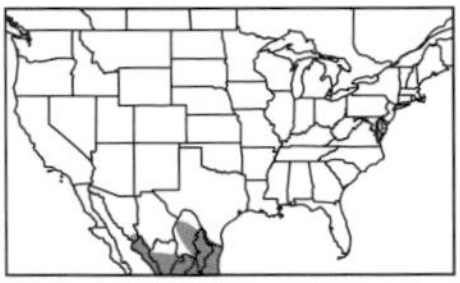

**Red Rim** *Biblis hyperia*
Swirled tan, dark brown and green with **tan V-shaped dorsal mark** toward the rear; head horns are twice as long as those on Common Mestra. *Habitat* Tropical woodlands and thorn scrub. *Found* Aug–Nov. in southern Texas, after flights. May not breed every year. *Host* Noseburns.

**Blue-eyed Sailor** *Dynamine dyonis*
Frosty pale, with green dorsal stripe and branched spines. *Habitat* Tropical woodland and scrub. *Found* All year, between flights. *Host* Euphorbiacaeae and noseburns. *Comments* Very aggressive, even cannibalistic, toward their own and other species.

**Gray Cracker** *Hamadryas februa*
Variable. Black, grayish, or orange, finely striated with white, yellow, orange, and black lines dorsally; **head red or orange with pair of long clubbed horns**; spines white or black, orange at the base. *Habitat* Tropical woodlands. *Found* All year. Breeding intermittent. *Host* Noseburns. *Comments* Young caterpillars rest on droppings and silk extensions of leaf veins.

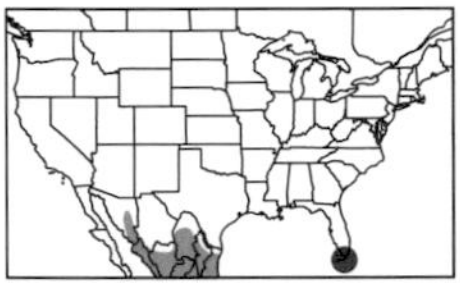

**Dingy Purplewing** *Eunica monima*
Brown with a pale lateral stripe and an orange and black head. *Habitat* Tropical woodland. *Found* All year. *Host* Gumbo-limbo. *Comments* Communal in a silken web.

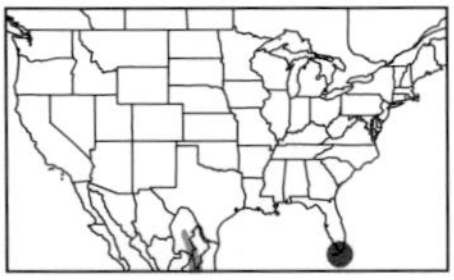

**Florida Purplewing** *Eunica tatila*
Dull orange to reddish-green, with black or white lateral band; spines small and black; head with pair of long sparsely branched horns, clubbed at the tips. *Habitat* Tropical woodland. *Found* All year. *Host* Crabwood.

**Mexican Bluewing** *Myscelia ethusa* Not shown.
Similar to purplewings and crackers. Head with pair of long branching horns, clubbed at the tips. *Habitat* Tropical woodland. *Found* All year. *Host* Adelias.

1. Ruddy Daggerwing

2. Dingy Purplewing adult

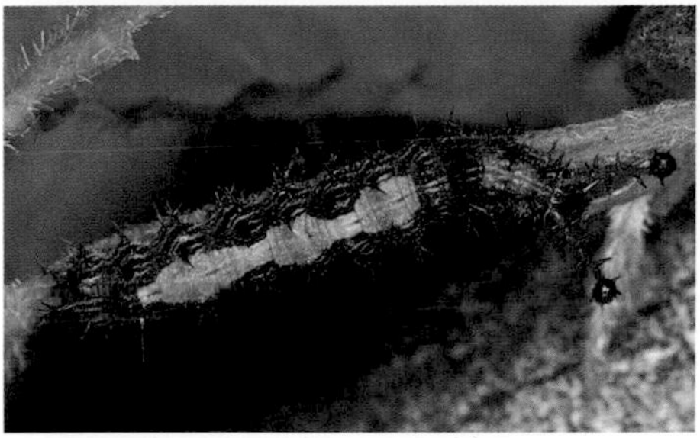

3. Common Mestra

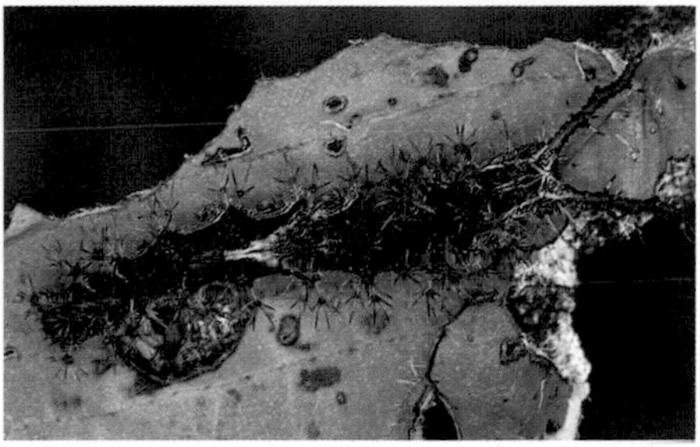

4. Red Rim

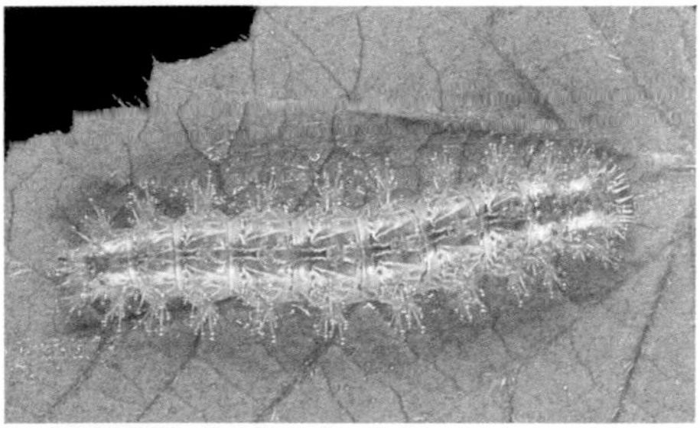

5. Blue-eyed Sailor

6. Gray Cracker

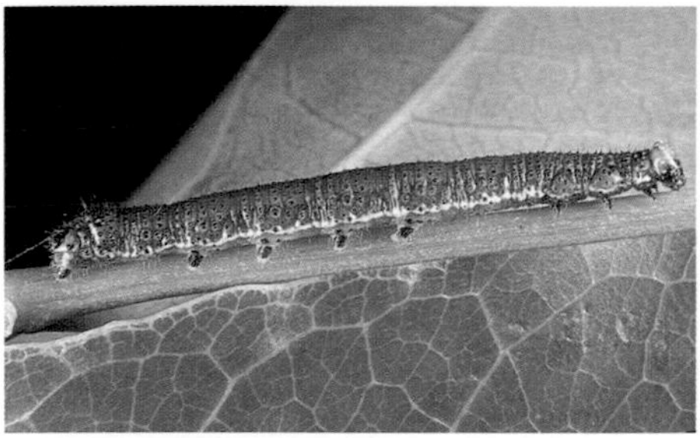

7. Dingy Purplewing

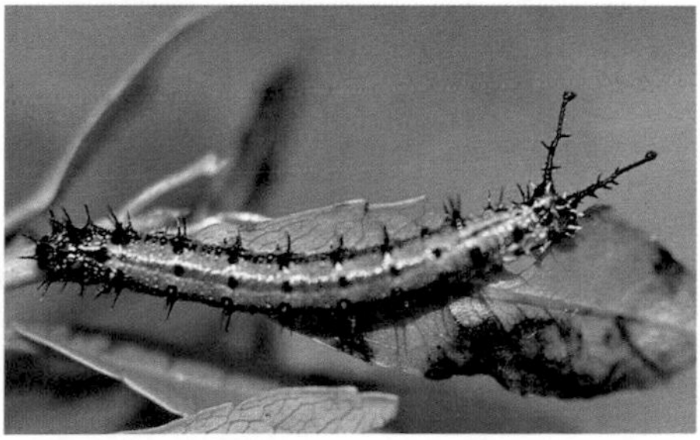

8. Florida Purplewing

## **Leafwings** (Charaxinae) and **Emperors** (Apaturinae)

Lack branching spines on the body. Leafwings have short knobs on head and other raised bumps. Older caterpillars usually are found in rolled leaf nest on the host. Emperors have long branching head horns, and the rear ends in a shallow fork. Eggs are laid in clusters, (small numbers by Empress Leilia and Hackberry Emperor), large numbers (over 100) by Tawny Emperor.

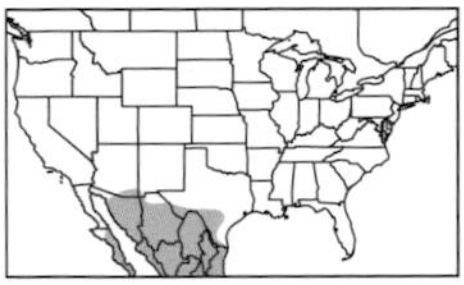

### Tropical Leafwing *Anaea aidea*

**Tan** with **prominent orange tubercles on head** and **red/brown blotches on side**. *Habitat* Tropical woodlands and thorn scrub. *Found* All year in southern Texas, Aug–Oct in southern Arizona. *Host* Crotons.

### Florida Leafwing *Anaea floridalis*

Tan (young) to green (mature) with **prominent orange tubercles on head** and **yellow lateral stripe; no side blotch**. *Habitat* Pine rockland. *Found* Most of year. *Host* **Pineland croton**.

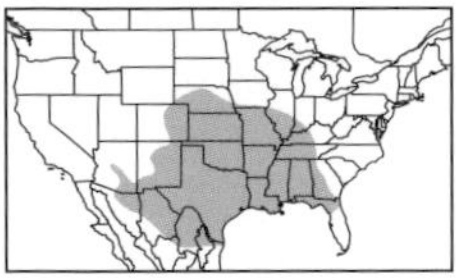

### Goatweed Leafwing *Anaea andria*

**Green** to gray-green, sometimes with black blotches on abdomen; **head gray-green without prominent orange tubercles**. *Habitat* Open woodlands and washes with its host, and adjacent areas. *Found* Most of year in some areas, April–Oct in others. *Host* Crotons.

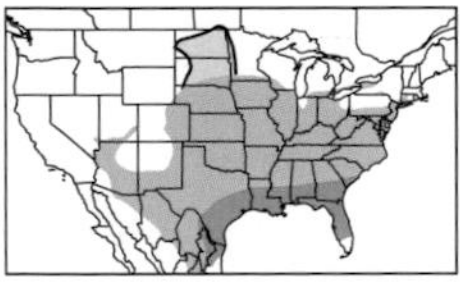

### *Hackberry Emperor *Asterocampa celtis*

Variable. Green, speckled with white dots, with yellow dorsal line extending along body from base of horn on head, and yellow lateral stripes. Can be confused with Tawny Emperor, which uses the same hosts. *Habitat* Closely tied to hackberry trees. *Found* April–May, and July–Sept, before, between, and after flights. *Host* Hackberries. *Comments* Half-grown caterpillars overwinter in groups (turning brown) inside curled or arched leaves of the host.

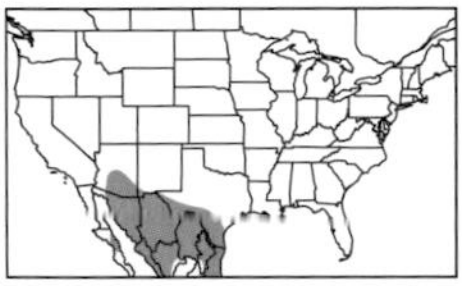

### Empress Leilia *Asterocampa leilia*

Green, speckled with tiny white dots, with pale yellow dorsal line extending along the body from base of horn on head, and a pale yellow sublateral line. *Habitat* Tropical and subtropical scrub, desert washes, and canyons with the host. *Found* March–Nov, before, between, and after flights. *Host* **Spiny or desert hackberry**.

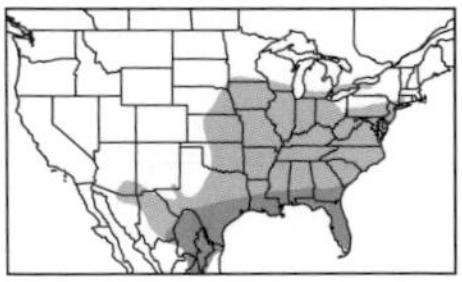

### *Tawny Emperor *Asterocampa clyton*

Variable. Light green with tiny white dots with alternating green and yellow bands (East) or green, speckled with tiny white dots, with yellow dorsal line extending along the body from base of horn on head, and yellow lateral stripe (Southwest). *Habitat* Hackberry woods. *Found* April–Sept, before, between, and after flights. *Host* Hackberries. *Comments* Young caterpillars feed and rest in large groups; older ones are solitary.

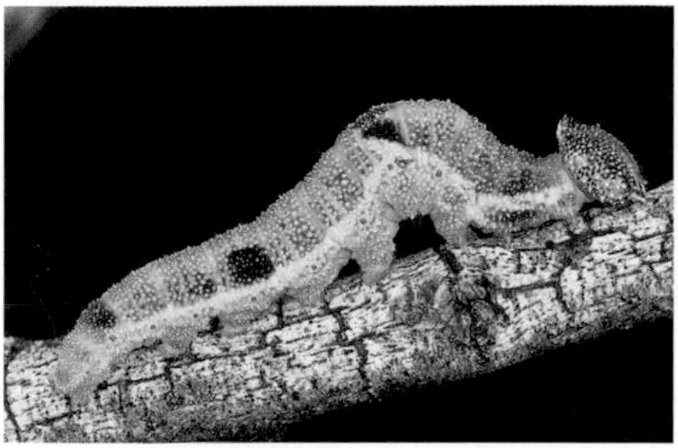

1. Tropical Leafwing

2. Empress Leilia adult

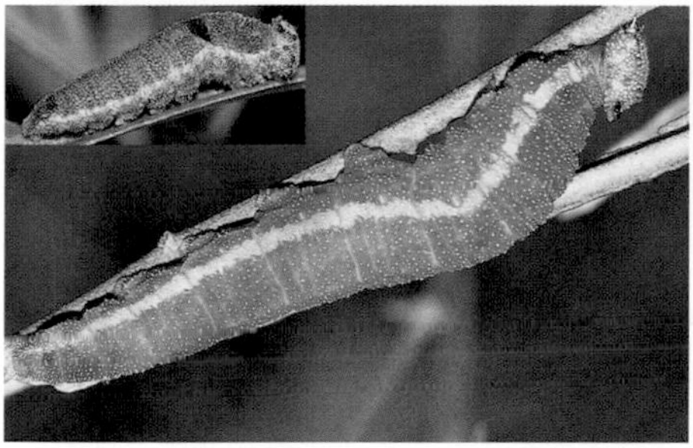

3. Florida Leafwing

4. Goatweed Leafwing

5. Hackberry Emperor

6. Empress Leilia

7. Tawny Emperor (Southwest)

8. Tawny Emperor (East)

## Satyrs (subfamily Satyrinae)

A distinctive group: Smooth with very short hairs, no spines, forked tails, and many with twin horn-like projections on the head. Most are green or brown, striped and somewhat flattened; overall an excellent design for blending with the leaves of grasses and sedges upon which they feed. Many rest at the base of the host, feeding primarily at night. Eggs can be scattered or laid on or off the host. Partially grown caterpillars overwinter.

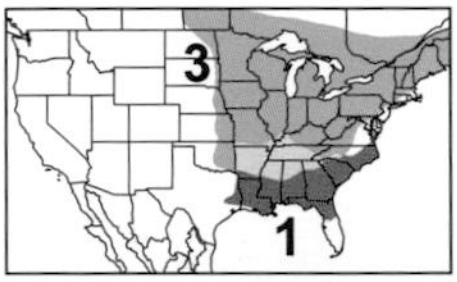


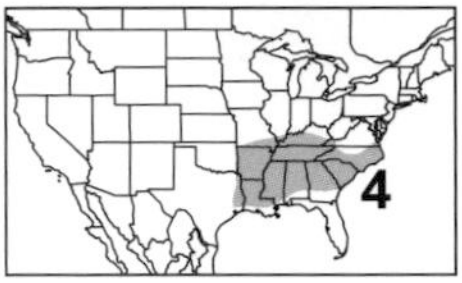


### 1. Southern Pearly-eye *Enodia portlandia*
### 3. Northern Pearly-eye *Enodia anthedon*
### 4. Creole Pearly-eye *Enodia creola*

Brown form with wavy dark lines and dots; green form with white or pale yellow stripes. Horns on head and tail are red-tipped. *Habitat* Southern in rich southern bottoms with cane; Northern prefers open rocky deciduous woods, especially near streams; Creole in canebrakes in dense woods. *Found* Mar–early May, June–Oct, between flights. *Host* Southern and Creole use cane. Northern uses grasses. *Garden Tips* Switch cane. *Comments* All three species of have both brown and green caterpillar forms, making reliable identification possible only by range in some cases and habitat in others.

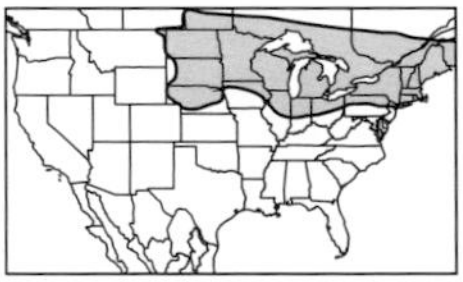

### Eyed Brown *Satyrodes eurydice*

Green or brown with white or pale yellow stripes, Horns on head and tail are red-tipped. Northern Pearly-eye uses different host. Probably not separable from Appalachian Brown where ranges overlap. *Habitat* Very wet meadows, marshes with sedges. *Found* April–May, July–Sept, before and after single flight. *Host* Sedges.

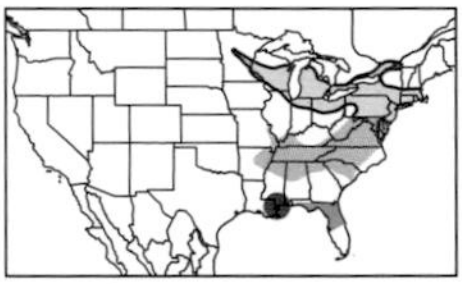

### Appalachian Brown *Satyrodes appalachia*

Green or brown with white or pale yellow stripes, Horns on head and tail are red-tipped. Pearly-eyes usually use different hosts. Probably not separable from Eyed Brown where ranges overlap. *Habitat* Wet, wooded situations adjacent to open areas. *Found* April–June, Aug–Sept, before, between, and after flights. *Host* Mainly sedges, also grasses.

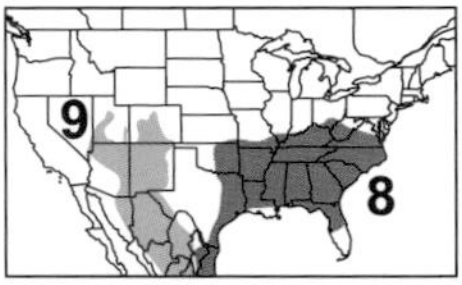


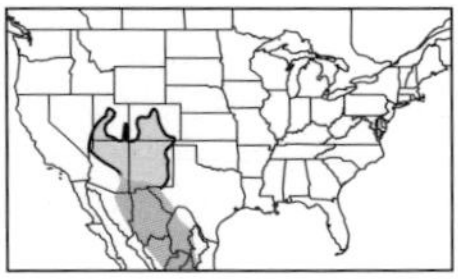

### 8. Gemmed Satyr *Cyllopsis gemma*
### 9. Nabokov's Satyr *Cyllopsis pyracmon*
### 10. Canyonland Satyr *Cyllopsis pertepida*

Not shown.

Gemmed and Nabokov's Satyrs green with yellow stripes or tan with yellow and brown stripes. Horns on head and tail are red-tipped, especially on individuals between broods. Canyonland Satyr caterpillar is unknown. *Habitat* Gemmed in moist, grassy areas within woodlands and, paradoxically, dry ridgetops (in the northern parts of its range). Nabokov's in oak woodland. Canyonland in mountain canyons and gulches, often in arid regions. *Found* March, June–July, and Aug–Sept, between flights. Nabokov's all year, but best in April and July. *Host* Grasses

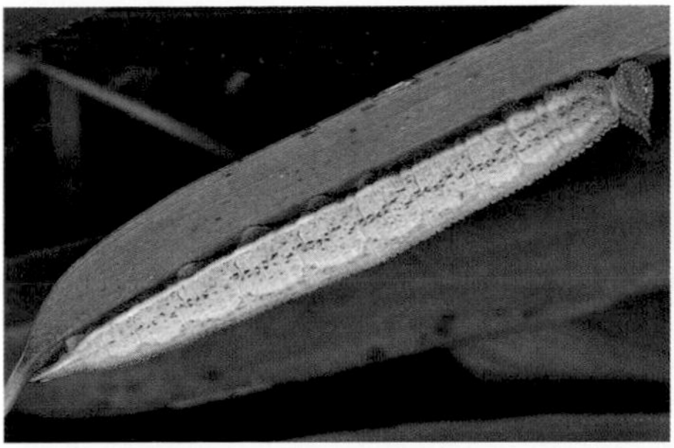
1. Southern Pearly-eye

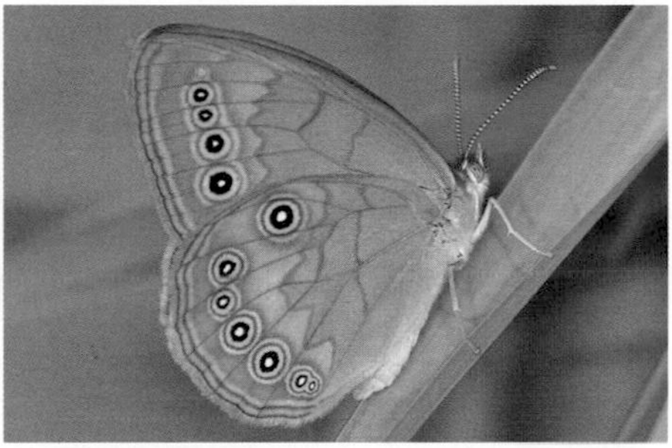
2. Eyed Brown adult

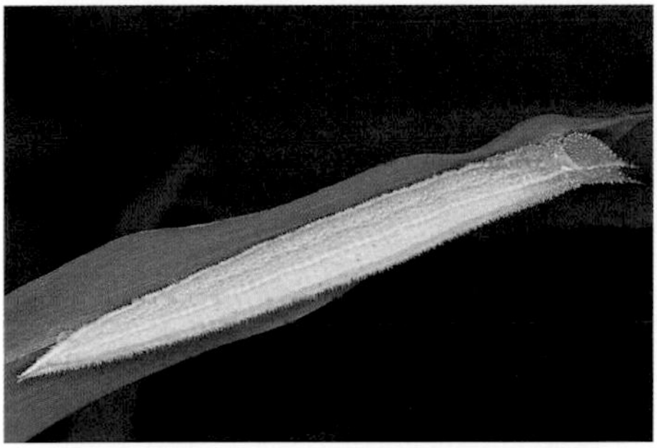
3. Northern Pearly-eye

4. Creole Pearly-eye

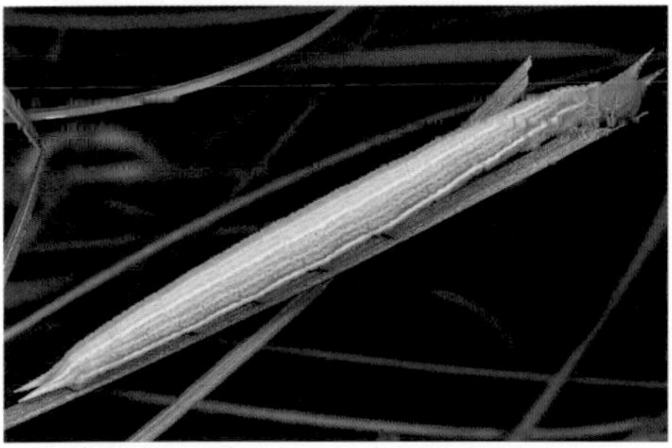
5. Eyed Brown

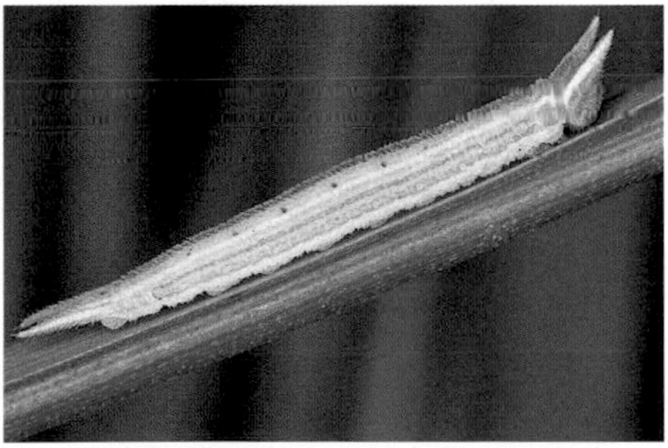
6. Eyed Brown

7. Appalachian Brown

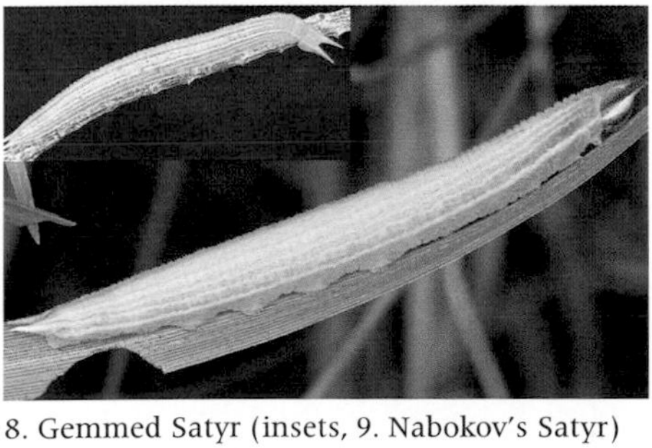
8. Gemmed Satyr (insets, 9. Nabokov's Satyr)

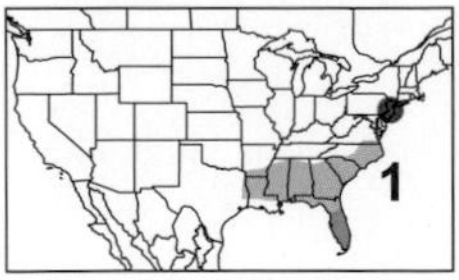


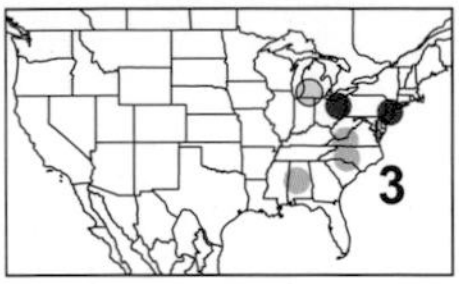


**1. Georgia Satyr** *Neonympha areolata*

**3. Mitchell's Satyr** *Neonympha mitchellii* Not shown.

Yellow-green with pale yellow stripes; **head green with a pair of tiny tan bumps**. Carolina Satyr is darker green. *Habitat* 1 in open pine-barrens in New Jersey, grass savannas in open pine woods further south; 3 in fens in Midwest, in dense semi-shaded swamps southward. *Found* 1 all year in the South, April–Sept northward, between flights. 3 April–July, depending on location. *Host* 1 uses sedges and grasses, 3 only sedges. *Comments* 3 is federally listed as endangered.

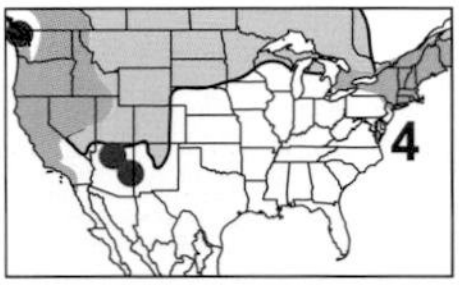


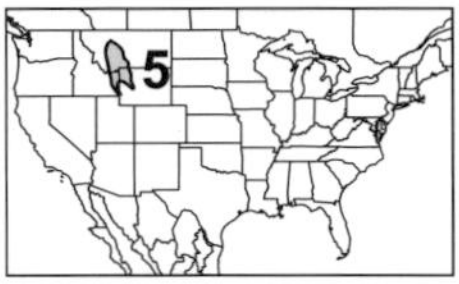


**4. Common Ringlet** *Coenonympha tullia*

**5. Hayden's Ringlet** *Coenonympha haydenii*
Not shown.

4 green with yellow to white stripes, a dark dorsal stripe, head rounded on top, no horns; tail with short horn tipped with red. 5: Last instar unknown. Wood-nymphs (pg. 111) are similar but bulkier and more tapered toward front and rear, and probably use different hosts where ranges overlap. *Habitat* 4 in meadows, dunes, marsh edges, and open grassy woodlands. 5 in high-elevation alpine and subalpine meadows. *Found* Mainly April–July, prior to flight. *Host* Grasses. *Comments* Also has brown forms.

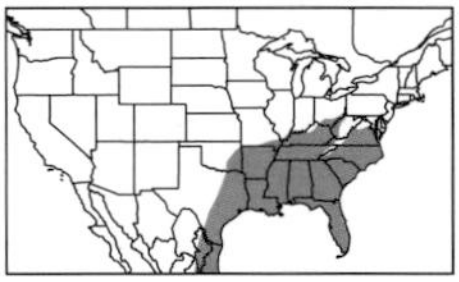

**Carolina Satyr** *Hermeuptychia sosybius*

Green, covered with short white hairs, head green and lacks pair of horns; tail horns short. *Habitat* A wide variety of woodland situations, especially moist forests. *Found* All year in the South, April–Sept northward, between flights. *Host* Grasses.

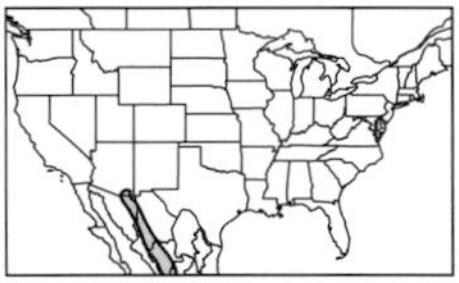

**Pine Satyr** *Paramacera allyni*

Yellow-green with numerous darker green stripes; head green with pair of squared-off horns. Rear horns short and red-tipped. *Habitat* High-elevation oak-pine forest. *Found* Probably May–June, prior to single flight with early instars Aug–Sept, after flight. *Host* Probably grasses. *Comments* Females lay their eggs on various substrates (including flowering plants and fallen pine needles) in extremely moist areas.

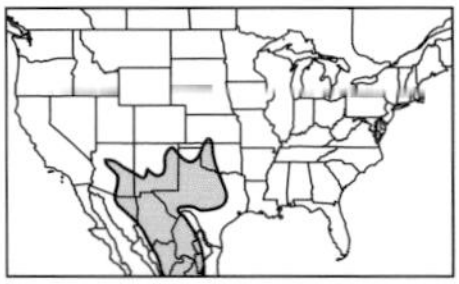

**Red Satyr** *Megisto rubricata*

Light brown with paler and darker stripes and bands; head tan or black with pair of very short knobs; rear horns short. *Habitat* Dry woodlands. *Found* Not known, undoubtedly before and between flights, April and again in July. *Host* Probably grasses. *Comments* Females prefer to lay their eggs on dead organic material (fallen pine needles, dead grass blades, etc.) rather than living grass, thus keeping their natural hosts a secret.

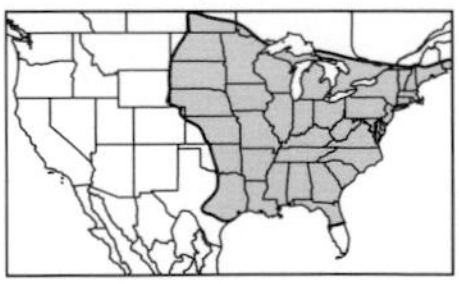

**Little Wood-Satyr** *Megisto cymela*

Light brown with white stripes and oblique dark lateral dashes; head brown with pair of very short knobs; rear horns short. *Habitat* Most often found along grassland and woodland interface; may also be found in open fields or deep woods. *Found* Feb–April and June–Sept, between flights. *Host* Grasses.

1. Georgia Satyr

2. Little Wood-Satyr adult

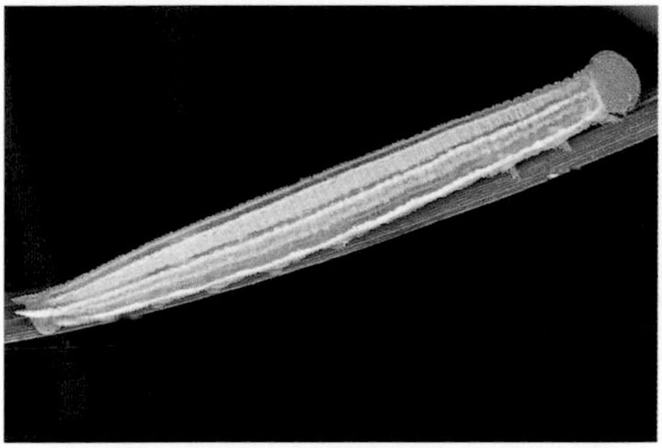

4. Common Ringlet (California)

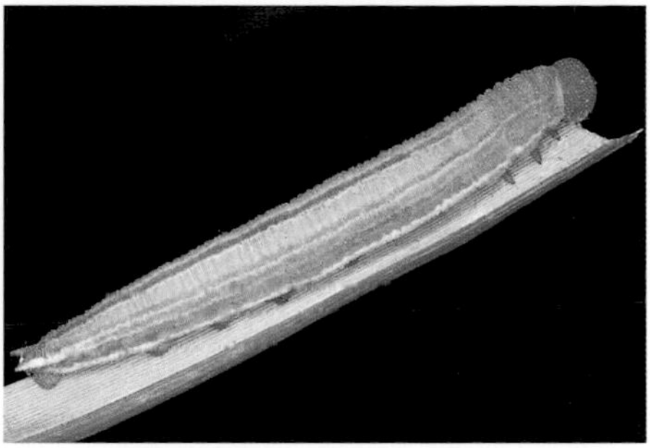

5. Common Ringlet

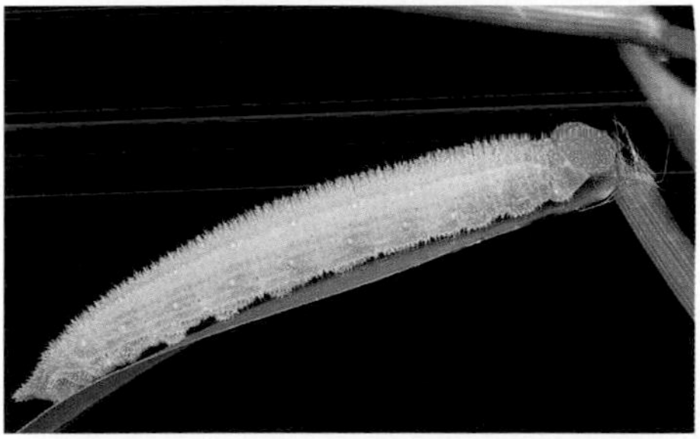

7. Carolina Satyr

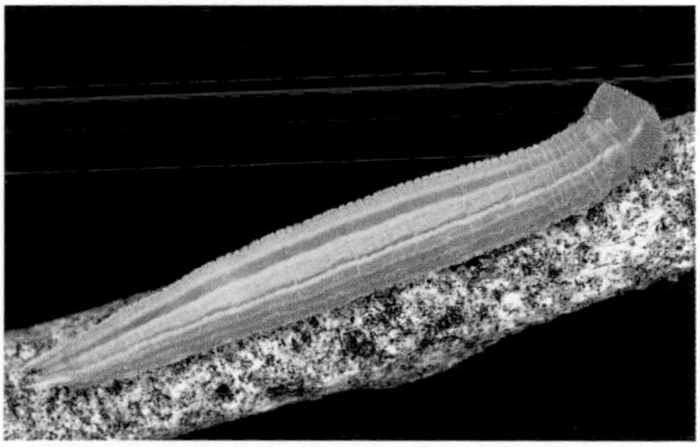

8. Pine Satyr

9. Red Satyr

10. Little Wood-Satyr

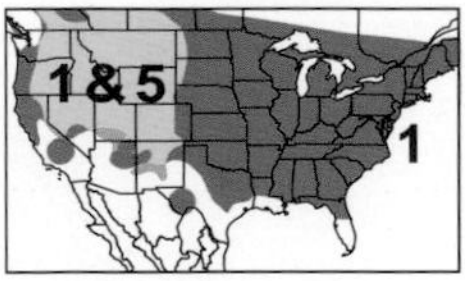


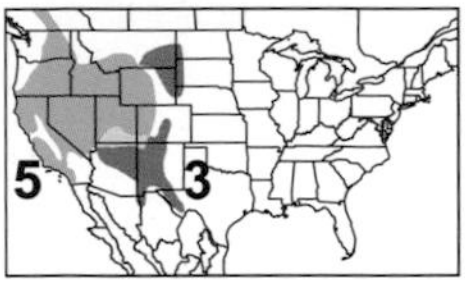


**1. Common Wood-Nymph** *Cercyonis pegala*

**3. Mead's Wood-Nymph** *Cercyonis meadi*

**4. Great Basin Wood-Nymph** *Cercyonis sthenele*
Not shown.

**5. Small Wood-Nymph** *Cercyonis oetus* Not shown.
Green with two white lateral stripes, short covering of light hair (Mead's with shorter hairs than Common); head green, without horns, rear horns short, tipped with red. See Common Ringlet (pg. 108). *Habitat* Open grassy areas. *Found* April–July. *Host* Grasses. *Comments* Overwinter as unfed first instar caterpillars so search for them prior to the single flight.

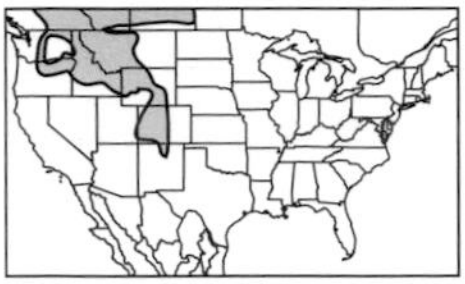

**Alpines** (genus *Erebia*)
Includes Common Alpine (*Erebia epipsodea*) and 6 species not shown. Brown striped caterpillars with brown, rounded heads, no horns; some lack forked tail. Most occupy arctic/alpine habitats. Many are biennial, requiring 2 years to complete their development to adult. They overwinter partially grown the first year, and nearly full-grown the next. Very little information is available about alpine caterpillars. All known species are very similar, making identification to species not possible at this time. These caterpillars are among the most difficult to locate in the field owing to the inaccessibility of their habitats and, to a lesser extent, their rarity. Thus, we illustrate one species, leaving an information gap for caterpillar enthusiasts to fill. See *BTB: West* and *BTB: East* for range maps of species. *Host* Grasses.

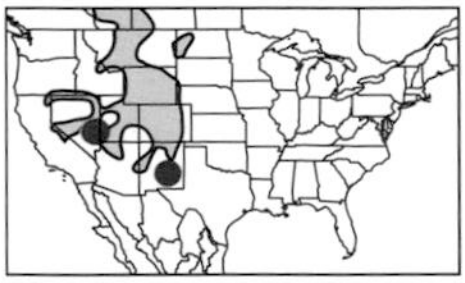

**Ridings' Satyr** *Neominois ridingsii*
Very similar to caterpillars of arctics. *Habitat* Open woodlands, prairies, sagebrush, and grassy mountain knolls and swales. *Found* Mainly April–June, prior to flight, biennial populations always present but still best sought prior to adults. *Host* Blue grama grass. *Comments* Bluish-white eggs and hatchling caterpillars are surprisingly large in size.

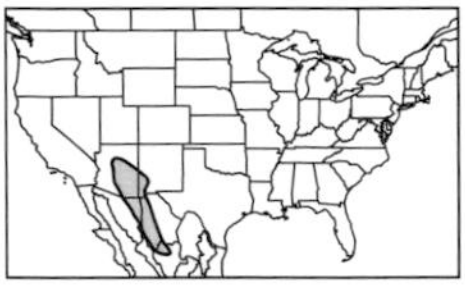

**Red-bordered Satyr** *Gyrocheilus patrobas*
Light brown with dark lateral stripe, head tan with short horns; tail long and forked. *Habitat* Coniferous and mixed woodland in mountain canyons. *Found* Mainly July–early Aug, prior to single flight. *Host* Bullgrass. *Comments* Rests during the day on rounded leaf bases near base of the host.

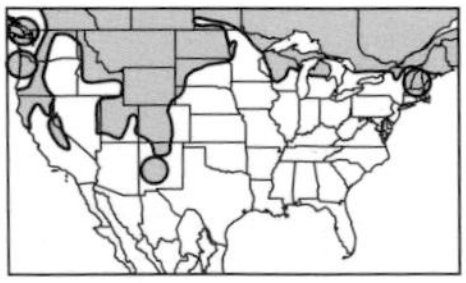

**Arctics** (genus *Oeneis*)
Includes Great Arctic (*Oeneis nevadensis*), Jutta Arctic (*Oeneis jutta*), and 6 species not shown. Like Alpines, arctics inhabit the far north or high elevations. Green, tan, or brown with many stripes, rounded heads no horns, and forked tails. Many are biennial, requiring 2 years to complete their development to adult. They overwinter partially grown the first year, and nearly full-grown the next. All are very similar. Like Alpines, they are extremely difficult to locate in the field except, perhaps, by people with extreme patience and devotion. See *BTB: East* and *BTB: West* for range maps of individual species. *Host* Grasses.

1. Common Wood-Nymph

2. Common Wood-Nymph adult

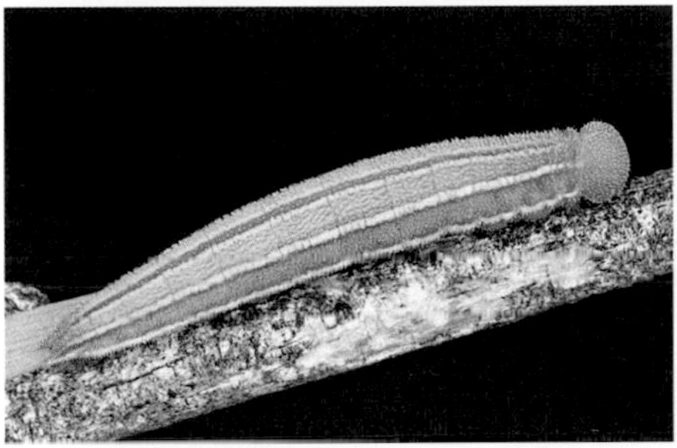

3. Mead's Wood-Nymph

6. Common Alpine

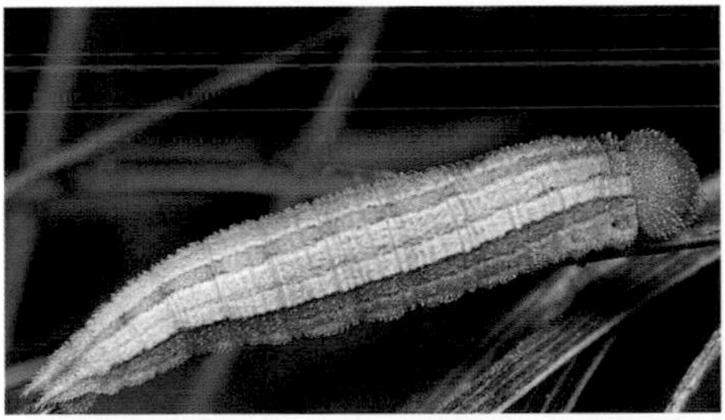

7. Ridings' Satyr

8. Red-bordered Satyr

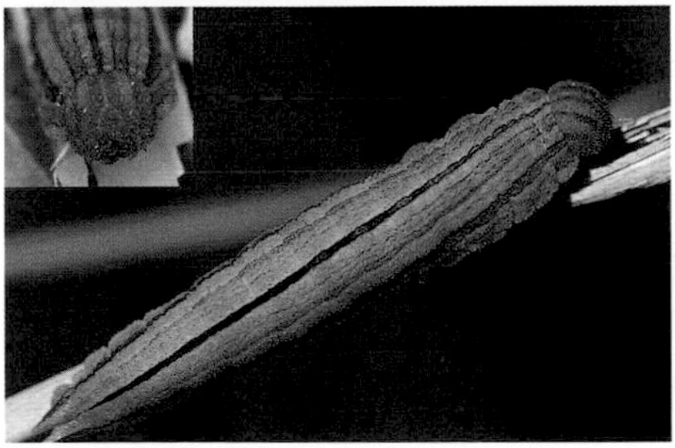

9. Great Arctic

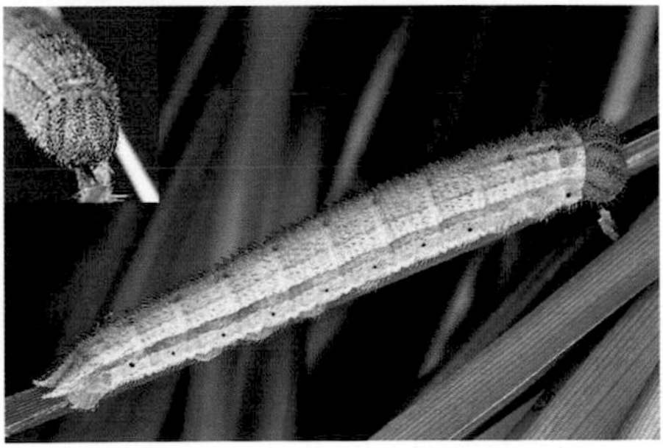

10. Jutta Arctic

## Monarchs or Milkweed Butterflies

This group includes some of our best known and largest butterflies (rivaling the swallowtails in size). Milkweed butterfly caterpillars are smooth with bright colors (warning predators of their toxicity) and have long fleshy filaments on their back. Feeding on milkweeds, they accumulate toxic chemicals (cardiac glycosides) that make them less palatable to predators. Because predators may avoid milkweed butterflies, Monarchs and Queens are mimicked by other species of butterflies. The whitish to pale green eggs are eliptical and pitted, and are usually laid one at a time on leaves or flowers. Caterpillars feed exposed on leaves, although older caterpillars sometimes hide at the base of the plant. The chrysalids are bright bluish-green, studded with gold spots.

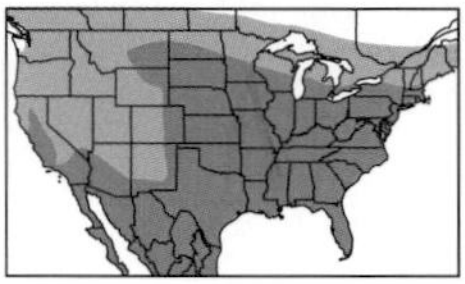

### *Monarch *Danus plexippus*

Transverse bands of yellow, black, and white with **two pairs of fleshy filaments**; head black-and-white striped. Queen and Soldier have three pairs of fleshy filaments. *Habitat* Open fields, roadsides, suburban areas; anywhere that milkweed occurs. *Found* March–Sept, between flights. *Host* Milkweeds and related plants. *Garden Tips* Orange milkweed, tropical milkweed, narrow-leaved milkweed, antelope horned milkweed, and swamp milkweed. *Comments* Not only our most renowned butterfly but perhaps our best-known caterpillar, as well. Females that have spent the winter in the Mexican mountains fly northward in early spring and lay eggs on milkweeds. Their offspring continue the journey north.

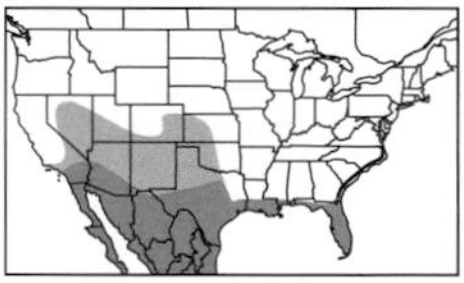

### *Queen *Danaus gilippus*

Transversely banded white and black, often with partial yellow bands or spots in black bands dorsally, often edged with maroon, with *three pairs of fleshy filaments* sometimes maroon or at least maroon at the base; head black-and-white striped. Soldier is similar but much rarer and has a series of tan spots or partial bands dorsally between white and yellow bands. *Habitat* Generally open areas, brushy fields, roadsides, deserts, etc. *Found* Most of the year in Florida and the Southwest; April–Sept along the Gulf Coast and in southern CA. *Host* Milkweeds and milkweed vines. *Garden Tips* Tropical milkweed, whitevine, climbing milkweed. *Comments* Caterpillars appear to tolerate mild freezes. As do Monarchs, Queens migrate south into Mexico in the fall, but no one knows where they go.

### *Soldier *Danaus eresimus*

Transversely banded white and black, **often with tan spots or partial bands dorsally between white and yellow bands**; three pairs of fleshy filaments dorsally. Queen has yellow spots or partial bands dorsally. *Habitat* Open areas and woodland edges. *Found* March–Sept, between flights. *Host* Milkweeds and milkweed vines. *Garden Tips* Vining milkweeds and whitevine.

1. Monarch egg on swamp milkweed

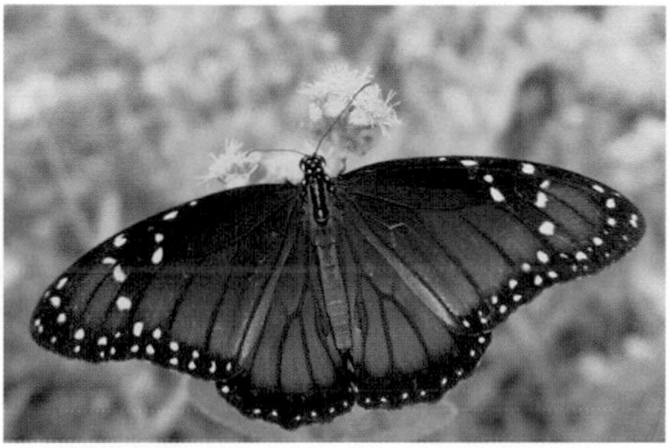

2. Soldier adult

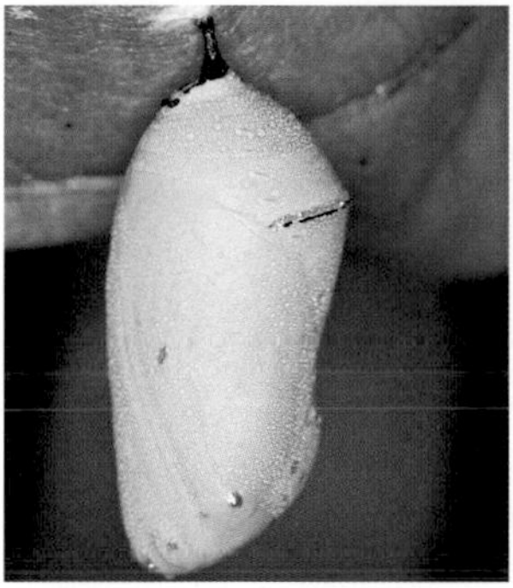

3. Monarch chrysalis

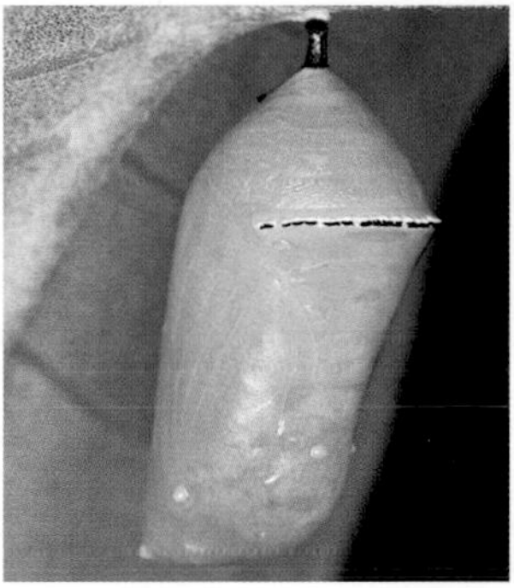

4. Queen chrysalis

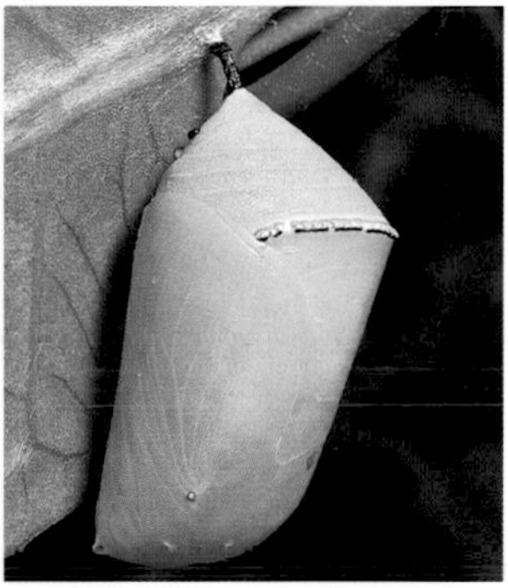

5. Soldier chrysalis

6. Monarch

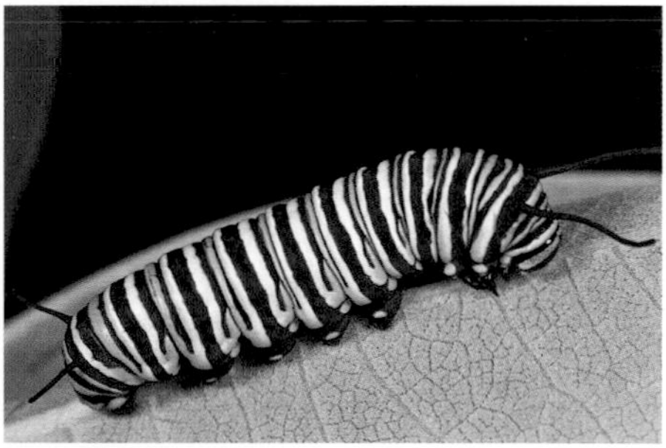

7. Monarch

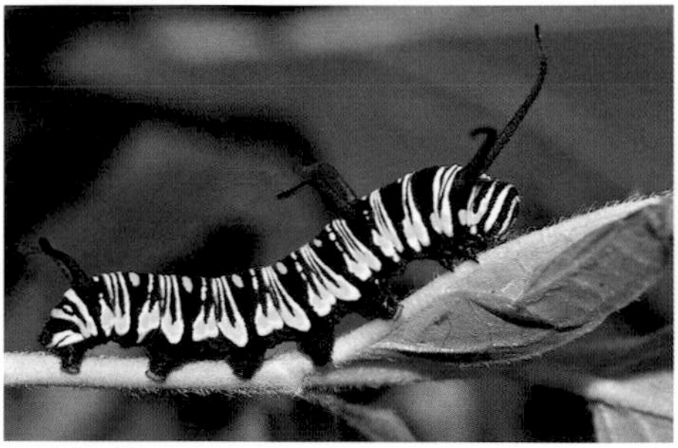

8. Queen

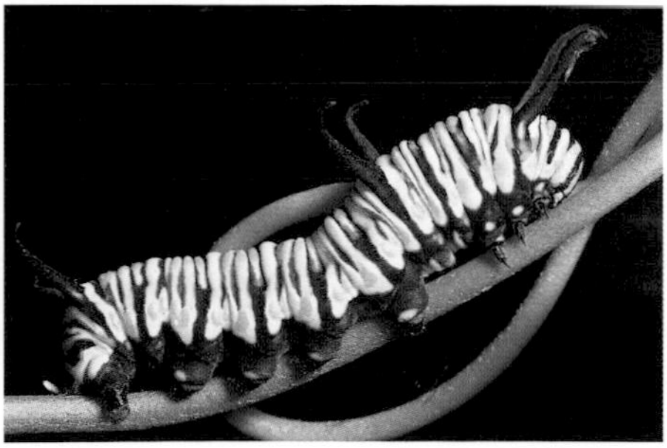

9. Soldier

## **Skippers** (family Hesperiidae)

See pg. 11 for general information about skippers.

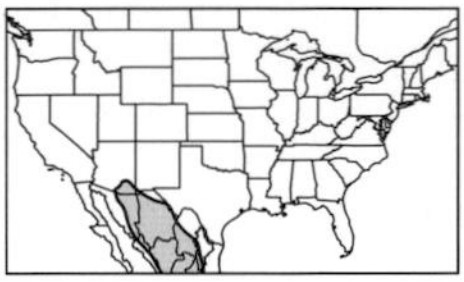

### **Dull Firetip** *Pyrrhopyge araxes*

Spectacular! Banded maroon and yellow, with sparse long white hairs; head covered with long white hairs. *Habitat* Oak woodlands, canyons, and hillsides. *Found* All year, but best April–June, prior to single flight. *Host* Emory oak, gray oak, Arizona white oak, and other oaks including hybrids. *Comments* Partially grown caterpillars overwinter in small nests on the host.

### ***3. Guava Skipper** *Phocides polybius*
### **4. Mangrove Skipper** *Phocides pigmalion*

Last instar white with light bluish-green cast (see insets to photos 3 and 4). Head brown with two yellow spots on lower face. *Early Instars* Maroon with yellow bands. Head is black and maroon with yellow lower facial spots. *Habitat* Guava Skipper in subtropical woodlands and suburban gardens. Mangrove Skipper in mangrove swamps and nearby areas. *Found* All year. *Host* Guava Skipper uses guava. Mangrove Skipper uses red mangrove.

### **Mottled Longtail** *Typhedanus undulatus*

Reddish-brown; head reddish-black with yellow-orange lower facial spots. *Habitat* Tropical and subtropical scrub and open woodlands. *Found* Rare stray to south Texas, should be sought after adult sightings. *Host* Sennas. *Comments* The only caterpillar that is red throughout development.

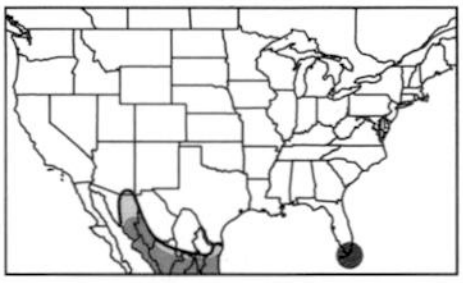

### **Hammock Skipper** *Polygonus leo*

Pale yellow-green with a yellow lateral stripe; **head whitish with a pair of black spots on top**. *Habitat* Tropical hardwood woodlands. *Found* All year. *Host* Jamaican dogwood and pongam.

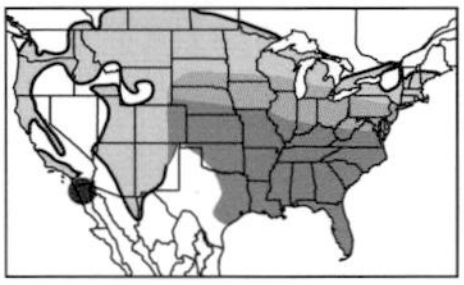

### ***Silver-spotted Skipper** *Epargyreus clarus*

Yellow-green with faint (photo 7) or strong black bands on each segment; head brown to black with yellow/orange lower facial spots. *Habitat* Wide ranging in open habitats, woodland borders and openings. *Found* May–Sept, after flights. *Host* Black locust, New Mexico locust, honey locust, hog peanut, indigos, milk-peas, beggar weeds, and many others. *Garden Tips* Black locust, New Mexico locust, Chinese wisteria. *Comments* A common and widespread garden butterfly. One of the easiest skipper caterpillars to locate, a good quest for beginners. Overwinters as a pupa.

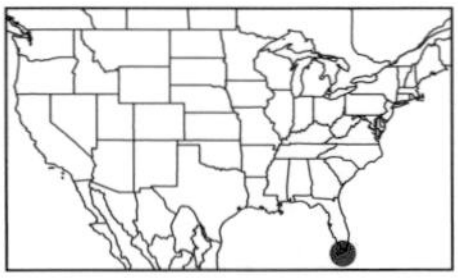

### **Zestos Skipper** *Epargyreus zestos*

Yellow-green with faint black stripes across each segment; head brown to dark brown with yellow lower facial spots. *Habitat* Tropical hardwood hammocks. *Found* All year. Florida Keys only. *Host* Legumes, especially striated milk-pea.

1. Dull Firetip

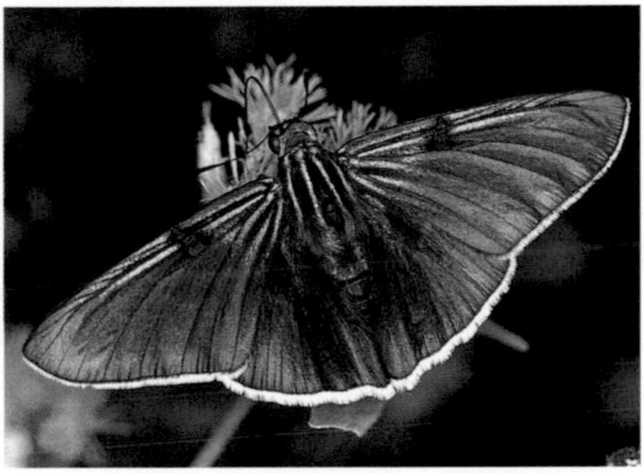

2. Guava Skipper adult

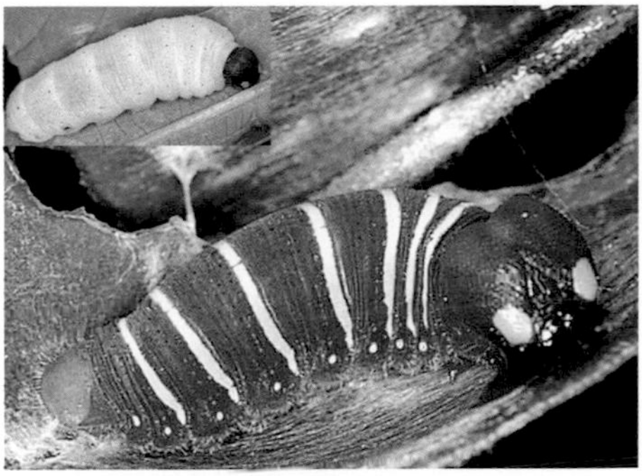

3. Guava Skipper (inset: final instar)

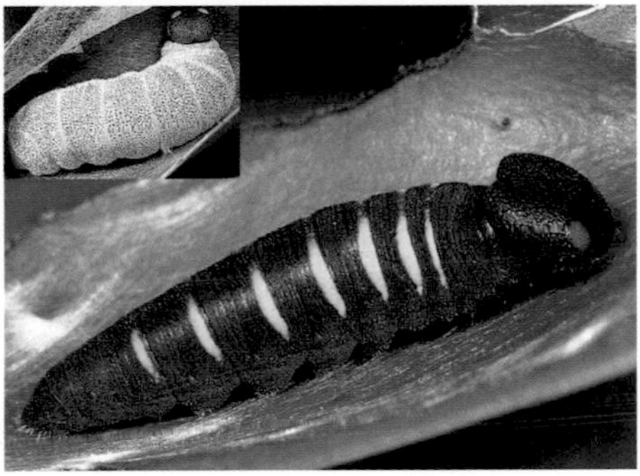

4. Mangrove Skipper (inset: final instar)

5. Mottled Longtail

6. Hammock Skipper (inset: early instar)

7. Silver-spotted Skipper

8. Zestos Skipper

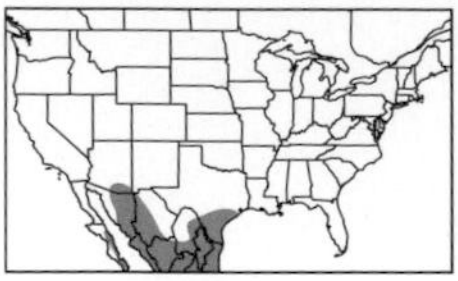

## White-striped Longtail *Chioides catillus*

Greenish turning pinkish-orange with tiny black dots and orange lateral lines; **head brownish-orange with two black spots on lower face**. *Habitat* Tropical and subtropical scrub, and open woodlands. *Found* All year. *Host* Wild lima beans, Wright's milk-pea and other vining legumes.

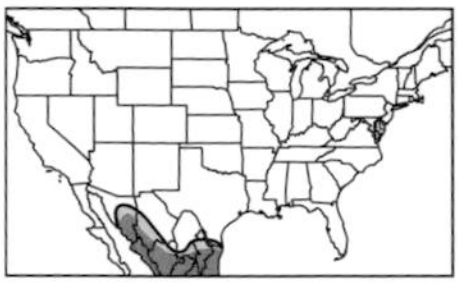

## Zilpa Longtail *Chioides zilpa*

Greenish or pinkish-orange with tiny black dots and orange lateral lines; head brown with orange spots. *Habitat* Tropical and subtropical scrub and open woodlands. *Found* Most of the year. *Host* Small-leafed *Nissolia* vines. *Comments* Older caterpillars outgrow the small leaves of the host and make nests of the larger leaves of an adjacent tree to conceal themselves.

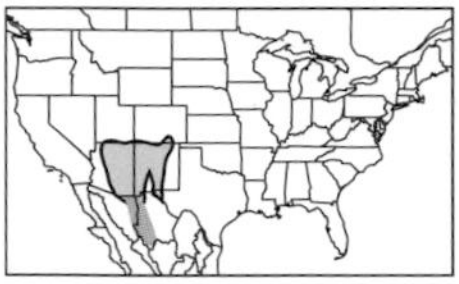

## Short-tailed Skipper *Zestusa dorus*

Pale green; **head red with cream spots, rounded on top**. Duskywings on same hosts have pointed foreheads. *Habitat* Oak and pine-oak woodlands. *Found* May–early July, late July–Sept, best after flights. *Host* Emory oak, Arizona white oak, probably other oaks. *Comments* Earlier instars have black heads with small orange spots on lower face.

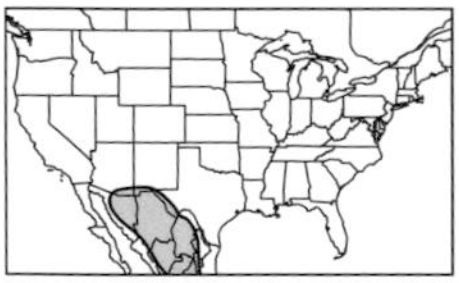

## Arizona Skipper *Codatractus arizonensis*

Green with **broken yellow lateral stripe**; **head gray-brown with yellow spots** on lower face. *Habitat* Rocky canyons near the host. *Found* June–Sept, after flights. *Host* Kidneywood.

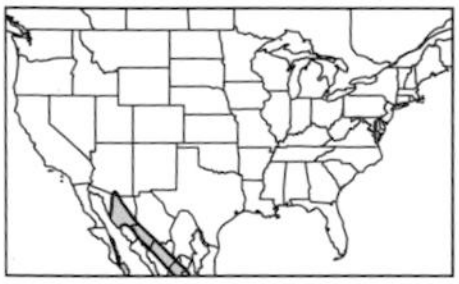

## Valeriana Skipper *Codatractus mysie*

Green with yellow-edged green dorsal stripe and yellow lateral stripe; head dark with **white spots on forehead** and orange spots on lower face. *Habitat* Red, rocky canyons and slopes near stands of the host. *Found* Late Aug–early Oct, after single flight. *Host Tephrosia leiocarpa*. *Comments* Hibernates full-grown on the ground in an orange silken cocoon, where pupation occurs the following July.

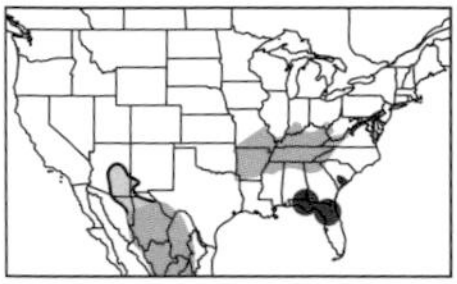

## Golden-banded Skipper *Autochton cellus*

Green with small yellow dots and a broad yellow lateral stripe; reddish collar. *Habitat* Woodland ravines along streams, riparian canyons, and gulches. *Found* June–Sept, after flight. *Host* Wild beans, butterfly pea, hog peanut, and other legumes. *Comments* Overwinters as pupa.

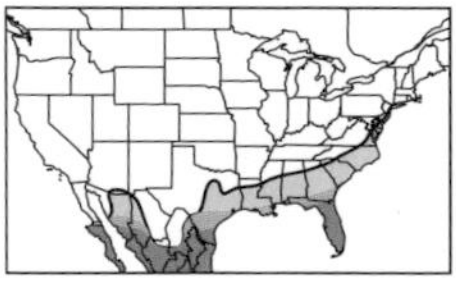

## *Long-tailed Skipper *Urbanus proteus*

Gray-green with pinkish cast; a yellow lateral stripe, **orange splash on rear segment;** head reddish with black markings in center. *Habitat* Open fields and woodland edges, especially brushy and disturbed situations. *Found* Most of the year in the South. *Host* Beggar weeds, wild and cultivated beans, milk-peas, butterfly peas, and many other shrubby and vining legumes in the pea family. *Garden Tips* Garden beans.

1. White-striped Longtail

2. Long-tailed Skipper adult

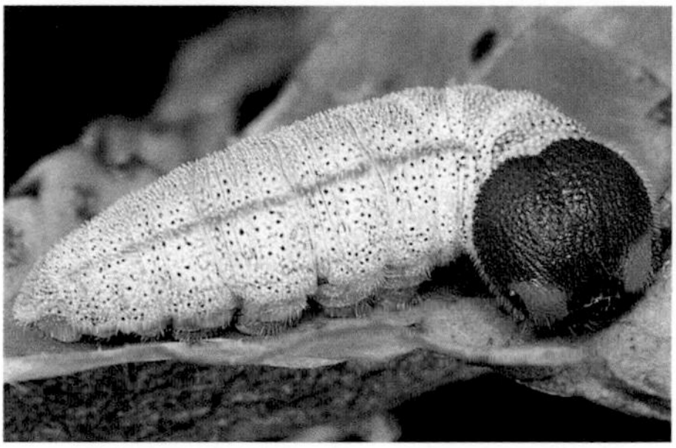

3. Zilpa Longtail

4. Short-tailed Skipper

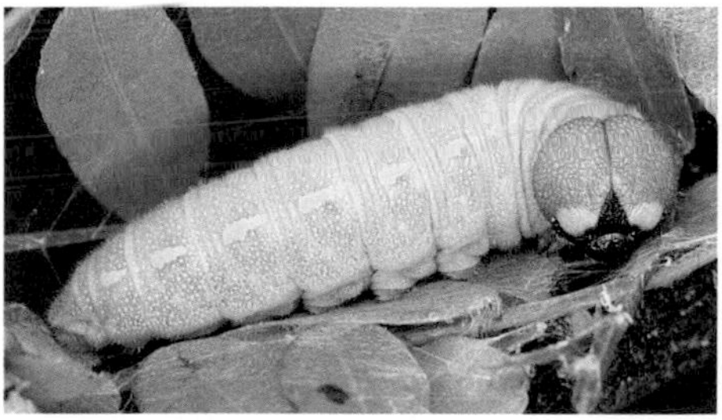

5. Arizona Skipper

6. Valeriana Skipper

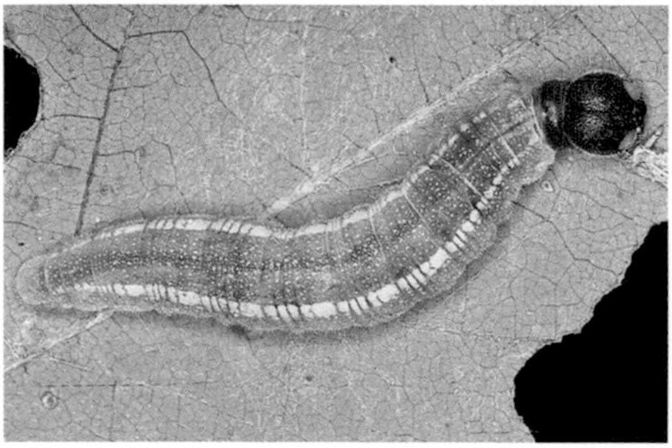

7. Golden Banded-Skipper

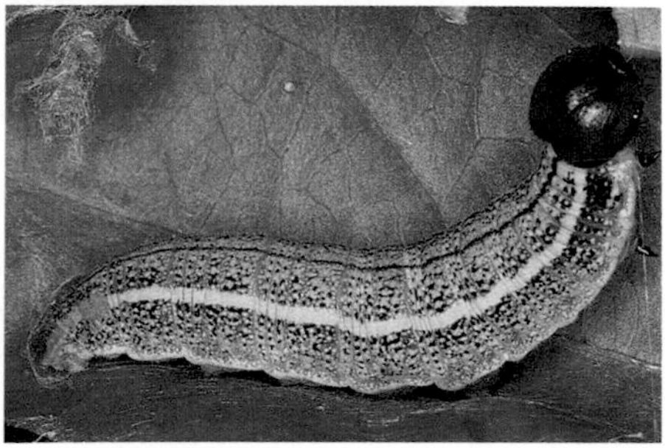

8. Long-tailed Skipper

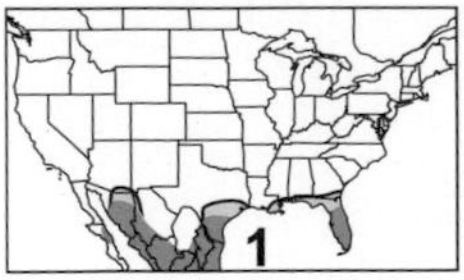


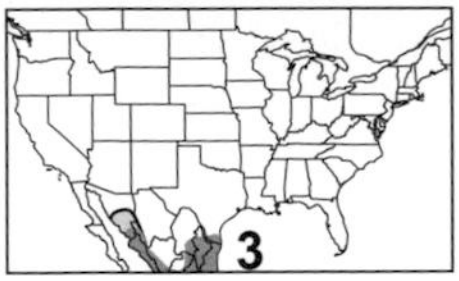


**1. Dorantes Longtail** *Urbanus dorantes*

**3. Brown Longtail** *Urbanus procne* Not shown.

Dorantes Longtail yellow, green, or orange with dark dorsal stripe and **lateral black, chain-like markings**; head black without facial spots. Cloudywings lack lateral chain-like markings. Brown Longtail caterpillar is unreported. *Habitat* Woodland edges, brushy fields, and gardens. *Found* All year in Florida and Texas, Sept–Oct in southern Arizona, after flights. Brown Longtail unknown. *Host* Dorantes Longtail uses beggarweeds, wild bean, and other legumes. Brown Longtail uses grasses. *Garden Tips* Garden beans.

## Cloudywings (genera *Achalarus* and *Thorybes*)

Like their adults, these caterpillars are difficult to distinguish from one another. With current information, reliable identification can't be made in areas where two or more species occur. Perhaps no two species share hosts in the same habitats, but much more study is needed.All Cloudywings hibernate as full-grown caterpillars, and they are best searched for after flights. All feed on legumes.

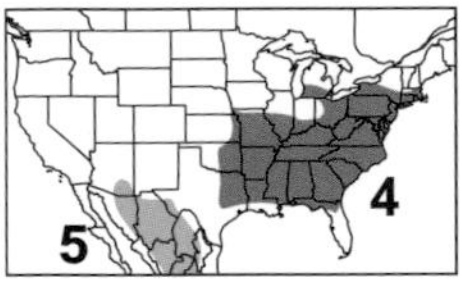


**4. Hoary Edge** *Achalarus lyciades*

**5. Desert Cloudywing** *Achalarus casica*

Light green turning pinkish with dark green speckling and a thin, pale orange lateral stripe; head black without facial spots. *Habitat* Hoary Edge widespread in open areas near woodlands. Desert Cloudywing in oak woods. *Found* June–July and Aug–Sept; Desert Cloudywing April–Oct. *Host* Hoary Edge on large-leaved legumes—beggar weeds, bush clovers, wild indigo, and others. Desert Cloudywing on butterfly pea and beggar weeds.

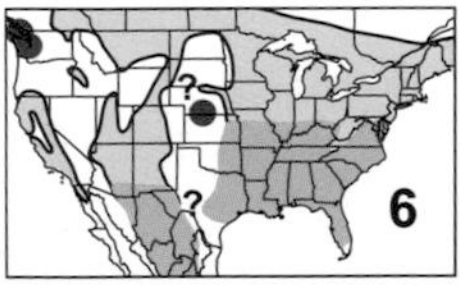


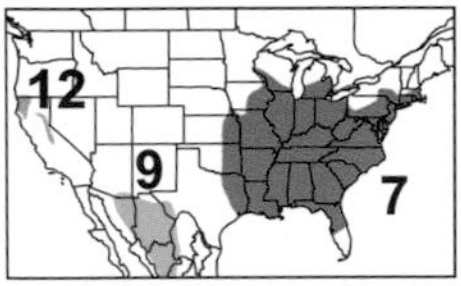


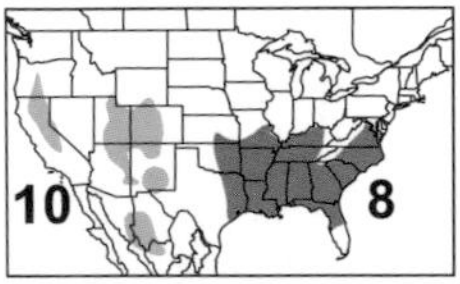


**6. Northern Cloudywing** *Thorybes pylades*

**7. Southern Cloudywing** *Thorybes bathyllus* Not shown.

**8. Confused Cloudywing** *Thorybes confusis* Not shown.

**9. Drusius Cloudywing** *Thorybes drusius*

**10. Mexican Cloudywing** *Thorybes mexicanus*

**12. Western Cloudywing** *Thorybes diversus* Not shown.

Light green turning pinkish with dark green speckling and a thin, pale orange lateral stripe; head black without facial spots. Sometimes dark (see photo 11). It is not known how to identify these caterpillars to species without raising them to adults. See page 162 for more information about these species. *Found* Northern and Southern are common and widespread. Northern found June–Sept; Southern May–Oct. Confused and Mexican can be locally common. Drusius and Western are the most limited. *Host* Legumes such as clovers and beggar weeds.

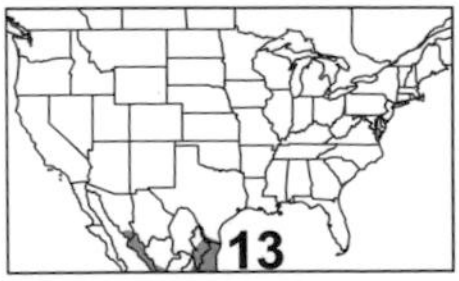


**13. Two-barred Flasher** *Astraptes fulgerator* Not shown.

Caterpillar is black with yellow bands or spots, and rusty facial marks. *Habitat* Tropical woodland. *Found* All year. *Host* Coyotillo.

1. Dorantes Longtail

2. Northern Cloudywing adult

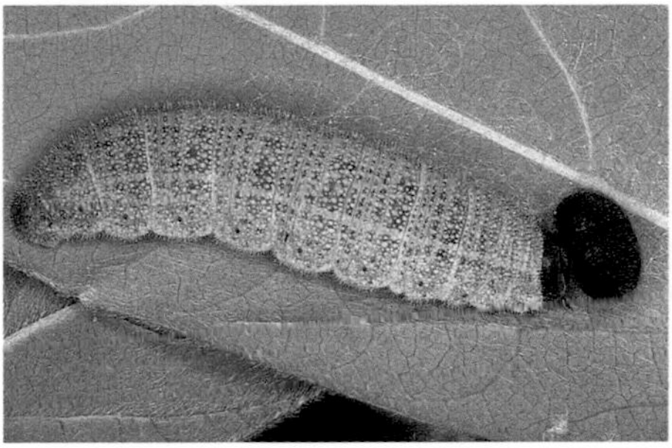

4. Hoary Edge

5. Desert Cloudywing

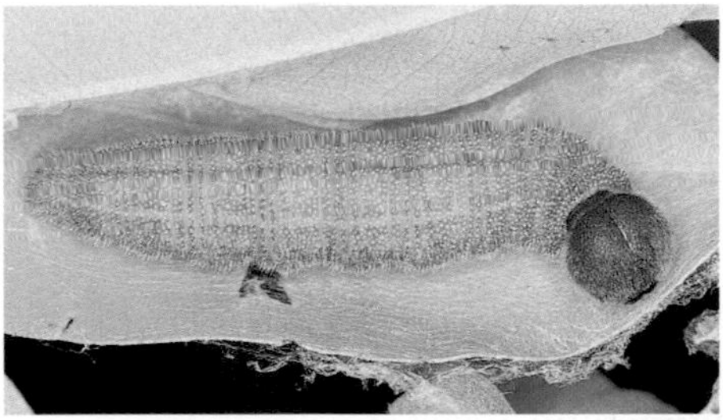

6. Northern Cloudywing

8. Drusius Cloudywing

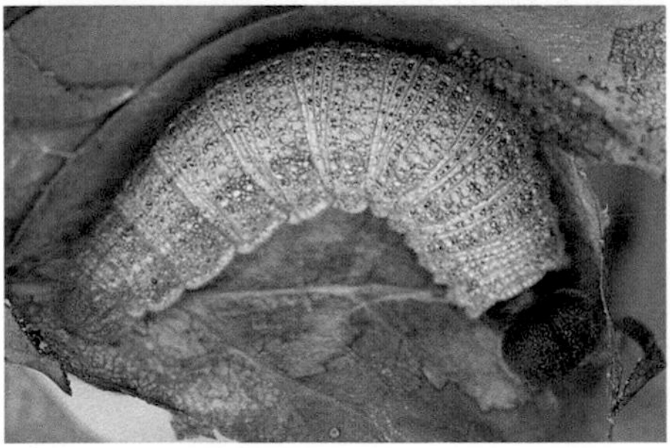

10. Mexican Cloudywing

11. Mexican Cloudywing

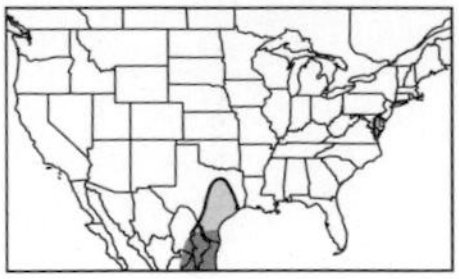

**Sickle-winged Skipper** *Achlyodes thraso*

Green with tiny pale yellow dots and a yellow lateral stripe, **heart-shaped head is gray rimmed with black**. *Habitat* Tropical woodlands and adjacent gardens. *Found* All year. *Host* Wild lime. *Garden Tips* Wild lime. *Comments* Pupates in nest on host.

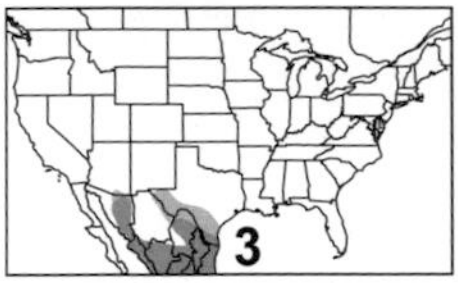


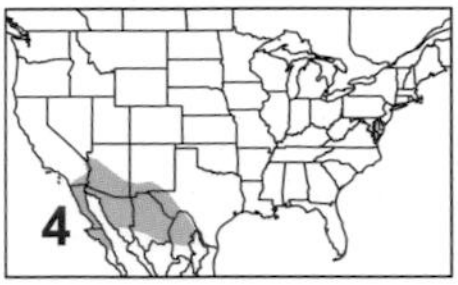


**3. Texas Powdered-Skipper** *Systasea pulverulenta*

**4. Arizona Powdered-Skipper** *Systasea zampa*

Light green, slightly transparent; lacks dorsal and lateral stripes; head tan or dark brown without facial spots. *Habitat* Lowland desert scrub and edges of subtropical and tropical woodlands. *Found* Most of year. *Host* Mallow family, including *Abutilon, Wissadula, Pseudabutilon,* and *Herrisantia*. *Garden Tips* Palmer's abutilon. *Comments* These are the only skipper caterpillars without an obvious constriction between head and thorax. Arizona Powdered-Skipper often makes nests in new unopened leaves of *Abutilon* plants.

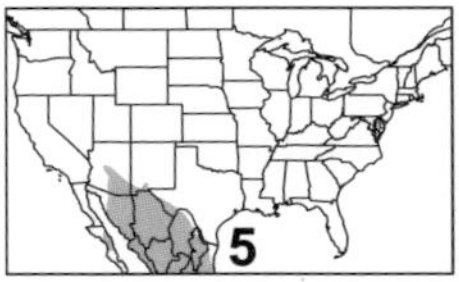


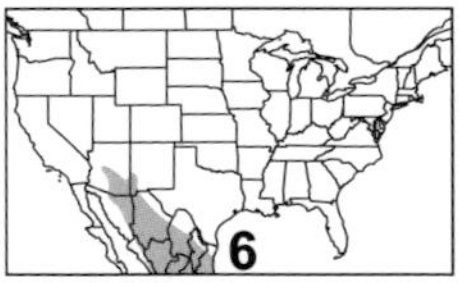


**5. Acacia Skipper** *Cogia hippalus*

**6. Gold-costa Skipper** *Cogia caicus*

Light bluish-green; lacks dorsal and lateral stripes; head brown with orange spots on lower face. Acacia Skipper with bolder spots than Gold-costa Skipper *Habitat* Desert grasslands, oak-juniper woodlands; rocky canyons and hillsides in arid situations. *Found* May–Oct. *Host* Fern acacia. *Garden Tips* Fern acacia. *Comments* An Acacia Skipper caterpillar and a Gold-costa Skipper caterpillar have been found feeding together on the same plant.

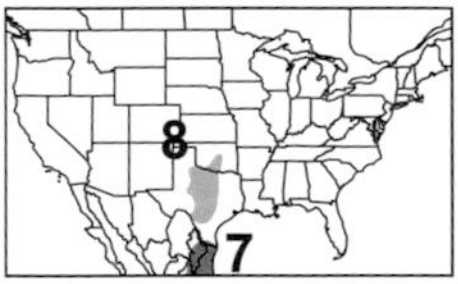


**7. Mimosa Skipper** *Cogia calchas* Not shown.

**8. Outis Skipper** *Cogia outis* Not shown.

Mimosa Skipper is pale green with tiny white dots, dark mid-dorsal line, and a faint pale yellow lateral stripe; head reddish-brown without facial spots. Outis Skipper is unreported. *Habitat* Thorn scrub. *Found* All year. *Host* Fern acacia.

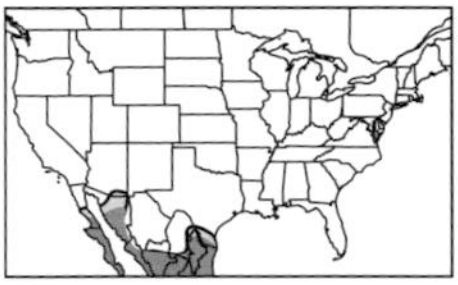

**White-patched Skipper** *Chiomara asychis*

Green with faint, lateral pale yellow stripe; head pale yellow with brown markings; **forehead pointed on either side with brown crescent mark**. *Habitat* Subtropical and tropical forest edges, desert scrub. *Found* Most of year in south Texas, April–Nov in southern Arizona. *Host* Barbados cherry in Texas; probably *Janusia gracilis* in Arizona. *Garden Tips* Barbados cherry, orchid vines.

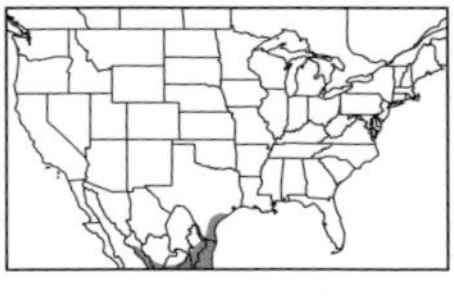

**False Duskywing** *Gesta gesta*

Light green with **yellow lateral spots; head orange with six black spots**. *Habitat* Open tropical scrub and disturbed situations. *Found* April–Nov. *Host* Indigos. *Comments* The caterpillars can defoliate host plants, leaving nothing visible but the conspicuous nests of the caterpillars.

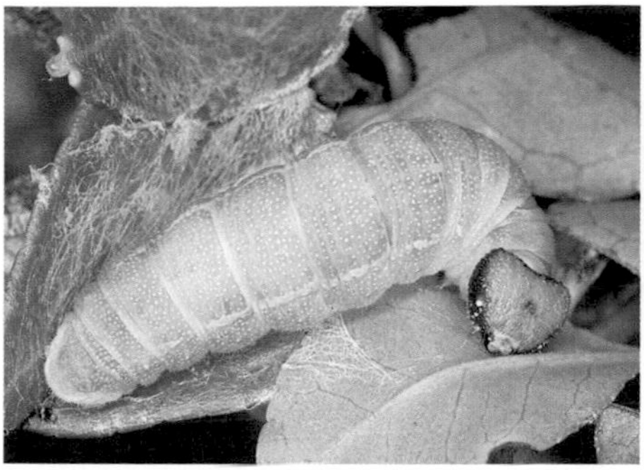

1. Sickle-winged Skipper

2. White-patched Skipper adult

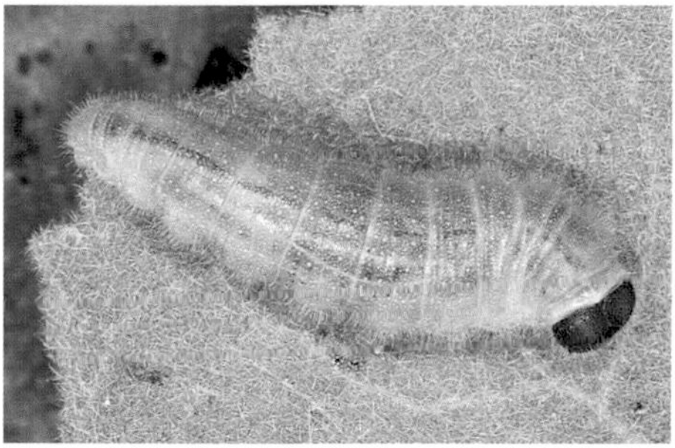

3. Texas Powdered-Skipper

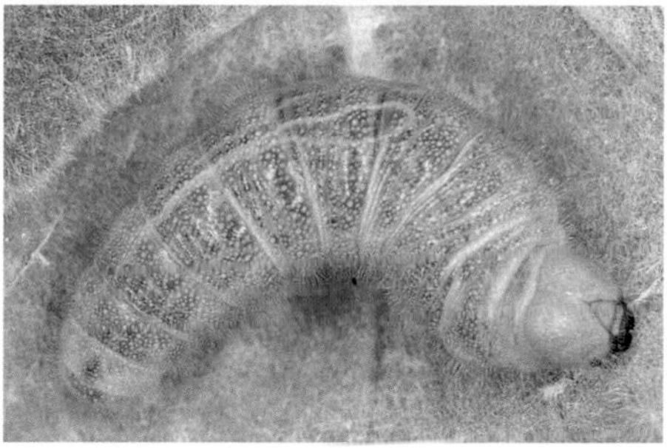

4. Arizona Powdered-Skipper

5. Acacia Skipper

8. Gold-costa Skipper

9. White-patched Skipper

10. False Duskywing

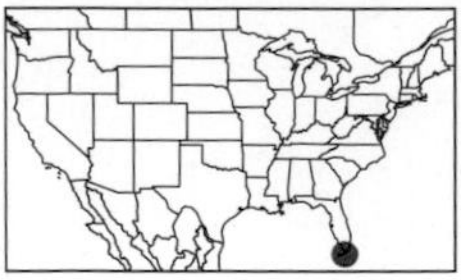

**1. Florida Duskywing** *Ephyriades brunneus*
Top of the head is more rounded than true duskywings. *Habitat* Pine rockland. *Found* All year. *Host* Locustberry.

**(True) Duskywings** (genus *Erynnis*)
See pg. 163 for more information about duskywings.

**3. Columbine Duskywing** *Erynnis lucilius*
Head strongly pointed at the corners, **lacks facial spots**. *Habitat* Rock outcrops with columbine. *Found* May–June; July–Aug/Sept. *Host* Wild columbine. *Garden Tips* If you live in northeastern rocky woodlands, plant native columbine and hope.

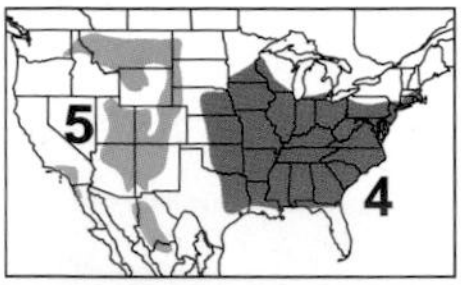


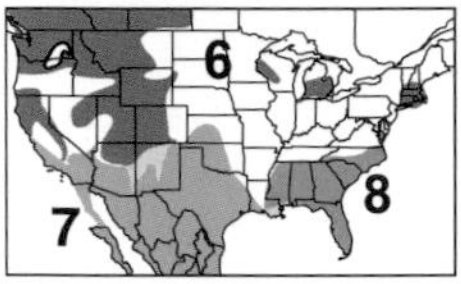


**Legume-feeding duskywings**

**4. Wild Indigo Duskywing** *Erynnis baptisiae*

**5. Afranius Duskywing** *Erynnis afranius*

**6. Persius Duskywing** *Erynnis persius* Not shown.

**7. Funereal Duskywing** *Erynnis funeralis* Not shown.

**8. Zarucco Duskywing** *Erynnis zarucco* Not shown.
Heads of Wild Indigo and Afranius are mottled with orange, that of Persius is mottled with tan or dark brown. Funereal and Zarucco heads are tan or brown and rimmed on either side with three yellow-orange patches. *Host* Wild Indigo mainly on crown vetch and wild indigo; other species on many legumes;

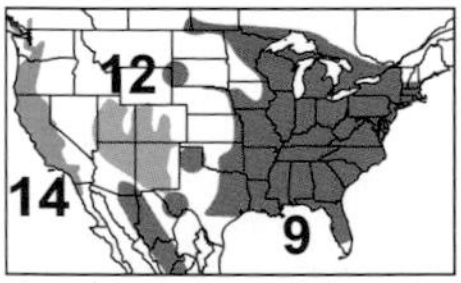


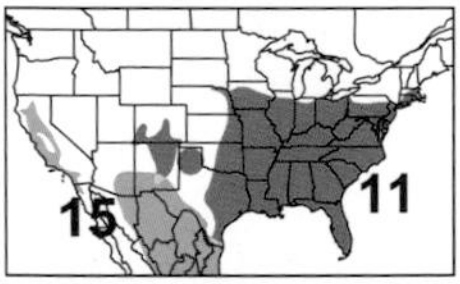


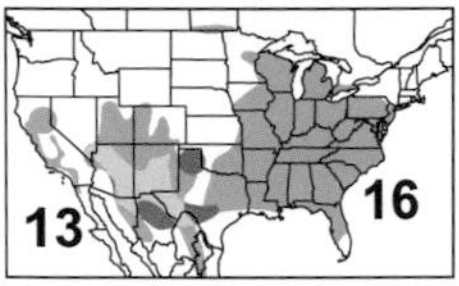


**Oak-feeding duskywings**

**9. Juvenal's Duskywing** *Erynnis junvenalis*

**10. Scudder's Duskywing** *Erynnis scudderi* Not shown.

**11. Horace's Duskywing** *Erynnis horatius* Not shown.

**12. Rocky Mountain Duskywing** *Erynnis telemachus* Not shown.

**13. Meridian Duskywing** *Erynnis meridianus* Not shown.

**14. Propertius Duskywing** *Erynnis propertius* Not shown.

**15. Mournful Duskywing** *Erynnis tristis* Not shown.

**16. Sleepy Duskywing** *Erynnis brizo* Not shown.
Green or bluish-green, covered with tiny white dots, with pale yellow lateral stripe; brown **head rimmed on either side with three yellow-orange patches**. Sleepy and Juvenal's have only a spring brood. Scudder's is found only in extreme southeastern Arizona. *Host* Oaks. See pg. 163 for more information.

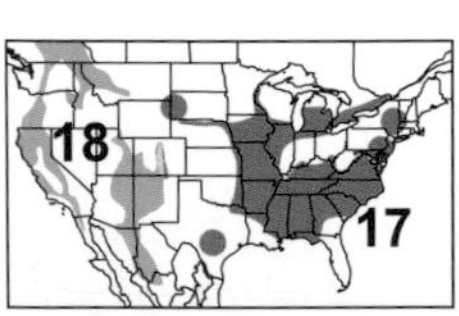


**Ceanothus-feeding duskywings**

**17. Mottled Duskywing** *Erynnis martialis* Not shown.

**18. Pacuvius Duskywing** *Erynnis pacuvius*
Similar to other duskywings. *Habitat* A wide range. *Host* New Jersey tea for Mottled. Other *Ceonothus* for Pacuvius.

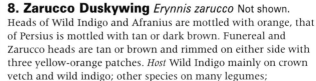

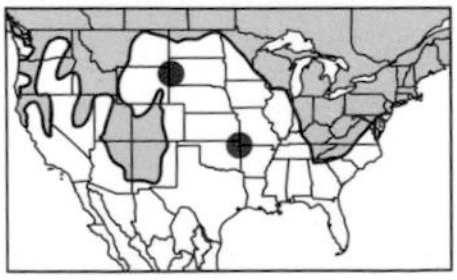

**19. Dreamy Duskywing** *Erynnis icelus*
*Habitat* Openings in moist forests. *Found* June–Aug. *Host* Willows, aspens, and poplars.

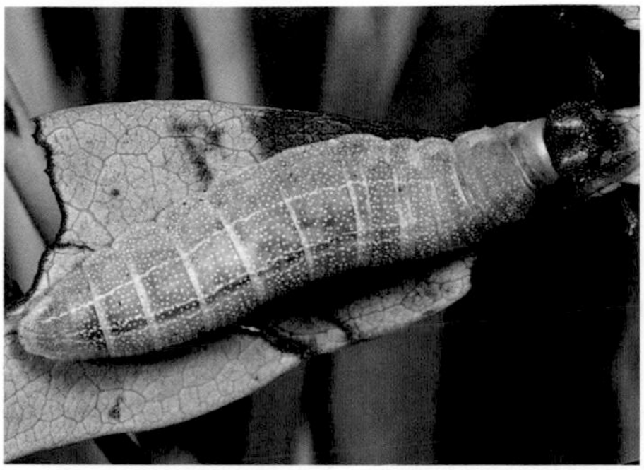

1. Florida Duskywing

2. Wild Indigo Duskywing adult

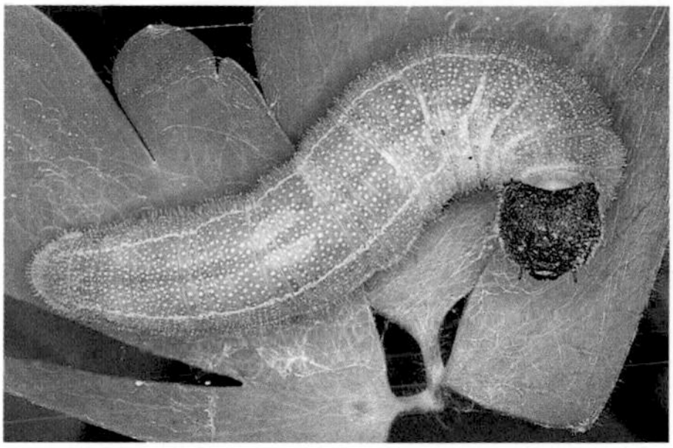

3. Columbine Duskywing

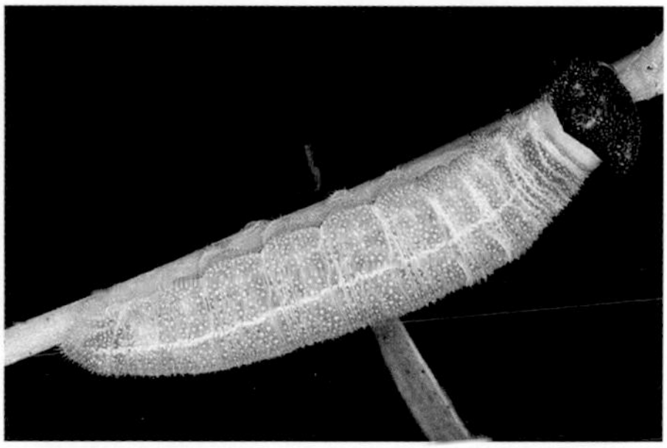

4. Wild Indigo Duskywing

5. Afranius Duskywing (inset 6. Perius)

8. Juvenal's Duskywing

17. Pacuvius Duskywing

18. Dreamy Duskywing

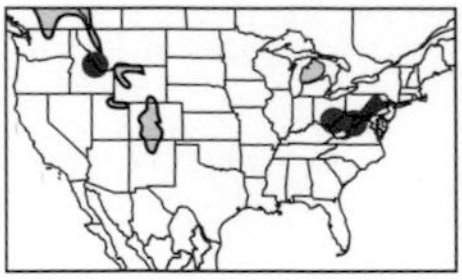

## 1. Grizzled Skipper *Pyrgus centaureae*

Appalachian Grizzled Skipper, possibly a distinct species, is pale green covered with fine white hairs. More northern and western populations are gray to pinkish. **Head is covered with fine reddish hairs**. Collar reddish. *Habitat* Open hilltops and grassy hillsides in wooded barrens in the East. Alpine meadows and talus slopes in the North and West. *Found* Late May through July. *Host* Cinquefoils, strawberries, and cloudberry.

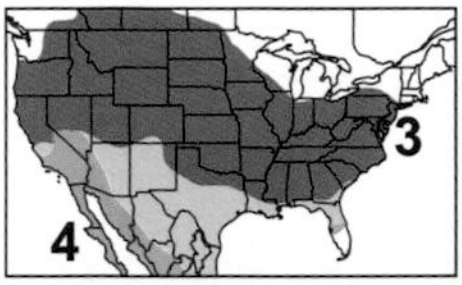


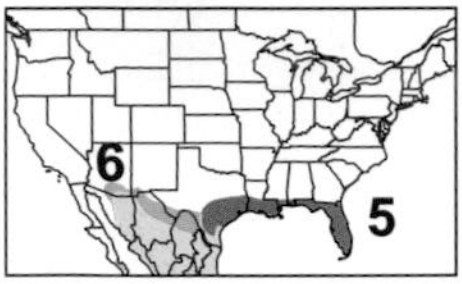


## 3. Common Checkered-Skipper *Pyrgus communis*

## 4. White Checkered-Skipper *Pyrgus albescens*

## 5. Tropical Checkered-Skipper *Pyrgus oileus*

Not shown.

## 6. Desert Checkered-Skipper *Pyrgus philetas*

Not shown.

Pale green, covered with fine white hairs. The **black head is sparsely covered with white hairs** and the **collar is reddish**. *Habitat* Desert is restricted to thorn scrub, the other species occur in a wide variety of open situations. *Found* All year in most of the range; Common Checkered-Skipper May–Oct northward. *Host* Mallows.

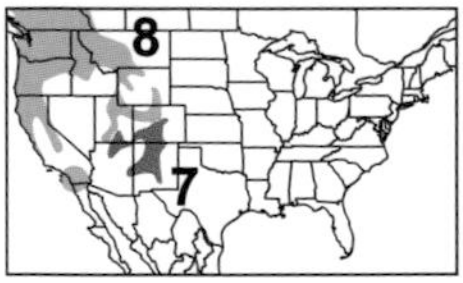


## 7. Mountain Checkered-Skipper *Pyrgus xanthus*

## 8. Two-banded Checkered-Skipper *P. ruralis*

Not shown.

Mountain Checkered-Skipper's **collar is black.** Two-banded green with black head. *Habitat* Mountain meadows and openings in cool coniferous forests for Two-banded. High, dry, mountain meadows and gullies for Mountain. *Found* June–Aug. *Host* Cinquefoils for Mountain. Cinquefoils and horkelias for Two-banded.

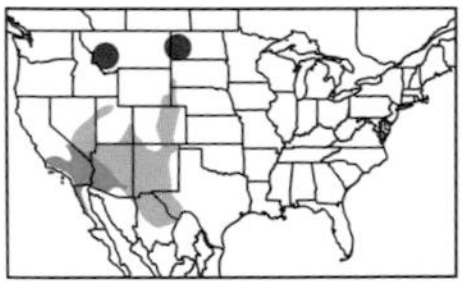

## 9. Small Checkered-Skipper *Pyrgus scriptura*

Pale green covered with fine white hairs. Black head is sparesely covered by tan hairs. **Collar is white**. *Habitat* Roadsides, gulches, alkali fields and disturbed open situations. *Found* April–Sept in 3 brood areas, late summer-early fall in 1 brood areas. *Host* Globe-mallows and other mallows.

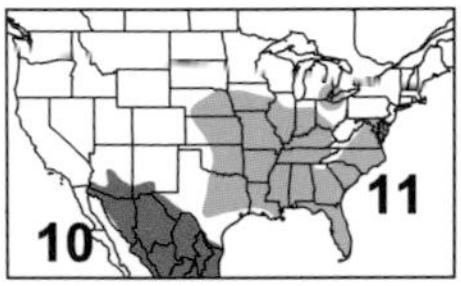


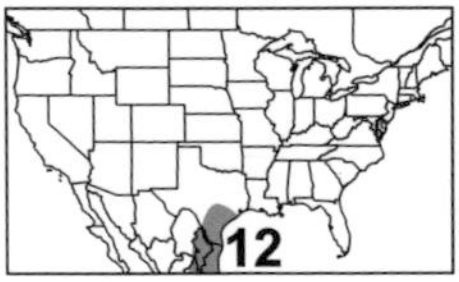


## 10. Golden-headed Scallopwing *Staphylus ceos*

## 11. Hayhurst's Scallopwing *Staphylus hayhurstii*

## 12. Mazans Scallopwing *Staphylus mazans* Not shown.

Pale greenish with white hairs; head dark, heart-shaped, and covered with fine white hairs; collar whitish. *Habitat* Riparian areas in arid regions for Golden-headed, open woodland for others. *Found* Most of year for Golden-headed; Mainly May–Sept for others. *Host* Mainly lambsquarters, but also pigweed for Mazans and bloodleaf for Hayhurst's. *Comments* Orange-brown eggs are usually laid on the upper sides of leaves. First instars cut and fold leaf flap over on top concealing themselves within. Full-grown caterpillars hibernate, turning pale yellow when feeding is complete.

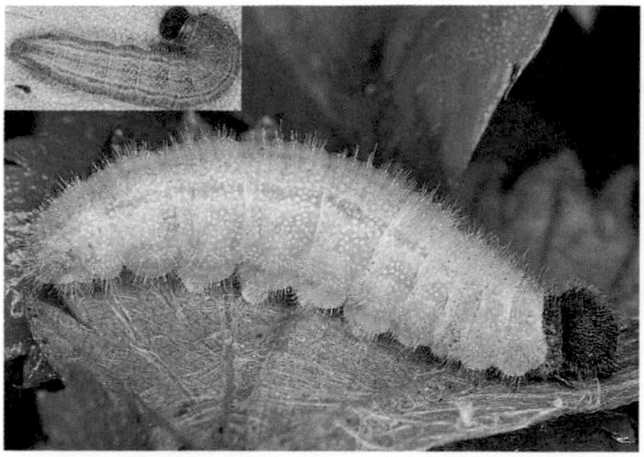

1. Grizzled Skipper

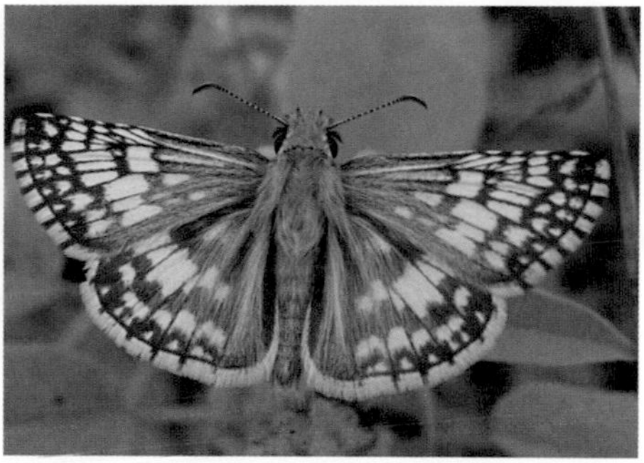

2. Common Checkered-Skipper adult

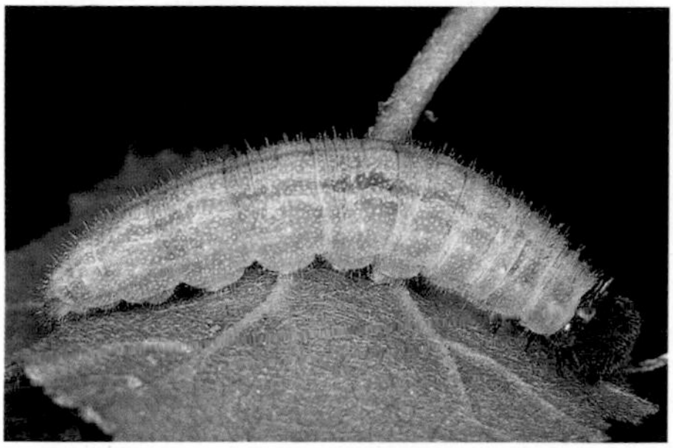

3. Common Checkered-Skipper

4. White Checkered-Skipper

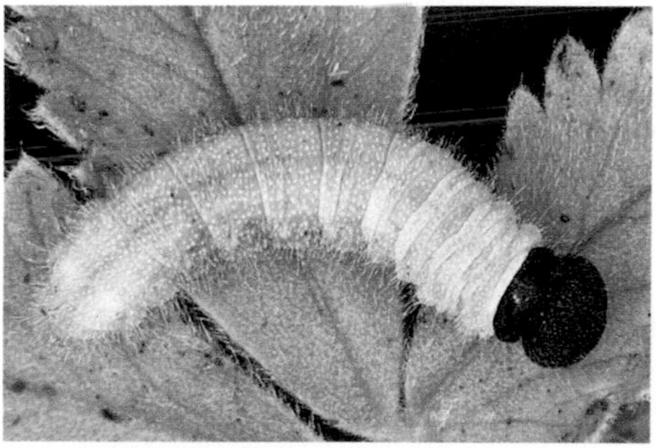

7. Mountain Checkered-Skipper

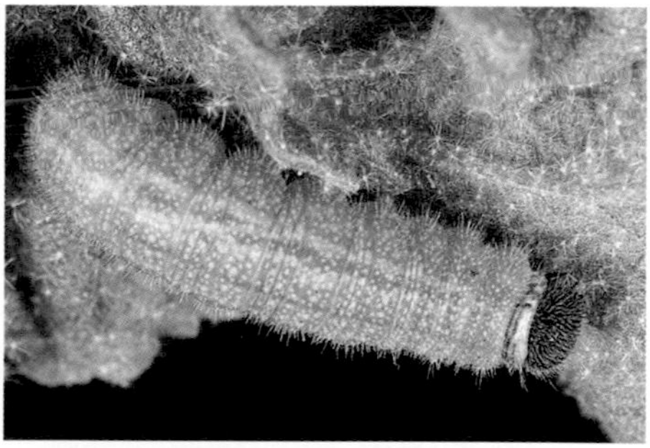

9. Small Checkered-Skipper

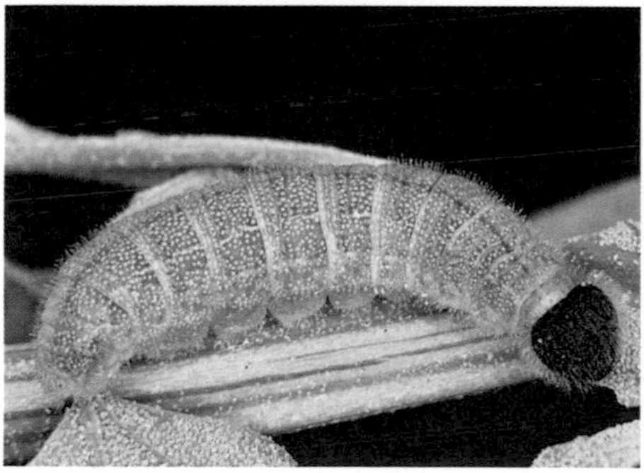

10. Golden-headed Scallopwing

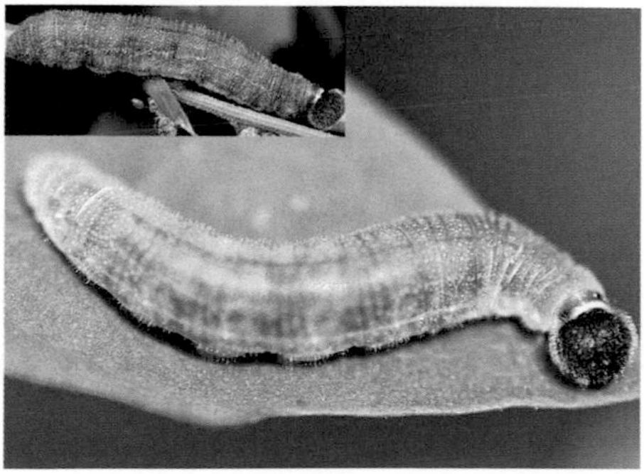

11. Hayhurst's Scallopwing

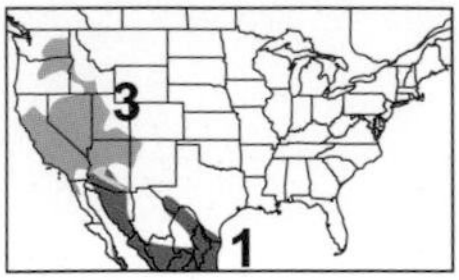


### 1. Erichson's White-Skipper *Heliopetes domicella*
### 3. Northern White-Skipper *Heliopetes ericetorum*

Pale green, covered with tiny, pale yellow dots and fine white hairs; pale yellow dorsal, subdorsal, and lateral stripes are faint or absent; **head black, densely covered with white hairs, not heart-shaped**. Late season caterpillars are pink-tinged (see inset to photo 4). *Habitat* 1. in lowland deserts and thorn scrub. 3. in chaparral and arid canyons. *Found* Most of year with little or no feeding during dry periods in the Southwest. *Host* Mallows. 1. prefers *Abutilon, Herrisantia*. 3 prefers *Sphaeralcea, Malacothamnus*. *Garden Tips* Globe mallows. *Comments* 1. Doesn't overwinter, feeding continues during winter months, at a much slower rate. 3. Overwinters as partially grown caterpillar.

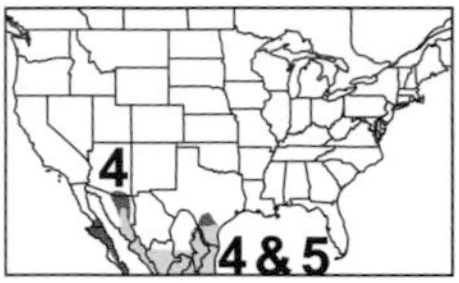


### 4. Laviana White-Skipper *Heliopetes laviana*
### 5. Turk's-cap White-Skipper *Heliopetes macaira*

Not shown.

Identical to Erichson's and Northern White-Skippers. *Habitat* Open tropical woodland, thorn scrub, and disturbed areas adjacent to fields. *Found* All year. *Host* Mallows.

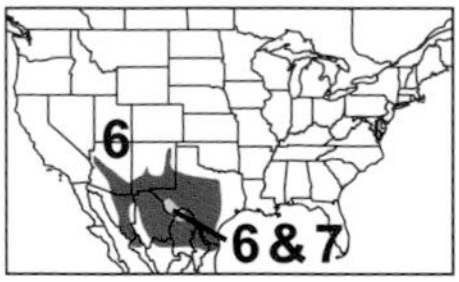


### 6. Common Streaky-Skipper *Celotes nessus*
### 7. Scarce Streaky-Skipper *Celotes limpia* Not shown.

6. Pale green, covered with tiny, pale yellow dots and fine white hairs; pale yellow dorsal, subdorsal, and lateral stripes may be present, faint or absent; **head black, flattened on top, densely covered with white hairs**. Sootywings use hosts in a different plant family. 7. Unreported. *Habitat* Thorn scrub, desert scrub and desert grassland. *Found* April–June, July–Sept 7. is rare and restricted to a small area in West Texas and adjacent Mexico. *Host* Mallows in Texas, ayenia in Arizona.

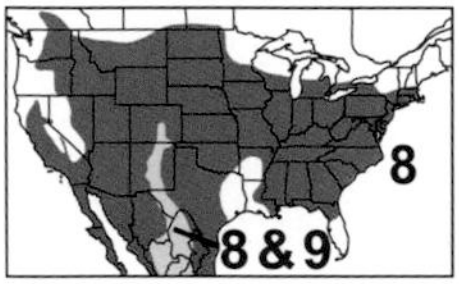


### 8. Common Sootywing *Pholisora catullus*
### 9. Mexican Sootywing *Pholisora mejicana* Not shown.

8. Pale green, covered with tiny, pale yellow dots and fine white hairs; pale yellow dorsal, subdorsal, and lateral stripes may be present, faint or absent; head black, densely covered with white hairs. Scallopwings are very similar. 9. Reported to be similar to 8 but with longer white body hairs. *Habitat* 8. Disturbed open areas, urban lots, railroad yards, etc. 9. Gulches and canyons. *Found* May–Sept. *Host* Lambsquarters and pigweeds.

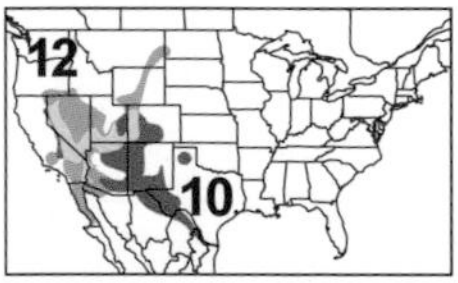


### 10. Saltbush Sootywing *Hesperopsis alpheus*
### 12. Mojave Sootywing *Hesperopsis libya* Not shown.

Pale blue-green or yellow-green, covered with tiny white dots; head black, heart-shaped and covered with short white (10) or tan (12) hair. *Habitat* Saltbush concentrations along desert washes and rivers, and in alkaline sage flats. *Found* May–July. *Host* 10. uses four-wing saltbush, quail saltbush (MacNeill's) and probably others. 12. Uses four-wing saltbush and shadscale saltbush.

1. Erichson's White-Skipper

2. Common Sootywing adult

3. Northern White-Skipper

4. Laviana White-Skipper

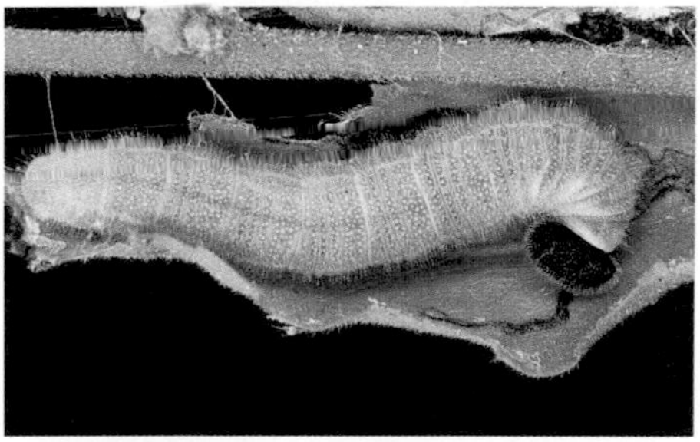

6. Common Streaky-Skipper

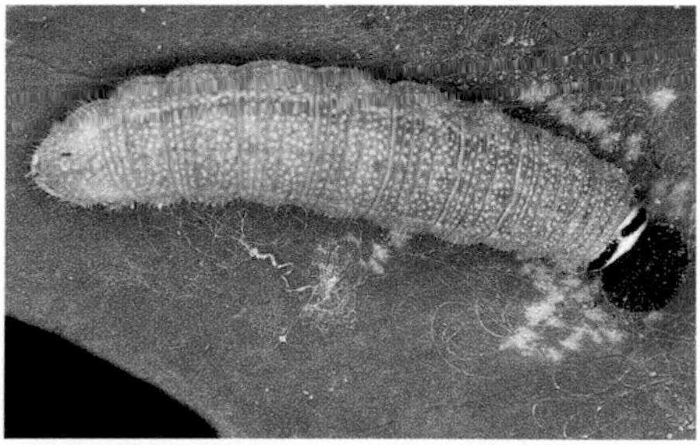

8. Common Sootywing

10. Saltbush Sootywing

11. 'MacNeil's' Saltbush Sootywing

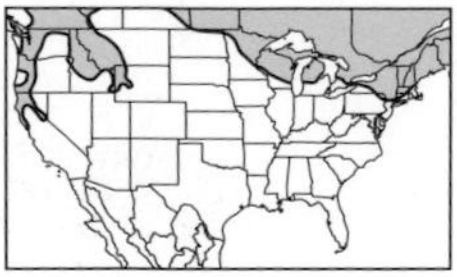

## Arctic Skipper *Carterocephalus palaemon*

Pale green with a white lateral stripe and 1–2 paler stripes below it; head green and round. *Habitat* Moist, grassy open areas within or adjacent to oak-pine transition forest. *Found* April–May, June–Sept, Oct. *Host* Grasses.

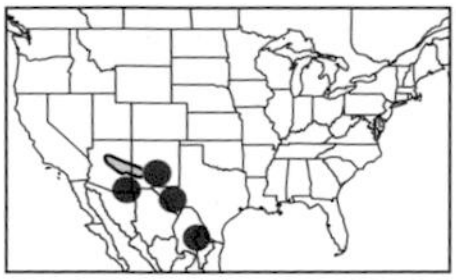

## Four-spotted Skipperling *Piruna polingi*

Note pair of outer **head stripes edged with black**. *Habitat* Very moist grass situations in high-elevation woodlands. *Found* All year, primarily late April–June; early instars after single flight, Aug–Oct. *Host* Smooth bromegrass, orchard grass, bent grass, and probably others. *Comments* Chooses grasses growing in moist shaded areas, often along or near watercourses.

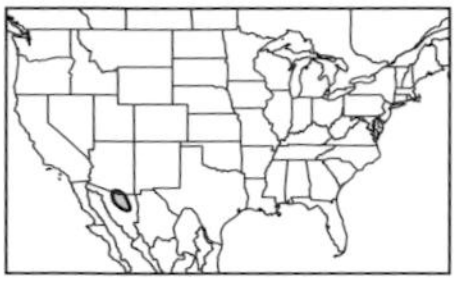

## Many-spotted Skipperling *Piruna cingo*

Head pale green; **face with brown Y.** Elissa Roadside-Skipper (pg. 146) has cream head and lacks Y. *Habitat* Mid-elevation grassy arroyos in arid oak-covered hillsides. *Found* All year, primarily before single flight July, and after flight Sept–Oct. *Host* Side-oats grama. *Comments* Early instars have more numerous white stripes. Chooses grasses in gully bottoms or under the shade of trees.

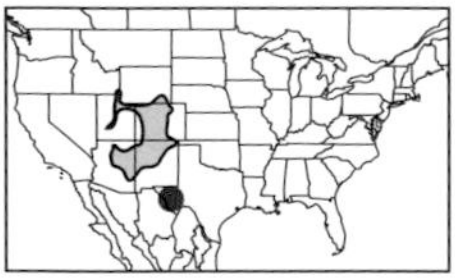

## Russet Skipperling *Piruna pirus*

Note pair of outer **head stripes edged with brown**. *Habitat* Moist grassy stream sides and other riparian situations in woodlands from prairie canyons to high elevations. *Found* Late April–May; early instars after single flight, July–Oct. *Host* Grasses. *Comments* Chooses grasses growing in shaded areas.

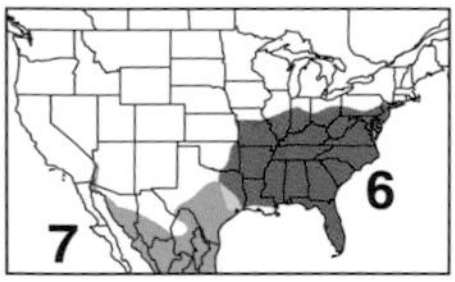

## 6. Swarthy Skipper *Nastra lherminier*

## 7. Julia's Skipper *Nastra julia* Not shown.

## 8. Neamathla Skipper *Nastra neamathla* Not shown.

6. & 7. Green with darker green mid-dorsal stripe; head tan edged in reddish-brown, with two broad, vertical reddish-brown stripes on face. Eufala Skipper (pg. 144) has less rounded head. 8. Unreported. *Habitat* Mainly low, open grassy situations. *Found* 6. April–June, and July–Sept. 7. All year. 8. April–Nov, after flights. *Host* 6. & 8. on little bluestem. 7. on Bermuda grass. *Comments* Lives in rolled leaf shelters.

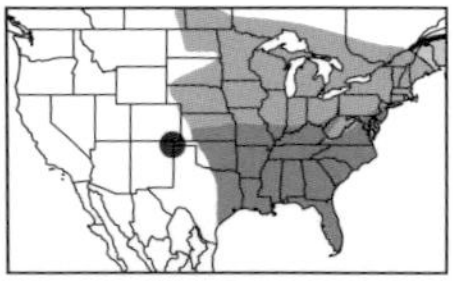

## Least Skipper *Ancyloxypha numitor*

Head roundish, tan to brown, darker near center, rimmed with white or tan. *Habitat* Wet grassy areas. *Found* May–June and Aug–Sept. *Host* Bluegrass, rice cutgrass, and other grasses.

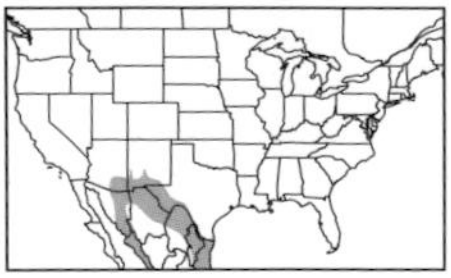

## Tropical Least Skipper *Anycloxypha arene*

Roundish **brown head with white markings**. *Habitat* Wet grassy areas. *Found* Most of year. *Host* Rabbitfoot grass, knotgrass, and other grasses found in wet soil at water edges. *Comments* Makes aerial nests, sometimes up to 3 feet from ground on taller grass clumps. Found Most of the year in southern areas.

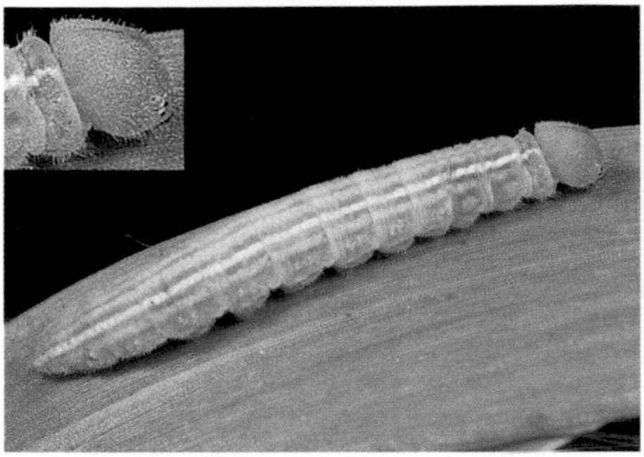

1. Arctic Skipper

2. Arctic Skipper adult

3. Four-spotted Skipperling

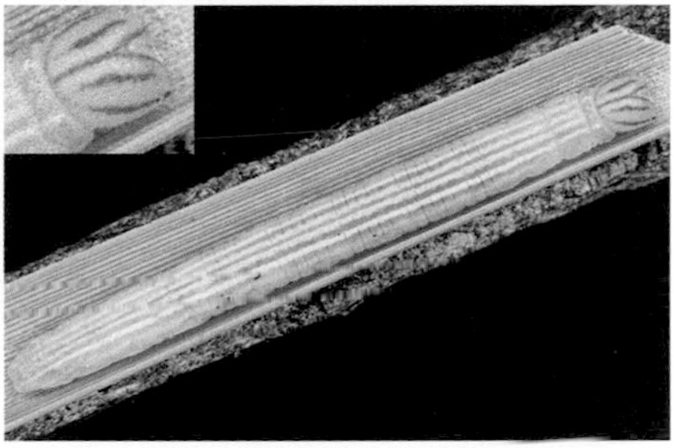

4. Many-spotted Skipperling

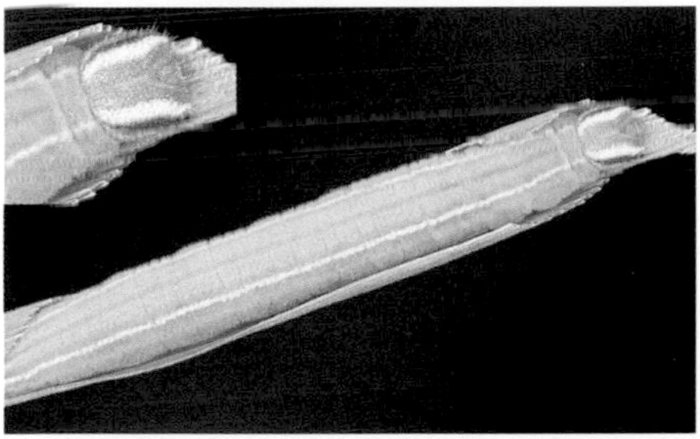

5. Russet Skipperling

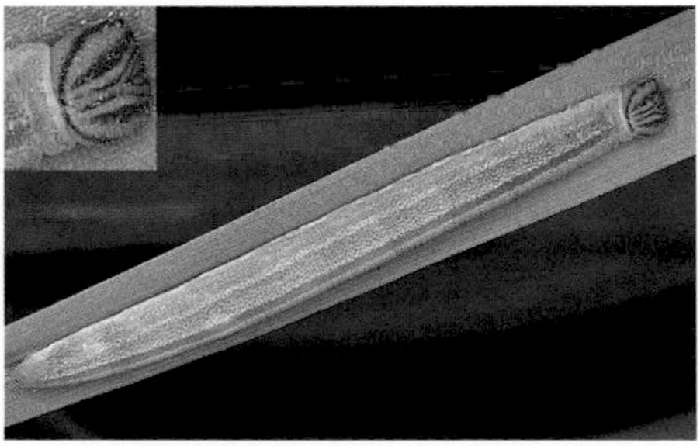

6. Swarthy Skipper

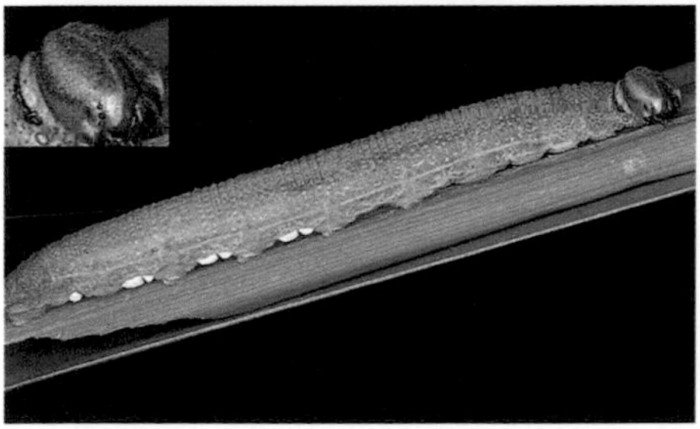

9. Least Skipper

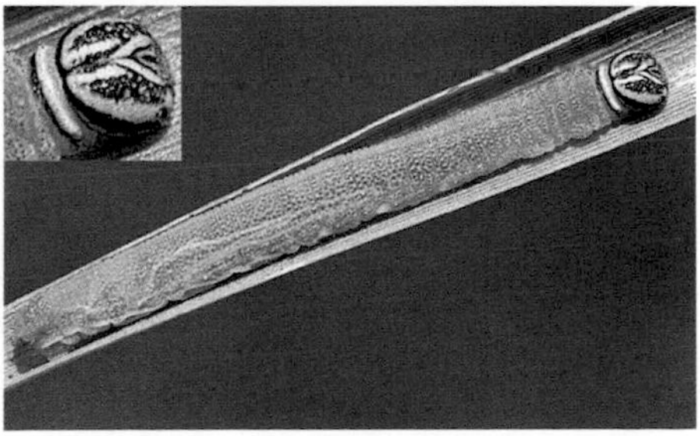

10. Tropical Least Skipper

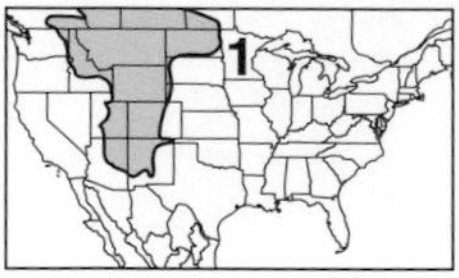


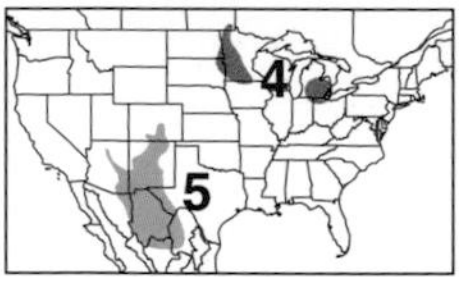


**1. Garita Skipperling** *Oarisma garita*

**4. Poweshiek Skipperling** *Oarisma poweshiek*
Not shown.

**5. Edwards' Skipperling** *Oarisma edwardsii* Not shown.
Garita and Poweshiek Skipperling have two forms: Green with numerous green and white stripes; head green, rounded; or tan to pinkish-tan with numerous red and tan stripes; head tan. Both forms have two paler streaks on the forehead and a pointed tail with a small notch. Edwards' Skipperling is unreported. See pg. 166 for more information about these species. *Host* Grasses.

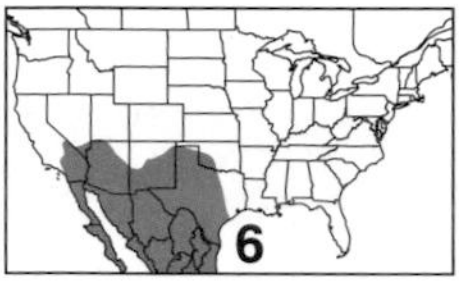


**6. Orange Skipperling** *Copaeodes aurantiacus*
Green, pale green, or tan with numerous green, red, or tan stripes, sometimes with purple or pink mid-dorsal stripe; head tan or pale green; **forehead pointed with a shallow notch**; **tail pointed and red tipped**. *Habitat* Arid regions, especially in canyons and gulches. *Found* Most of the year. *Host* Bermuda grass, side-oats grama, and other grasses. *Garden Tips* Side-oats grama. *Comments* Does not make a nest at any time during development. Can be located by searching for the distinctive triangular feeding notches cut in the host leaves. Caterpillars shake while they walk (perhaps resembling a piece of grass blowing in the wind).

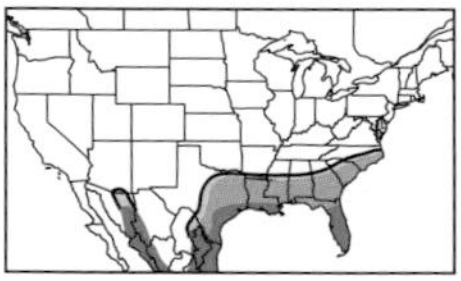

**7. Southern Skipperling** *Copaeodes minimus*
Green with darker mid-dorsal stripe divided by a lighter line, with numerous paler stripes; **head green, not pointed**; **tail pointed and red-tipped**. We do not know if there is a tan form. *Habitat* Open grassy habitats, but usually not in very wet or dry situations. *Found* All year. *Host* Bermuda grass. *Comments* Does not make a nest at any time during development.

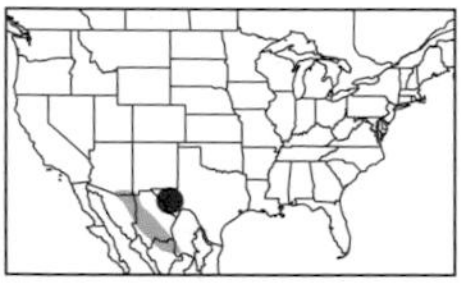

**8. Sunrise Skipper** *Adopaeoides prittwitzi*
Green with darker green mid-dorsal stripe and green spots; head mostly reddish-brown with two paler vertical streaks arising from forehead. *Habitat* Cienegas between 4000′ and 6000′. *Found* May–Oct, before and after flights. *Host* Knotgrass.

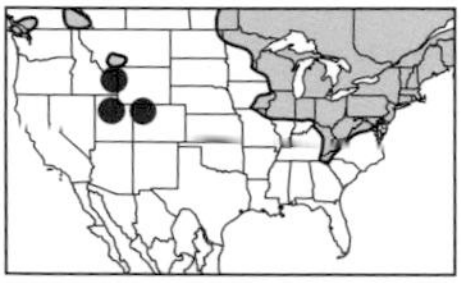

**9. European Skipper** *Thymelicus lineola*
Green with dark dorsal stripe and two lateral yellow lines; **head greenish-tan with two white or yellow vertical stripes**. *Habitat* Dry, grassy fields, especially those with tall grasses. *Found* April–June, prior to flight. *Host* Timothy and other tall broad-leaf species. *Comments* Eggs overwinter; feeds in spring from a rolled leaf shelter high up on a grass blade.

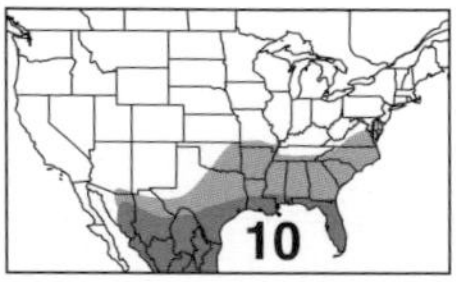


**10. Clouded Skipper** *Lerema accius*

**11. Three-spotted Skipper** *Cymaenes tripunctus*

**12. Fawn-spotted Skipper** *Cymaenes odilia* Not shown.
See pg. 166 for more information about these species.

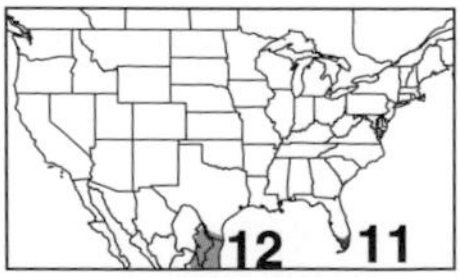

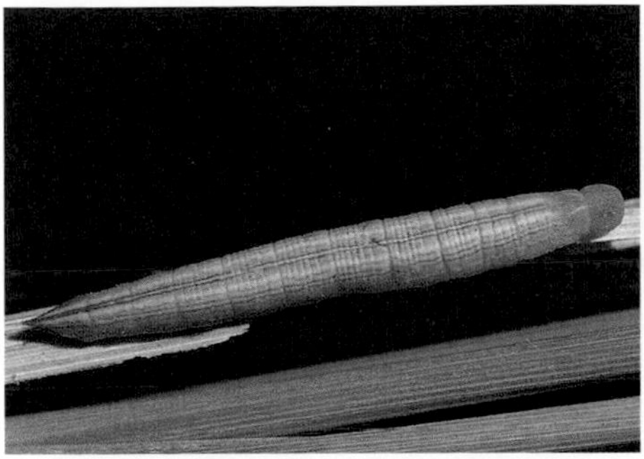
1. Garita Skipperling

2. Southern Skipperling adult

3. Garita Skipperling

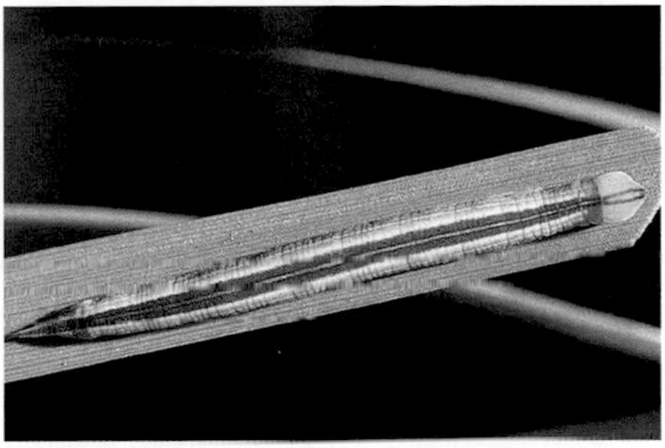
6. Orange Skipperling

7. Southern Skipperling

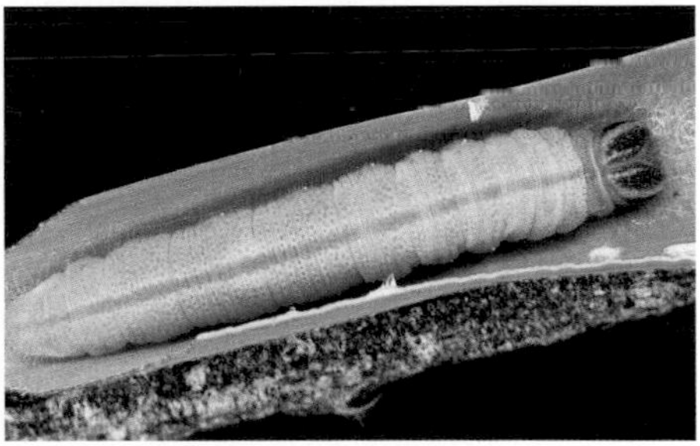
8. Sunrise Skipper

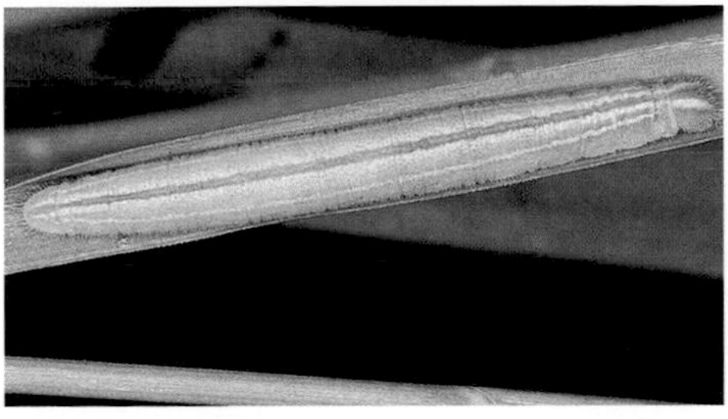
9. European Skipper

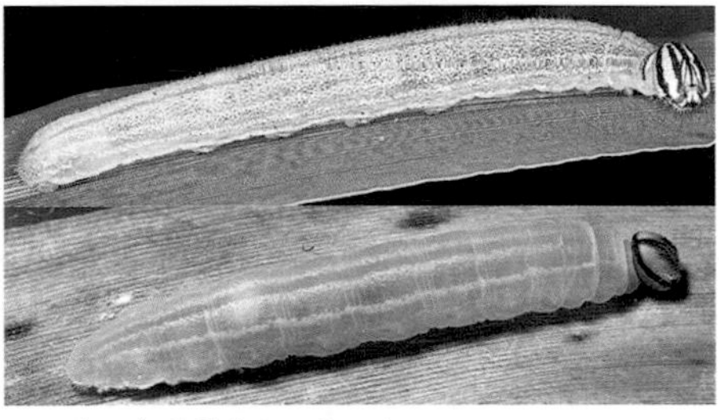
10. Clouded Skipper (top)
11. Three-spotted Skipper (bottom)

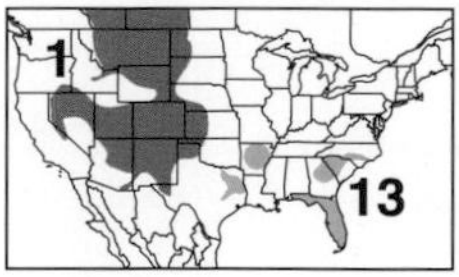


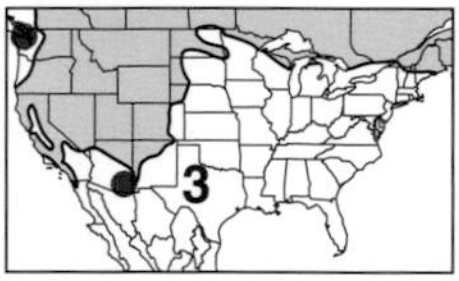


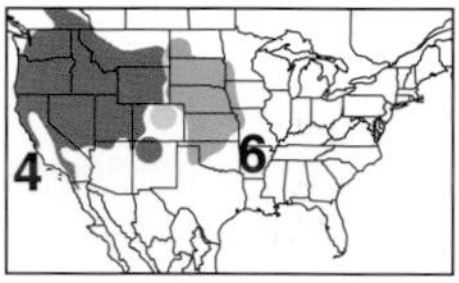


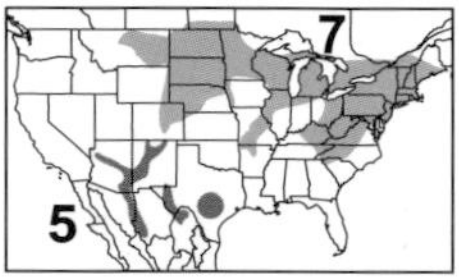


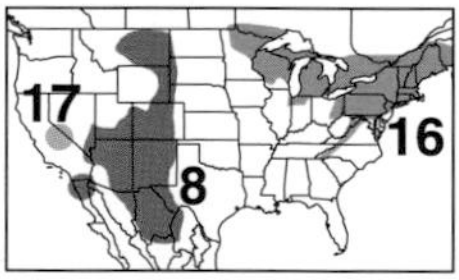


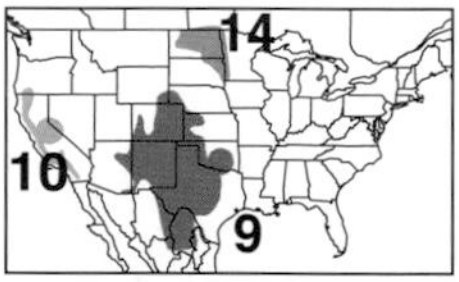


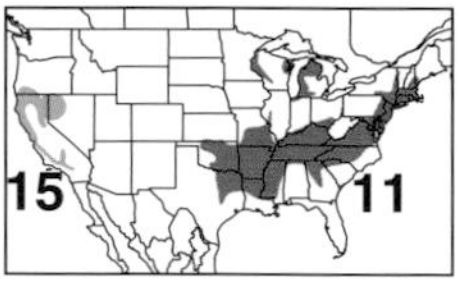


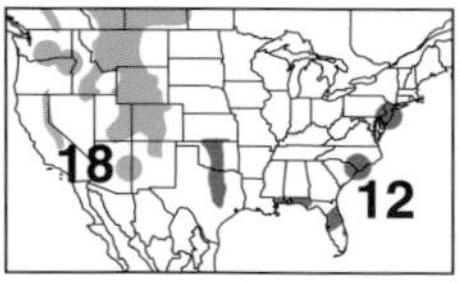


## **Hesperia Skippers** (genus *Hesperia*)

These caterpillars are light to dark brown with black heads that may or may not have two light vertical stripes on the forehead. Nests are constructed in or near the base of grass clumps (sometimes partially subterranean), making them nearly impossible to locate in the field. All species of hesperia skippers that we have seen are similar enough and show enough specific variation to preclude reliable identification based on color and pattern. In some cases, habitat, feeding period, and range will help to determine your hesperia caterpillar should you "win the lottery" and actually locate one in the field. The overwintering stage depends on the species. Some overwinter as eggs, others partially grown (summer flyers), and a few as early instars (fall flyers). All hesperia skippers feed on grasses, with many using mainly bluestem grasses or grama grasses.
See pg. 167 for more information about hesperia skippers.

### 1. Uncas Skipper *Hesperia uncas*

Individuals we've seen have been very dark brown to almost black, but they may be more variable. *Habitat* Short grass, grassland, and grassy alkaline flats. *Found* Probably prior to flights, April–June; Aug–Oct. *Host* Blue grama grass and others.

### 3. Common Branded Skipper *Hesperia comma*
### 4. Juba Skipper *Hesperia juba*
### 5. Apache Skipper *Hesperia woodgatei* Not shown.
### 6. Ottoe Skipper *Hesperia ottoe* Not shown.
### 7. Leonard's Skipper *Hesperia leonardus* Not shown.
### 8. Pahaska Skipper *Hesperia pahaska* Not shown.
### 9. Green Skipper *Hesperia viridis* Not shown.
### 10. Columbian Skipper *Hesperia columbia* Not shown.
### 11.Cobweb Skipper *Hesperia metea*
### 12. Dotted Skipper *Hesperia attalus*
### 13. Meske's Skipper *Hesperia meskei*
### 14. Dakota Skipper *Hesperia dacotae* Not shown.
### 15. Lindsey's Skipper *Hesperia lindseyi* Not shown.
### 16. Indian Skipper *Hesperia sassacus*

See pg. 167 for more information about these species.

### 17 Sierra Skipper *Hesperia miriamae* Not shown.

Caterpillar is unknown. *Habitat* Above treeline (over 10,500′) in the central Sierra Nevada Mountains of California and adjacent White Mountains of California and Nevada. *Found* Probably prior to single flight, May–July. *Host* Grasses. *Comments* An oxygen mask is recommended but not required.

### 18. Nevada Skipper *Hesperia nevada* Not shown.

Caterpillar is unknown. *Habitat* High-elevation grasslands, mountain meadows, and northern prairies. *Found* Probably prior to single flight, April–July. *Host* Grasses.

1. Uncas Skipper

2. Dottted Skipper adult

3. Common Branded Skipper

4. Juba Skipper

11. Cobweb Skipper

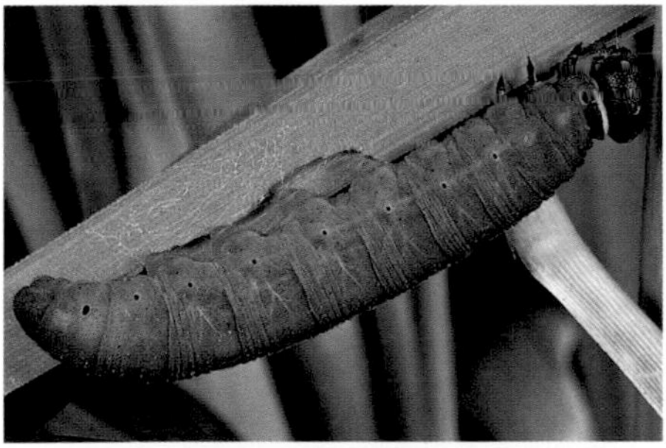

12. Dotted Skipper

13. Meske's Skipper

16. Indian Skipper

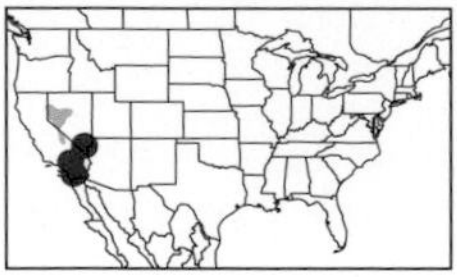

**Alkali Skipper** *Pseudocopaeodes eunus*
Light brown to tan; head black with two vertical white stripes. Many other species have similar face pattern but they use different hosts. *Habitat* Grassy, alkali flats, desert seeps, and springs. *Found* June–Sept. *Host* Saltgrass. *Comments* Nests down in clump of host and nearly impossible to find.

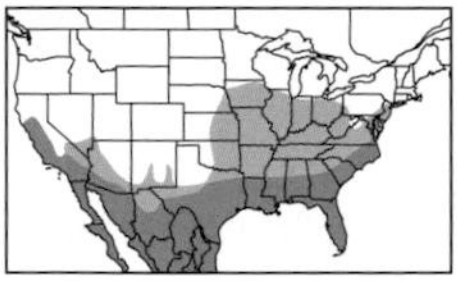

***Fiery Skipper** *Hylephila phyleus*
Variable. Yellow-brown, gray, greenish-brown, or dull green with brown dorsal stripe; head dark brown with black vertical stripe bordered by two lighter lines. Sachem may be the only other caterpillar with same host, habitat, and behavior. *Habitat* Lawns and other low grassy areas such as dry fields and roadsides. *Found* March–Oct. *Host* Bermuda grass and other grasses. *Garden Tips* Lantana (for adult). *Comments* Nest is horizontal, at the base of the host and difficult to locate; a good way to avoid caterpillar hunters and lawnmowers!

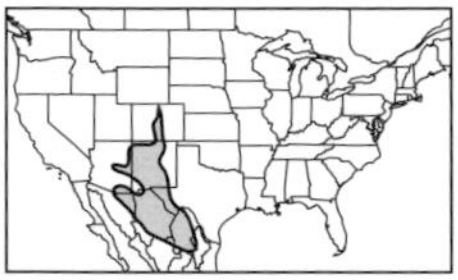

**Morrison's Skipper** *Stinga morrisoni*
Dull greenish-tan, slightly transparent; head light brown with dark center stripe. *Habitat* Open pine or pine-juniper-oak woodlands. *Found* Last instar May, just prior to flight; early-middle instars, July–Sept. *Host* Needlegrass, side-oats grama, and other grasses. *Comments* Nests are usually vertical, within the grass.

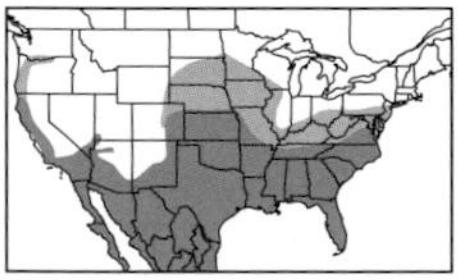

**Sachem** *Atalopedes campestris*
Speckled dark green; head black. *Habitat* Open disturbed fields, roadsides, suburban and urban lots, barrens. *Found* All year in the South; July–Oct, northward. *Host* Bermuda grass, crabgrasses. *Garden Tips* Lantana (for adult). *Comments* Nests at the base of the host like other dark skippers with black heads.

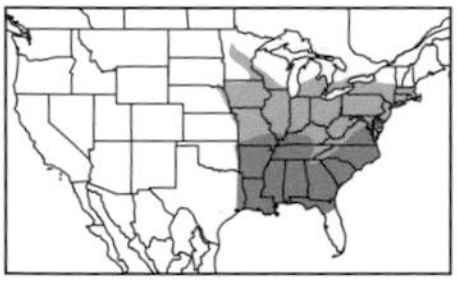

**Little Glassywing** *Pompeius verna*
Greenish with darker mid-dorsal line and a pair of spots at the rear; head black and smooth. *Habitat* Moist brushy fields near woodlands; rarer in poor soil areas. *Found* March–July, prior to and after flights. *Host* Purple-top grass. *Comments* Northern populations are single brooded, and partially grown caterpillars overwinter.

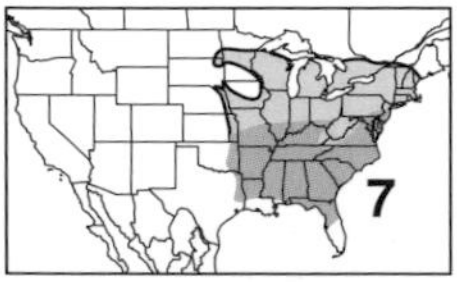


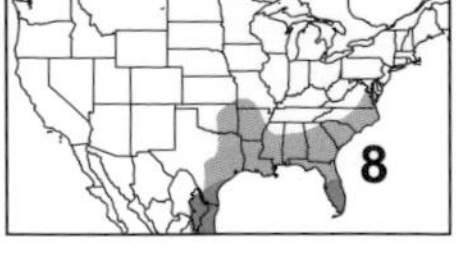


**7. Northern Broken-Dash** *Wallengrenia egeremet*
**8. Southern Broken-Dash** *Wallengrenia otho*
Speckled brown, sometimes with a pinkish hue; head black with or without two lighter vertical stripes on forehead. Not separable from *Polites* skippers that share range. *Habitat* 7. Open fields and meadows; common in moist, but not wet, situations. 8. Moist woodland edges and trails, and open areas. *Found* 7. April–May, Sept–Oct. 8. Most of the year in the South; March–April, July–Aug, and Oct–Nov northward. *Host* 7. Panic grasses. 8. Crabgrasses and others. *Comments* 8. Carries its shelter with it as it moves. Partially grown caterpillars overwinter in a small tube cut from a grass blade.

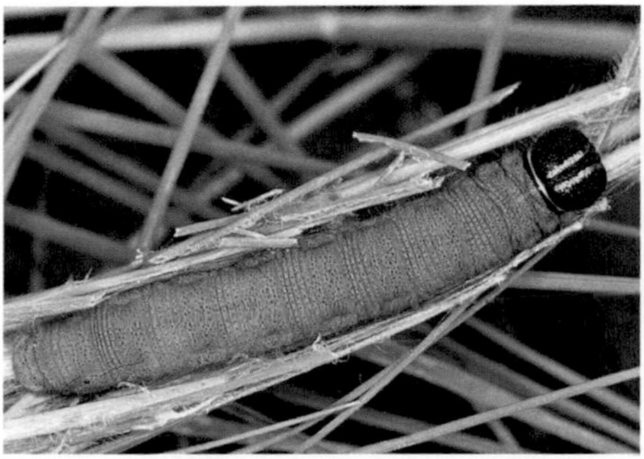

1. Alkali Skipper

2. Northern Broken-Dash adult

3. Fiery Skipper

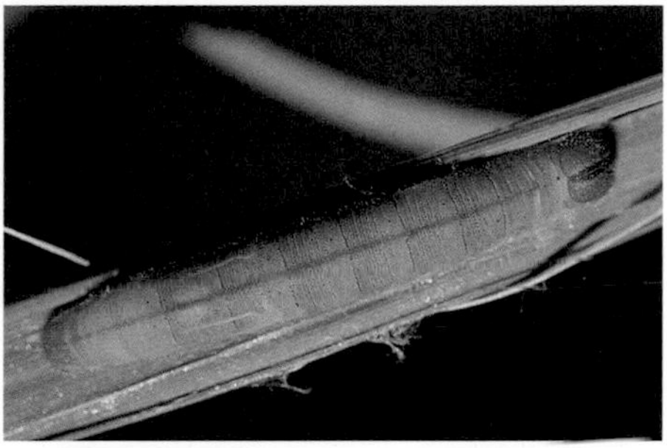

4. Morrison's Skipper

5. Sachem

6. Little Glassywing

7. Northern Broken-Dash

8. Southern Broken-Dash

See page 169 for more information about *polites* skippers.

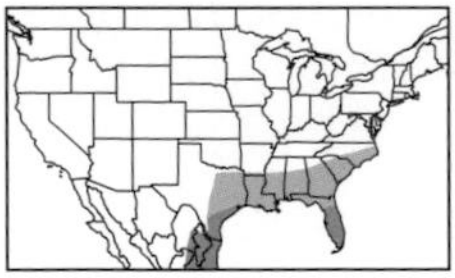

**Whirlabout** *Polites vibex*
Green to purplish; head dark brown to black with two light vertical stripes on forehead and light spots at margins. *Habitat* Disturbed grassy fields, roadsides, and woodland edges. *Found* Most of the year in the South. March–April and July–Sept where double-brooded. *Host* Bermuda grass, crabgrass, and others.

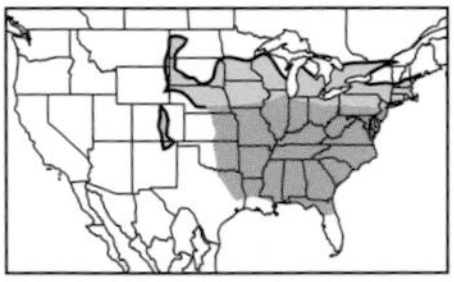

**Crossline Skipper** *Polites origenes*
Speckled brown, sometimes with a pinkish hue; head black. *Habitat* Dry, grassy fields, prairies, power-line cuts, especially in poor soil areas. *Found* July–Aug, Sept–Oct, between flights, and April. *Host* Big bluestem and other grasses. *Comments* The only *Polites* that makes aerial nests.

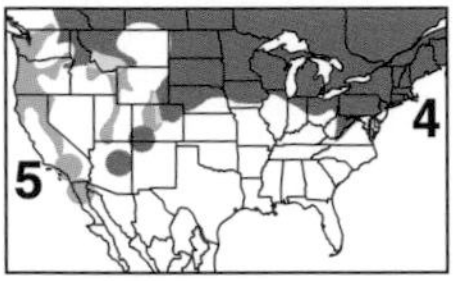


**4. Long Dash** *Polites mystic*
**5. Sonoran Skipper** *Polites sonora* Not shown.
Speckled brown, sometimes with a pinkish hue; head black. *Habitat* Wet meadows and marshes southward; more tolerant of drier sites northward. *Found* April–Sept. *Host* Blue grasses and giant bent grass for Long Dash. Unreported for Sonoran Skipper.

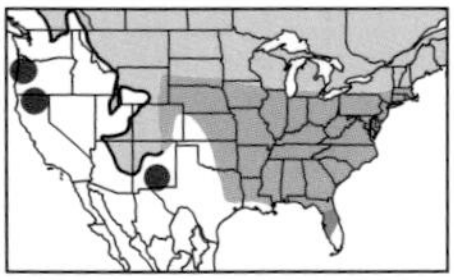

**Tawny-edged Skipper** *Polites themistocles*
Speckled brown; head black. *Habitat* Open grassy areas including suburban habitats, roadsides, and mountain meadows. *Found* May–Oct; varies with region. *Host* Many grasses. *Comments* Usually lays eggs on flowering plants adjacent to host grasses.

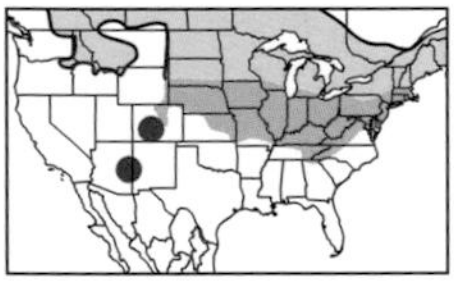

**Peck's Skipper** *Polites peckius*
Speckled dark brown; head black. *Habitat* Any open grassy area, including meadows, power-line cuts, suburban lawns, and roadsides. *Found* June–Oct, where double brooded; July–Sept where single brooded. *Host* Many grasses.

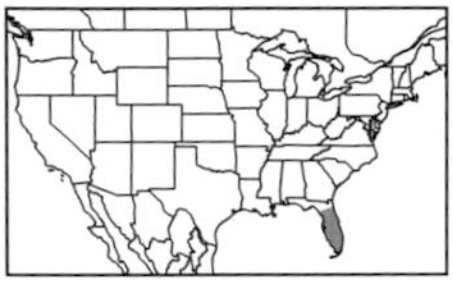

**Baracoa Skipper** *Polites baracoa*
Speckled dark brown; may have black dorsal and lateral lines; head black with or without two white vertical stripes on forehead. Tawny-edged Skipper uses different grass. *Habitat* Open situations with low grasses, especially lawns. *Found* All year. *Host* Centipedegrass and probably other low grasses.

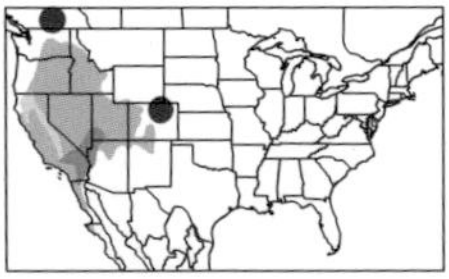

**Sandhill Skipper** *Polites sabuleti*
Dark greenish-brown to brown or maroon; head black with two white vertical stripes. *Habitat* Many grassy open situations. *Found* March–Oct; varies with region. *Host* Saltgrass, alkali sacaton, Bermuda grass, and others.

**Mardon Skipper** *Polites mardon*
Speckled dark brown; head black with two white vertical stripes on forehead. *Habitat* In Washington State, lowland pastures and grassy slopes. In northwestern California, mainly grassy areas in rhododendron/conifer forest in the fog belt. *Found* July–Sept. *Host* Grasses. Those used in nature are unknown.

1. Whirlabout

2. Crossline Skipper adult

3. Crossline Skipper

4. Long Dash

6. Tawny-edged Skipper

7. Peck's Skipper

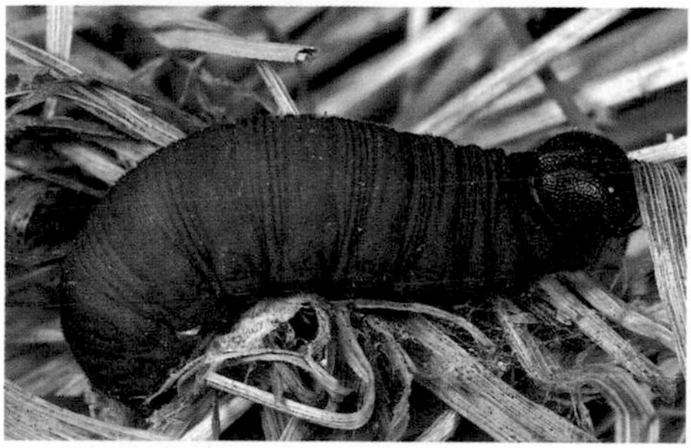

8. Baracoa Skipper

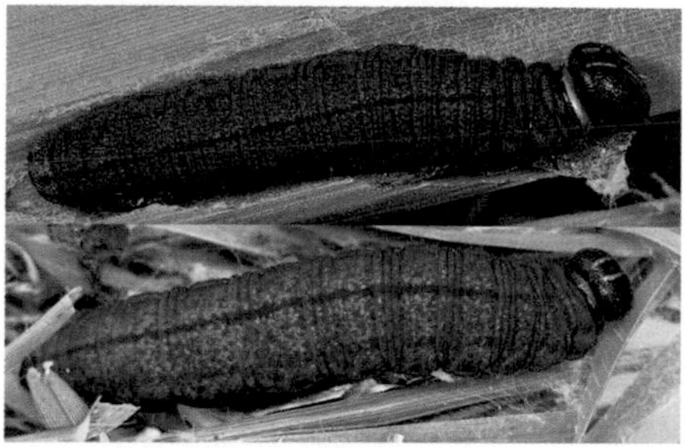

9. Sandhill Skipper (top)
10. Mardon Skipper (bottom)

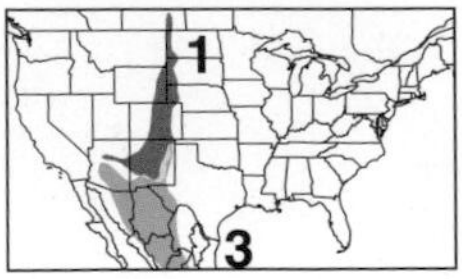


**1. Rhesus Skipper** *Polites rhesus*

**3. Carus Skipper** *Polites carus* Not shown.

Rhesus Skipper tan or brownish-green with **rear segments reddish**, the red sometimes extends farther along the body; head black with two white stripes on forehead. Carus Skipper is unknown. *Habitat* Rhesus Skipper in high elevation short-grass prairie, often on high-quality blue grama grassland. Carus Skipper in arid open oak-grassland. *Found* Possibly March–April, probably late June–Sept. *Host* Rhesus Skipper on blue grama grass. Carus Skipper probably uses grasses.

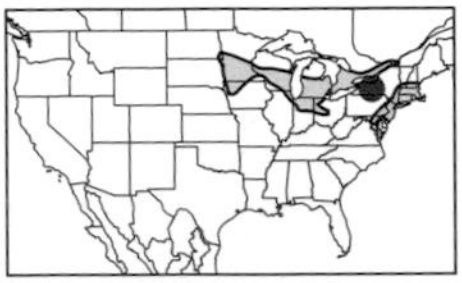

**Mulberry Wing** *Poanes massasoit*

Tan with dark mid-dorsal stripe and faint lateral stripes, speckled with green, and covered with fine hair, giving it a velvety appearance; head brown, pubescent. *Habitat* Wet meadows, open freshwater marshes, fens, or bogs. *Found* Prior to flight April–June, and after flight, Aug–Sept. *Host* Sedges.

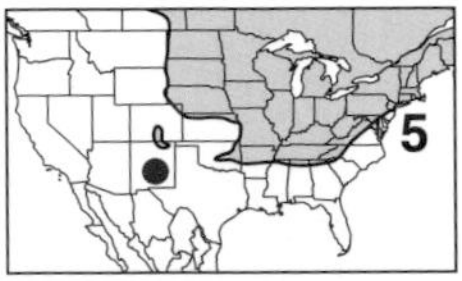


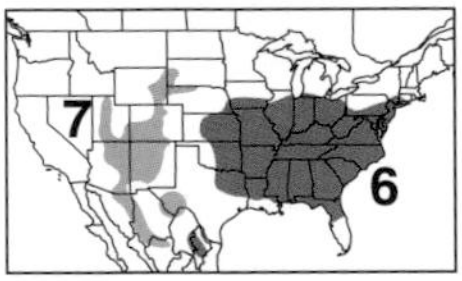


**5. Hobomok Skipper** *Poanes hobomok*

**6. Zabulon Skipper** *Poanes zabulon*

**7. Taxiles Skipper** *Poanes taxiles* Not shown.

Brownish green, sometimes with a pinkish hue; head brown, pubescent. *Habitat* Deciduous woodland edges and openings, parks, and gardens. Taxiles also in small patches of woods in prairie. *Found* June–Oct. *Host* Many grasses. Hobomok often on panic grasses. Taxiles often uses wheat grasses or brome grasses.

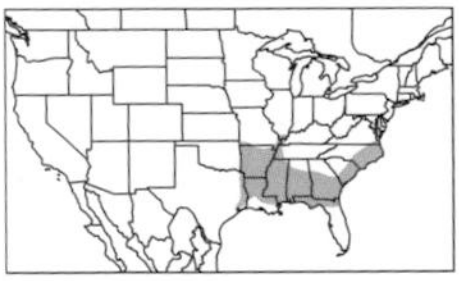

**Yehl Skipper** *Poanes yehl*

Tan with dark mid-dorsal stripe and faint lateral stripes, speckled with green and covered with fine hair, giving it a velvety appearance; head brown, pubescent. *Habitat* Wooded swamps and adjacent areas. *Found* July–Aug, Sept–Nov. *Host* Giant cane.

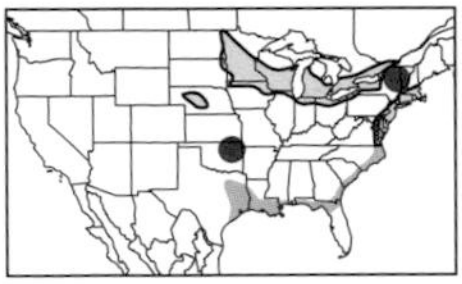

**Broad-winged Skipper** *Poanes viator*

Tan with faint whitish lateral stripes; head tan with a black vertical dash. *Habitat* Tidal and freshwater marshes with tall grasses. *Found* June–July, Oct–Nov, and April. *Host* Common reed and wild rice. *Comments* Overwinters partially grown.

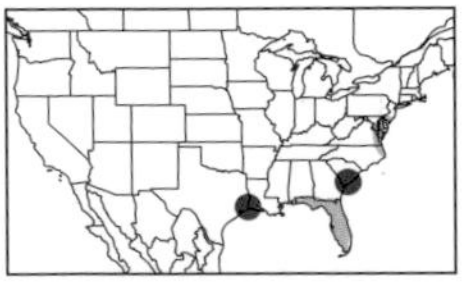

**Aaron's Skipper** *Poanes aaroni*

Tan with faint whitish lateral stripes; head tan with a black vertical dash. *Habitat* Marshes and roadside ditches. *Found* March–April, and June–July, between flights. *Host* Saltgrass and possibly smooth cordgrass in salt marshes; sedges in freshwater marshes. *Comments* Lives in tube shelters on host.

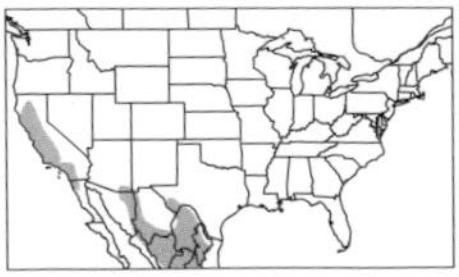

**Umber Skipper** *Poanes melane* Not shown.

Yellow-green; head brown. *Habitat* Woodland trails and edges, lowland and foothill canyons; often near water; gardens along California coast. *Found* Most of year along coast, a rarity elsewhere, March–Oct. *Host* Grasses.

1. Rhesus Skipper

2. Mulberry Wing adult

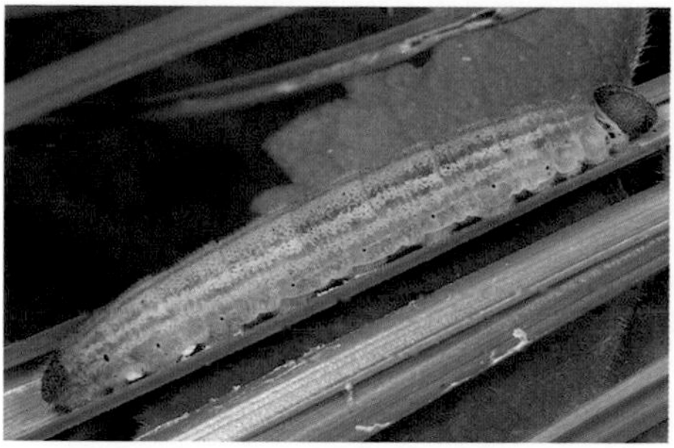

4. Mulberry Wing

5. Hobomok Skipper

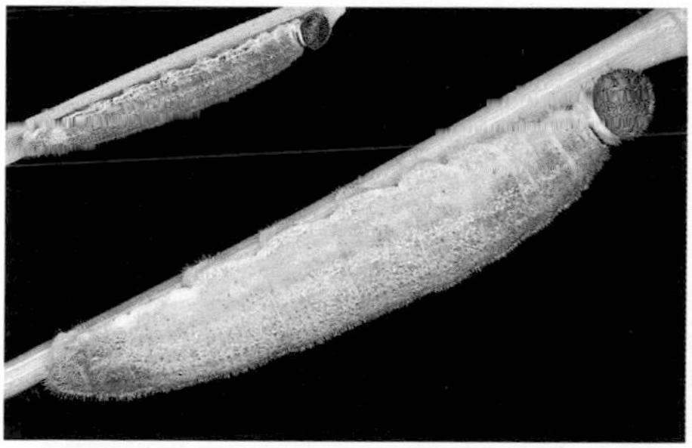

6. Zabulon Skipper

8. Yehl Skipper

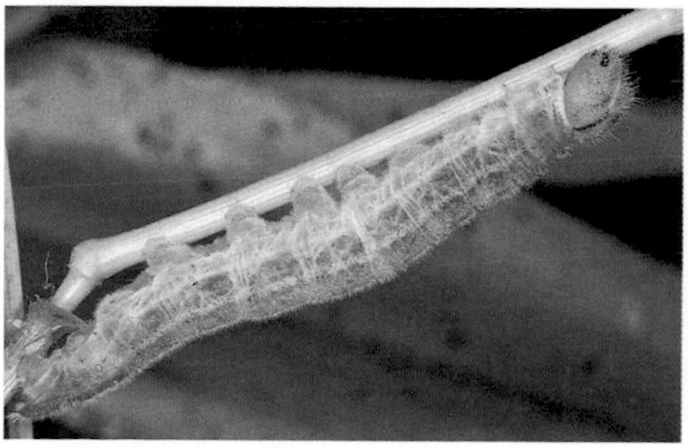

9. Broad-winged Skipper

10. Aaron's Skipper

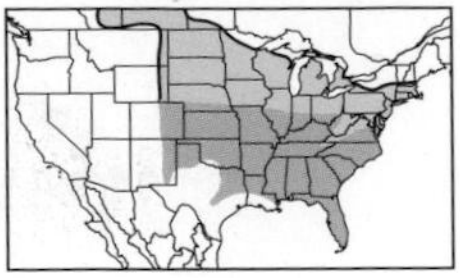

**Delaware Skipper** *Anatrytone logan*
Bluish-green with **black crescent on rear**; head white, rimmed brown or black with three vertical brown or black stripes. *Habitat* Open brushy fields, moist meadows, prairies, sedge marshes, and coastal marshes. *Found* June–July, Sept–Oct, in 2-brood areas; April–June, in 1-brood areas. *Host* Smooth bromegrass and other grasses. *Comments* First instar has black head, contrasting with first instar of Arogos Skipper, which had a red-brown head.

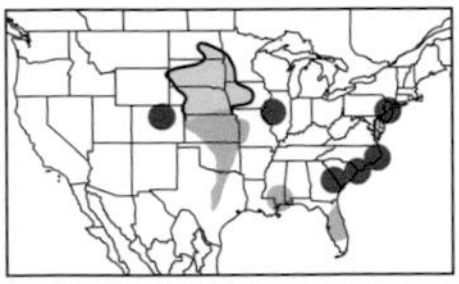

**Arogos Skipper** *Atrytone arogos*
Bluish-green; head tan rimmed with red and face with three vertical reddish-brown stripes. *Habitat* Tall grass prairie in the Midwest, grassy barrens in the East. *Found* Varies with locality, but usually June–July and April–May. *Host* Big bluestem (Midwest), little bluestem (north New Jersey), pine barrens, reed grass (south New Jersey), and lopsided Indian grass (Florida).

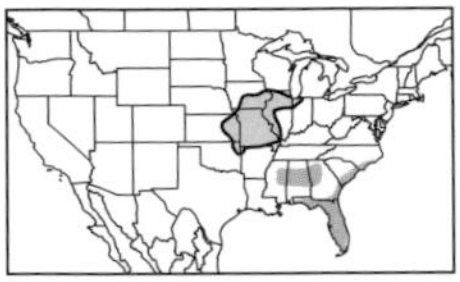

**Byssus Skipper** *Problema byssus*
Bluish-green; head white rimmed with black, with three black vertical facial stripes. *Habitat* Prefers edges of wooded wetlands, savannas, and marshes along the Atlantic coast; tall-grass prairies farther west. *Found* March–early May, July–Aug where double-brooded. July–Oct; and April where single-brooded. *Host* Plume grasses, eastern gama grass, and probably others.

**Rare Skipper** *Problema bulenta*
Bluish-green; head gray, rimmed with black, with faint white vertical stripes. *Habitat* Brackish river tidal marshes near the coast. *Found* July and April–May in 2 brood areas, Aug–Sept in single brood areas. *Host* Intertidal cordgrass in New Jersey, wild rice southward.

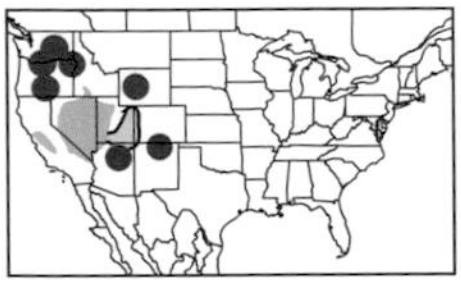

**Yuma Skipper** *Ochlodes yuma*
Pale greenish to tan; head cream to tan, bisected on the forehead by a darker brown vertical stripe. *Habitat* Arid-country seeps, marshes and streams with giant reed colonies. *Found* July–Sept where single-brooded, June–Oct where double-brooded. *Host* **Giant reed grass**. *Comments* Perhaps the only skipper restricted to one host.

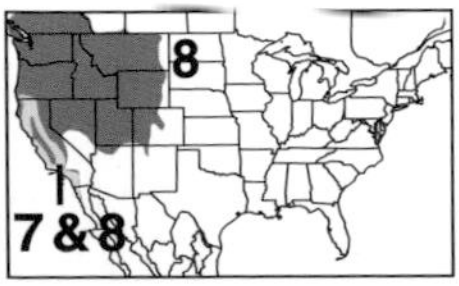


**7. Rural Skipper** *Ochlodes agricola*
**8. Woodland Skipper** *Ochlodes sylvanoides*
Variable. Tan or greenish-tan with mid-dorsal stripe, speckled with tiny brown or green spots that fade when full-grown. *Habitat* Woodland edges, chaparral, roadsides, and riparian habitats, all at low to moderate elevations. *Found* March–July, prior to flight. *Host* 7. on melic grass, probably others. 8. on many grasses. *Comments* Early instars in aerial nests; last instars nest in center of host. Overwinters as unfed first instar.

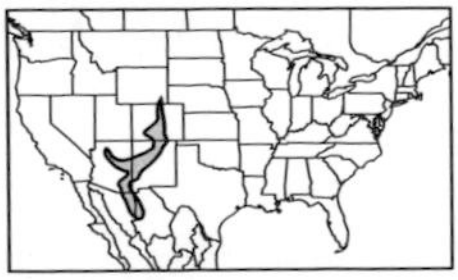

**Snow's Skipper** *Paratrytone snowi* Not shown.
Unreported. *Habitat* High-elevation openings in pine forest.

1. Delaware Skipper

2. Woodland Skipper adult

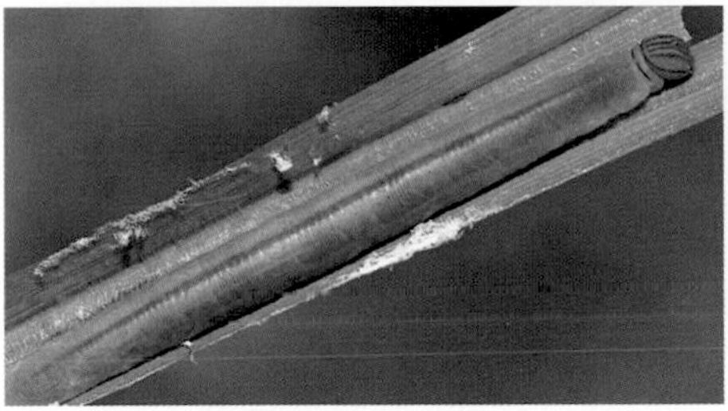

3. Arogos Skipper

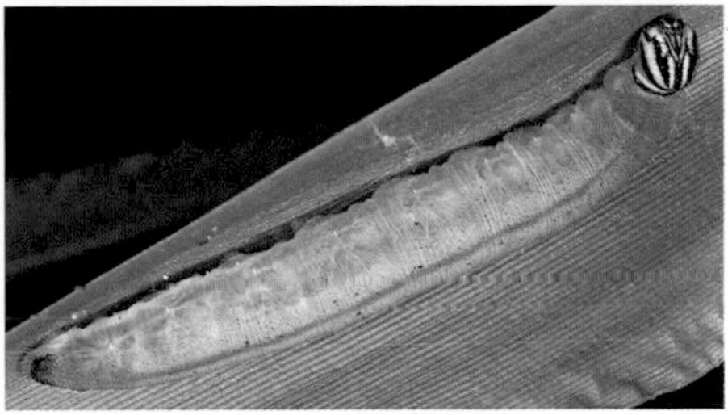

4. Byssus Skipper

5. Rare Skipper

6. Yuma Skipper

7. Rural Skipper

8. Woodland Skipper

**Euphyes Skippers** have a unique white-and-brown striped facial pattern that always has a **black oval on the upper forehead** edged with white or cream. Like the Hesperia the caterpillars of this genus of skippers are essentially identical and not distinguishable by color pattern or markings. However, some can be separated by host, range, or habitat. Most are sedge feeders and occur in wet areas. They may be single- or double-brooded. Third or fourth instar caterpillars overwinter in aerial nests made by constructing a tube shelter by tying several leaf blades together.

## Palmetto Skipper *Euphyes arpa*

*Habitat* Open pine flats and saw palmetto. *Found* Most of the year, between flights. *Host* **Saw palmetto**. *Comments* Nests are located at the base of the fan-like leaves. Monk Skipper (pg. 144) is the only other species to use saw palmetto.

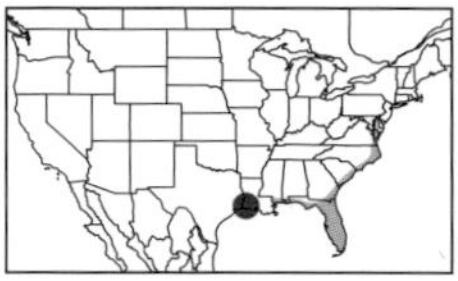

## Palatka Skipper *Euphyes pilatka*

*Habitat* Brackish marshes and adjacent areas. *Found* Most of year in Florida, March–May and July–Aug northward. *Host* Sawgrass.

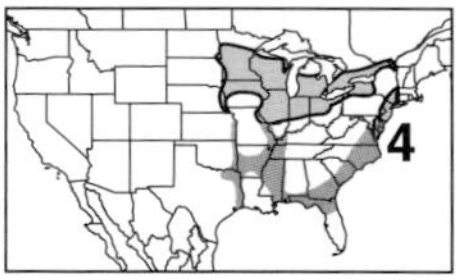


## 4. Dion Skipper *Euphyes dion* Not shown.

## 5. Bay Skipper *Euphyes bayensis* Not shown.

*Habitat* Dion in calcareous fens and other alkaline to neutral wetlands from northern New Jersey northward; bogs, roadside ditches, and other acid wetlands from southern NJ south. Bay Skipper in brackish marshes on MS and TX coast. *Found* Varies with locality, March–June, Sept–Nov. *Host* Sedges.

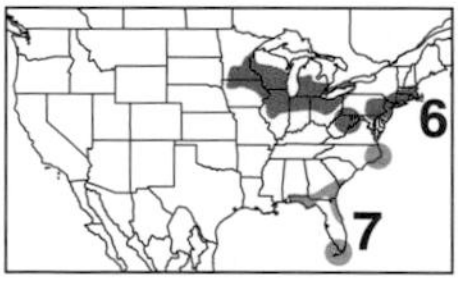


## 6. Black Dash *Euphyes conspicua*

*Habitat* Wet meadows and freshwater marshes. *Found* April–June, Aug–Oct, prior to and after flights. *Host* Sedges.

## 7. Berry's Skipper *Euphyes berryi*

*Habitat* Swamps and swamp edges. *Found* June–July, Oct–Nov, and Feb–March. *Host* Sedges.

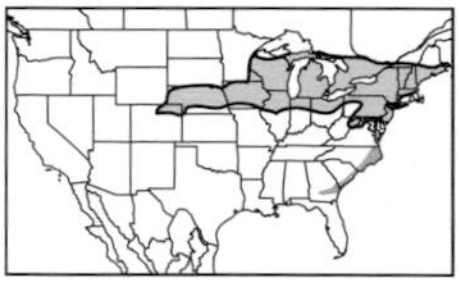

## Two-spotted Skipper *Euphyes bimacula*

*Habitat* Wet acid soil areas such as bogs, acid marshes, and meadows with sedges. *Found* April–May and June, Aug–Sept. *Host* Sedges.

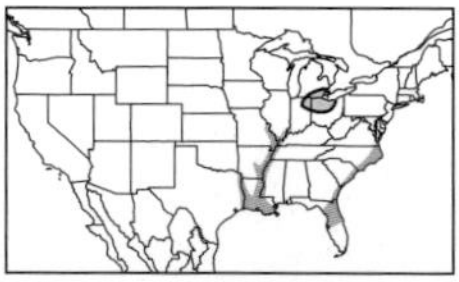

## Dukes' Skipper *Euphyes dukesi*

*Habitat* Shady, freshwater swamps and roadside ditches. *Found* March–Sept, prior to and after flights. *Host* Sedges.

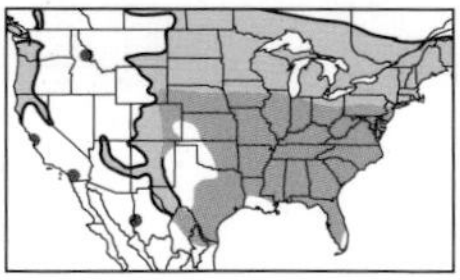

## Dun Skipper *Euphyes vestris*

*Habitat* Moist open situations near deciduous woodlands, prairies, and roadsides; seeps and springs in southern California. *Found* Varies with region, March–May, Aug–Oct, and Nov. *Host* Sedges. *Comments* Flights are earlier further south.

1. Palmetto Skipper

2. Black Dash adult

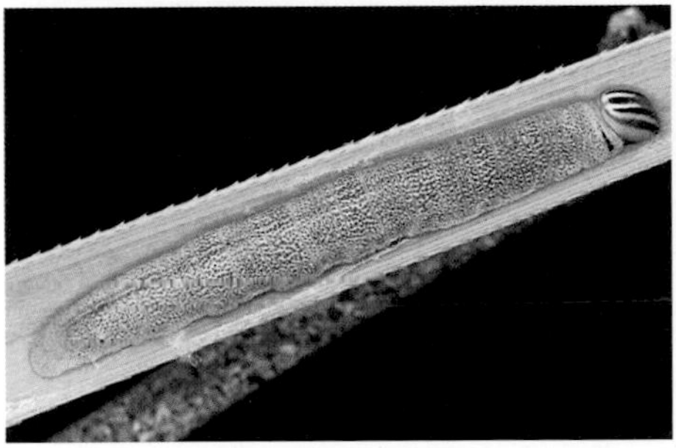

3. Palatka Skipper

6. Black Dash

7. Berry's Skipper

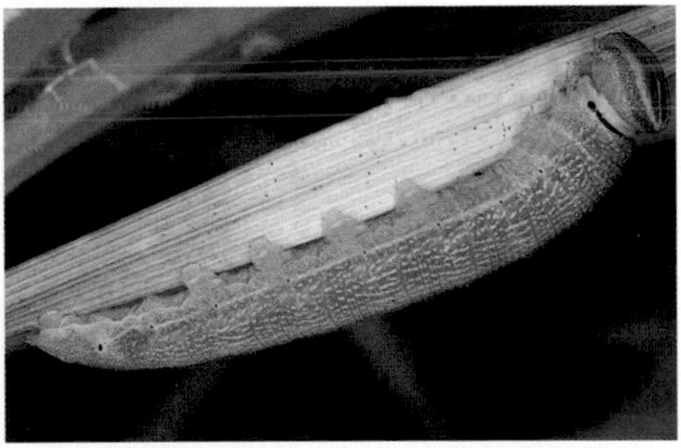

8. Two-spotted Skipper

9. Dukes' Skipper

10. Dun Skipper

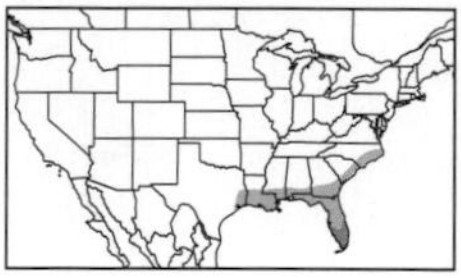

**Twin-spot Skipper** *Oligoria maculata*
Light brown to reddish. *Early Instars* Pale green; darker at rear. Hobomok Skipper occupies a different range. Might be confused with Zabulon Skipper where two occur together. *Habitat* Pinelands and wetlands. *Found* All year. *Host* Bluestem grasses.

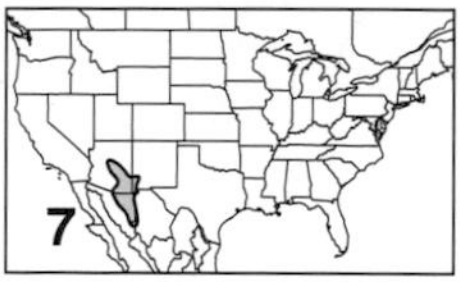


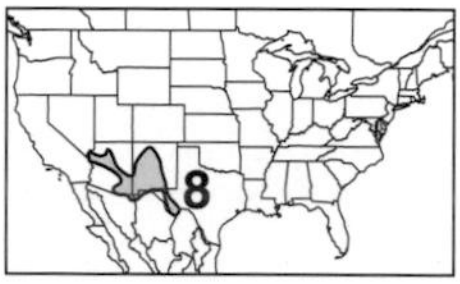


**Dusted-Skippers** (genus *Atrytonopsis*) are primarily southwestern. Based on the small sample observed so far, we see no distinct differences between species. Full-grown, most are nearly $1\frac{1}{4}$ inches long. All species we've seen are dull green to tan, sometimes with a pinkish hue, and covered with fine short hairs, creating a velvety appearance; the rear segments are more tan than green; head yellow-brown, pubescent. Aerial nests are constructed by those that we are familiar with, at least during the early part of their development. Partially grown or full-grown caterpillars overwinter. See pg. 169–71 for more information about individual species of dusted-skippers.

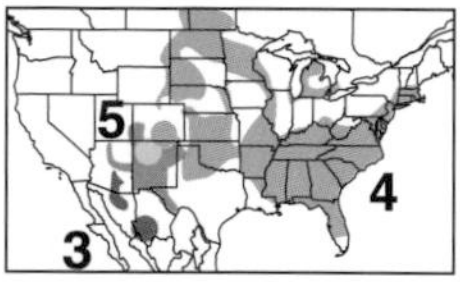


**3. Cestus Skipper** *Atrytonopsis cestus*
*Host* Bamboo muhly.

**4. Dusted Skipper** *Atrytonopsis hianna*
*Host* Bluestem grasses.

**5. Viereck's Skipper** *Atrytonopsis vierecki* Not shown.
*Host* Unreported.

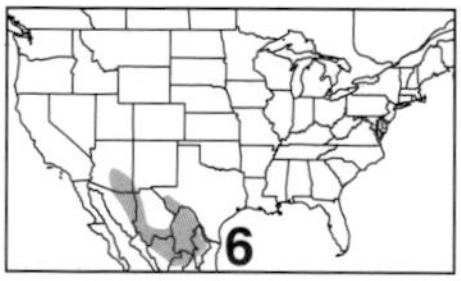


**6. Sheep Skipper** *Atrytonopsis edwardsii*
*Host* Side-oats grama and probably other grasses.

**7. Moon-marked Skipper** *Atrytonopsis lunus* Not shown.
*Host* Bull grass.

**8. Python Skipper** *Atrytonopsis python* Not shown.
*Host* Unreported; grass suspected.

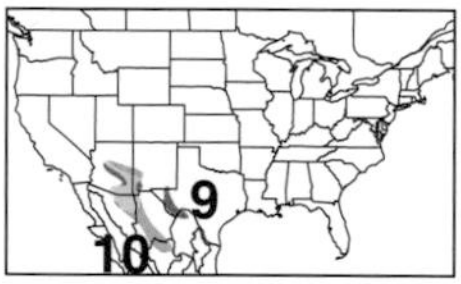


**9. White-barred Skipper** *Atrytonopsis pittacus*
*Host* Side-oats grama and probably other grasses.

**10. Deva Skipper** *Atrytonopsis deva* Not shown.
*Host* Unreported; grasses suspected.

**Monk Skipper** *Asbolis capucinus*
Yellow-green; rear plate may or may not have a black crescent; **head orange**. *Habitat* Gardens and woodland edges near palms. *Found* All year. *Host* Palms and saw palmetto. *Garden Tips* Coconut palm, Christmas palm. *Comments* Coats inside of last leaf nest with heavy wax covering.

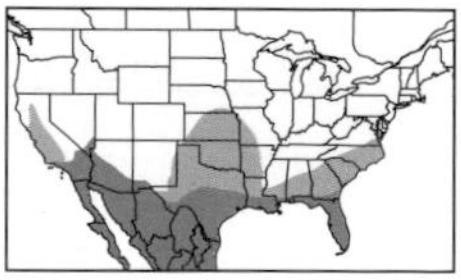

**Eufala Skipper** *Lerodea eufala*
Green; head white rimmed with brown, with brown vertical stripe in center, bordered by broad, brown inverted V-shaped marks (not well-defined stripes) that do not reach the top of the forehead. *Habitat* A wide variety of open situations. *Found* All year in southern regions, March–Nov, northward. *Host* Bermuda grass, Johnson grass, and others. *Comments* Does not overwinter, instead continually feeds at a slower rate.

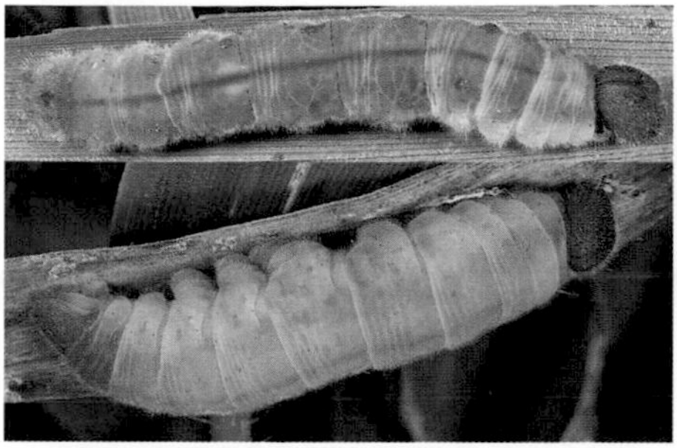

1. Twin-spot Skipper

2. Dusted Skipper adult

3. Cestus Skipper

4. Dusted Skipper

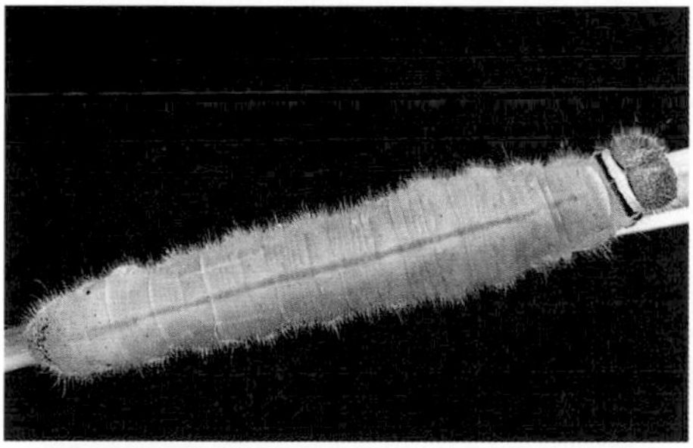

6. Sheep Skipper

9. White-barred Skipper

11. Monk Skipper

12. Eufala Skipper

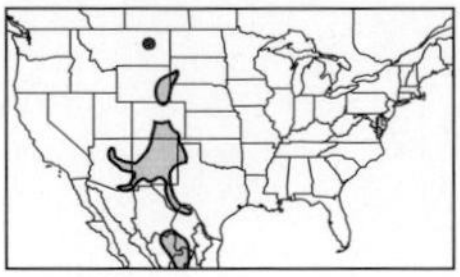

**Simius Roadside-Skipper** *"A." simius* Not shown.
Light bluish-green, head tan. *Habitat* High, short-grass prairie. *Found* April–June, prior to flight. *Host* Blue grama grass. *Comments* Not a true roadside-skipper. Overwinters as first instar and does not make aerial nest, unlike true roadside-skippers.

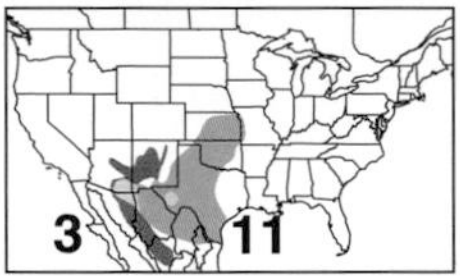


**Roadside-Skippers** (genus *Amblyscirtes*)
Green or whitish-green with cream, white, or tan head rimmed with brown or black and with a vertical stripe that splits into an inverted V above the mouth. Some species have a pair of vertical stripes arising on either side of the V mark but not touching the forehead (like stalagmites). The arrangement, intensity, and/or presence or absence of these vertical stripes is useful for identifying some species. All known feeding stages construct aerial nests. *Early Instars* Black head. Rolls a single grass blade into a tube or fold the tip over onto itself. Later instars develop their characteristic facial markings and may tie two or more leaf blades together. All overwinter as full-grown caterpillars in a nest, usually on the ground. These nests are usually constructed on the host by cutting and tying together grass blades, sealing the rear end with silk, then chewing through the leaf upon which it rests and dropping to the ground. The caterpillar then uses a pair of pointed tooth-like appendages near its mouth to crawl to a secure location before sealing itself in the nest, where it will remain until the following spring or summer. Roadside-skipper caterpillars are best found 2–8 weeks after the adult flights. See pg. 171 for more information.

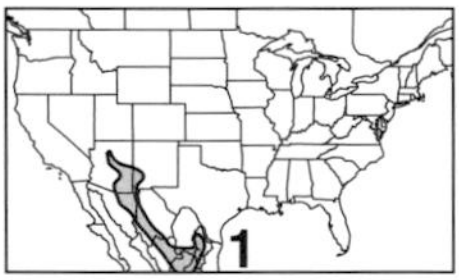


**1. Large Roadside-Skipper** *Amblyscirtes exoteria*
Dark brown stripes lighter at back of head. *Host* Bull grass.

**3. Cassus Roadside-Skipper** *Amblyscirtes cassus*
Yellow-green body; prominent side-stripes on head. *Host* Bulb panic grass and probably other grasses.

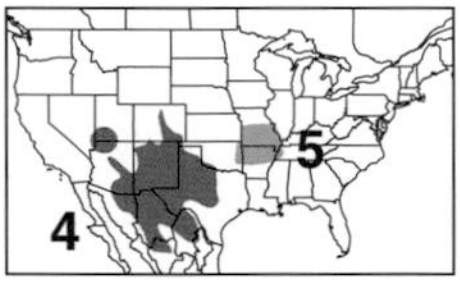


**4. Bronze Roadside-Skipper** *Amblyscirtes aenus*
Blue-green; short side-stripes on head. *Host* Many grasses.

**5. Linda's Roadside-Skipper** *Amblyscirtes linda*
Not shown.
Very similar to Bronze. *Host* Broad-leaved uniola.

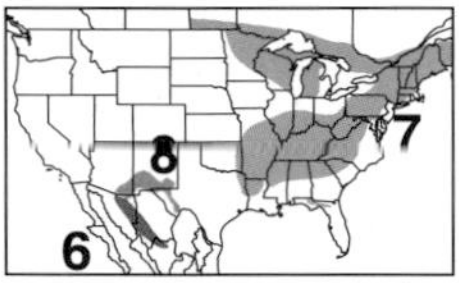


**6. Elissa Roadside-Skipper** *Amblyscirtes elissa*
Facial stripes narrow. *Host* Side-oats grama.

**7. Pepper and Salt Skipper** *Amblyscirtes hegon*
Triangular central facial stripe. *Host* Fowl manna grass, Indian grass, uniolas, and others.

**8. Texas Roadside-Skipper** *Amblyscirtes texanae*
Thick central facial stripe. *Host* Bulb panic grass in Arizona; unreported in Texas.

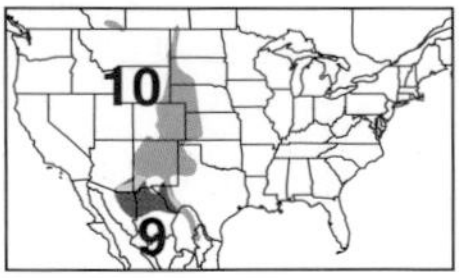


**9. Slaty Roadside-Skipper** *Amblyscirtes nereus*
Outer two facial stripes wide. *Host* Common beardgrass.

**10. Oslar's Roadside-Skipper** *Amblyscirtes oslari*
Not shown.
Similar to Slaty Roadside-Skipper. *Host* Side-oats grama.

**11. Nysa Roadside-Skipper** *Amblyscirtes nysa* Not shown.
Very similar to Bronze. *Host* Many grasses.

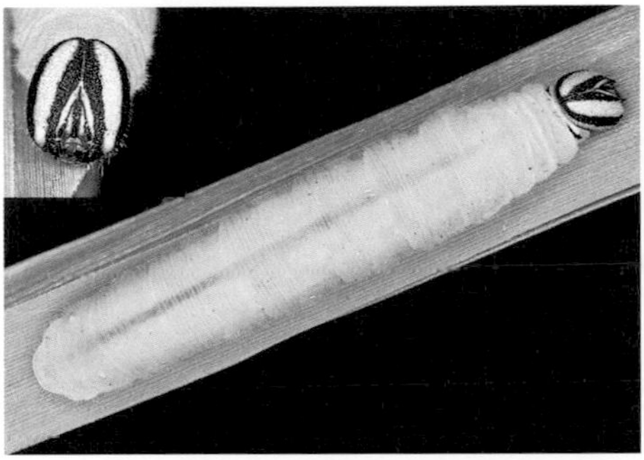

1. Large Roadside-Skipper

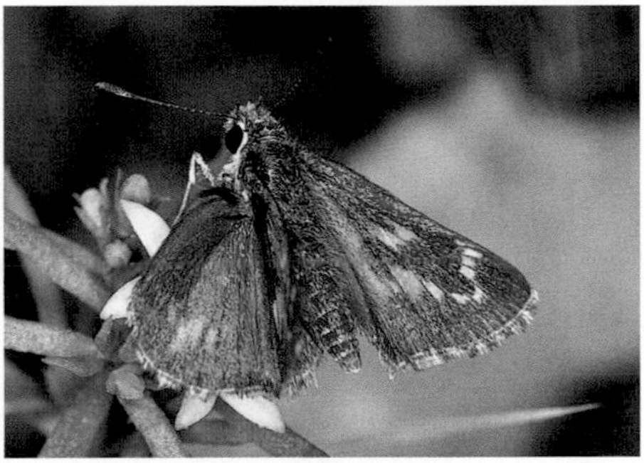

2. Cassus Roadside-Skipper adult

3. Cassus Roadside-Skipper

4. Bronze Roadside-Skipper

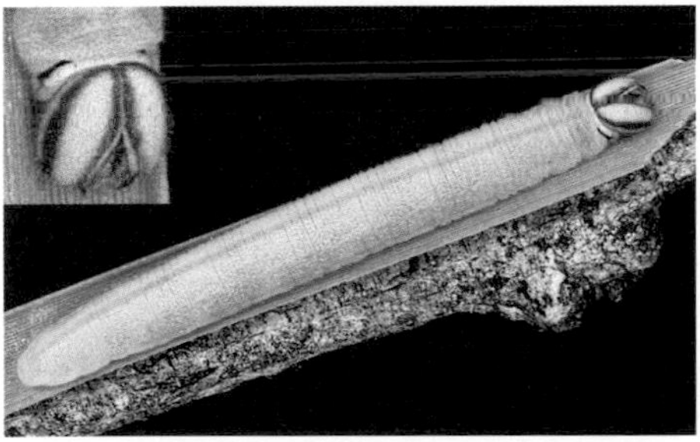

6. Elissa Roadside-Skipper

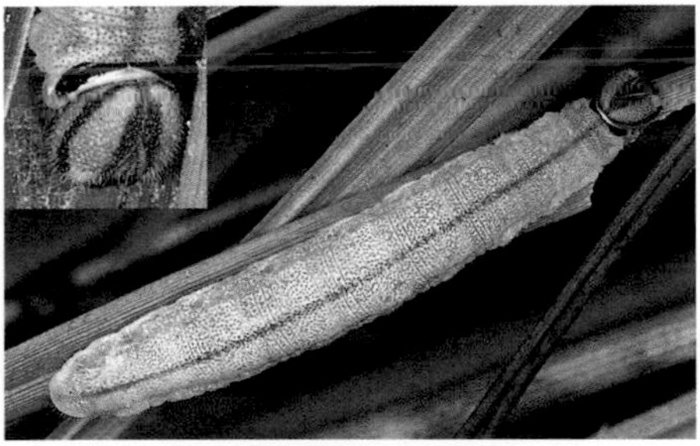

7. Pepper and Salt Skipper

8. Texas Roadside-Skipper

9. Slaty Roadside-Skipper

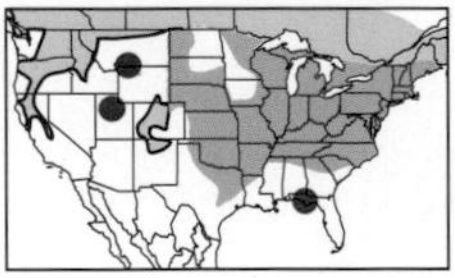

**1. Common Roadside-Skipper** *Amblyscirtes vialis*
Whitish powdery bloom; **head light gray** with gray facial stripe in center, bordered by two vertical dashes; **covered by flattened hairs**. No other roadside-skipper has as much facial or body hair. *Habitat* Roadsides and other edge areas where woodlands meet grasslands. Barrens. *Found* May–June, and Aug–Sept. *Host* Broad-leaved uniola, bromegrasses, little bluestem, and others.

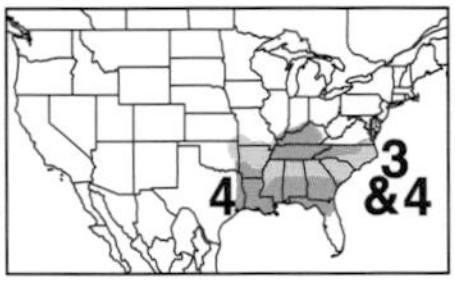


**3. Carolina Roadside-Skipper** *Amblyscirtes carolina*
**4. Lace-winged Roadside-Skipper** *A.aesculapius*
**5. Reversed Roadside-Skipper** *A. reversa* Not shown.
Dark gray-green with pale middorsal stripe; head tan or whitish with brown central vertical stripe. Brown vertical stripe broader and more solid on Reversed. *Habitat* Moist woodlands (mainly deciduous) with cane. *Found* March–Oct. *Host* Cane.

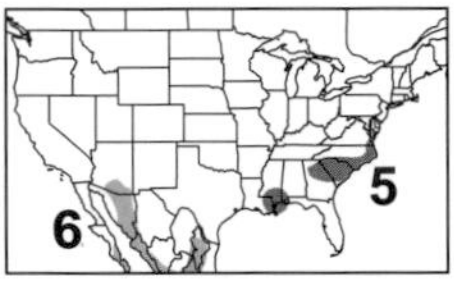


**6. Toltec Roadside-Skipper** *Amblyscirtes tolteca*
See Slaty Roadside-Skipper, which occurs in higher habitats. *Habitat* Canyons and washes at low to mid-elevations in desert ranges. *Found* June–early Oct. *Host* Many grasses. *Comments* A rarity in the U.S., most common roadside-skipper in NW Mexico.

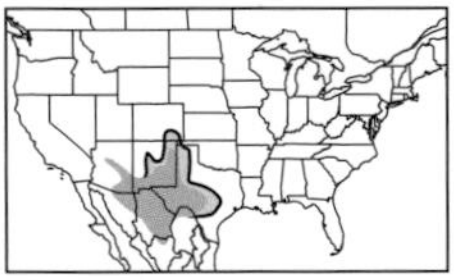

**7. Dotted Roadside-Skipper** *Amblyscirtes eos*
Bluish-green; head white with brown central vertical stripe, fairly narrow just above an inverted V mark on the face. *Habitat* Grassy areas, depressions, and wash edges in sparse oak woodlands with the host. *Found* April–Oct. *Host* Obtuse panic grass preferred, other grasses suspected *Comments* Possibly overwinters on the host and not on the ground.

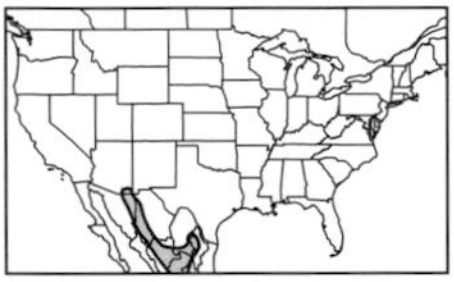

**8. Orange-edged Roadside-Skipper** *A. fimbriata*
Yellow-green; head white or cream with thin central brown vertical stripe. **Head with white covering much of the face**. *Habitat* Moist grassy openings in high-elevation pine forest. Found July–Sept. *Host* Brome, wild rye, and others. *Comments* Last instar nest often is a single grass blade rolled into a tube.

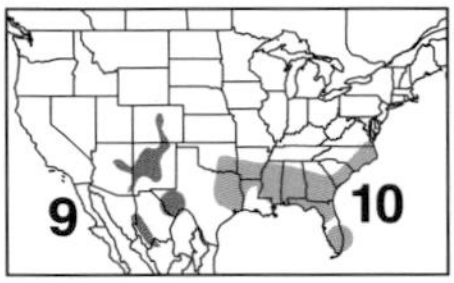


**9. Orange-headed Roadside-Skipper** *A. phylace*
Green; head white to slightly tan with central brown vertical stripe. *Habitat* High prairie gulches and canyons in sparsely wooded regions. *Found* July–Sept. *Host* Prefers big bluestem. *Comments* Although adult is similar to Orange-edged Roadside-Skipper, the caterpillars are quite distinctive.

See pg. 173 for more information about the following species.

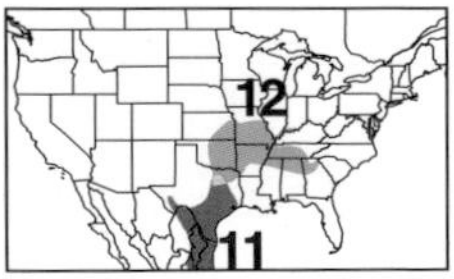


**10. Dusky Roadside-Skipper** *A. alternata* Not shown.
*Host* Bluestem grasses.

**11. Celia's Roadside-Skipper** *A. celia* Not shown.
*Host* Guinea grass and others.

**12. Bell's Roadside-Skipper** *A. belli* Not shown.
*Host* Broad-leaved uniola, Johnson grass, and others.

1. Common Roadside-Skipper

2. Lace-winged Roadside-Skipper adult

3. Carolina Roadside-Skipper

4. Lace-winged Roadside-Skipper

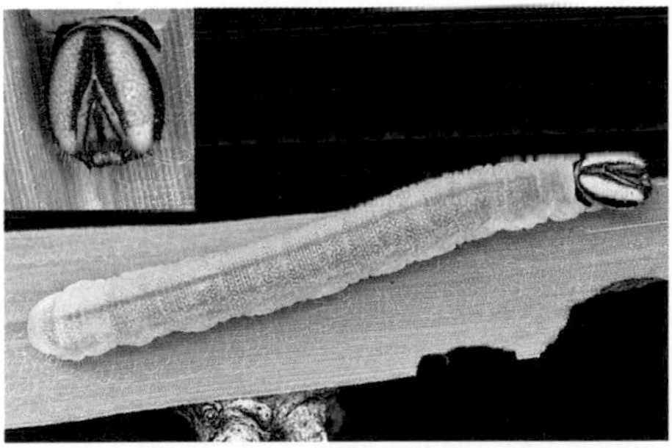

6. Toltec Roadside-Skipper

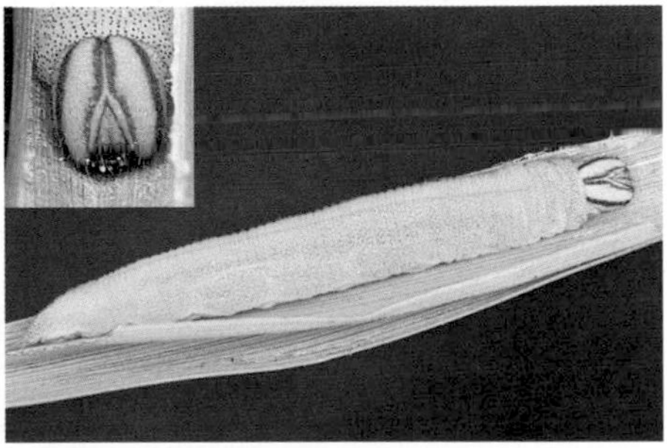

7. Dotted Roadside-Skipper

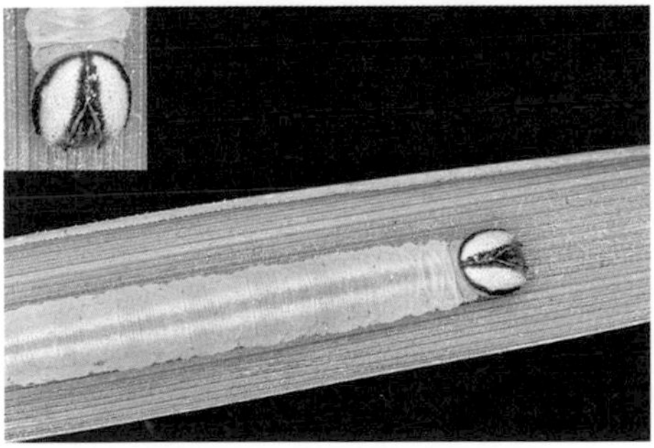

8. Orange-edged Roadside-Skipper

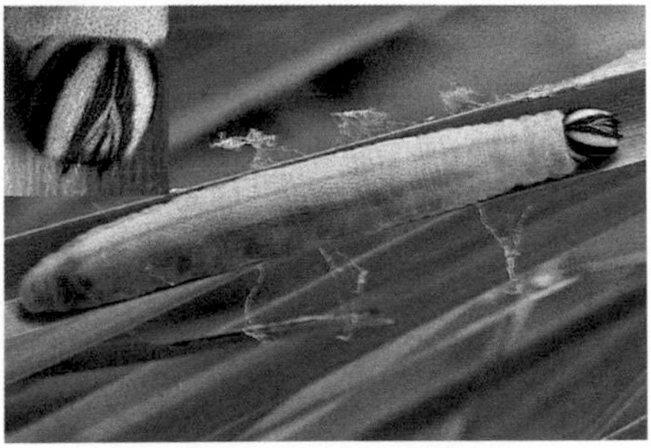

9. Oranged-headed Roadside-Skipper

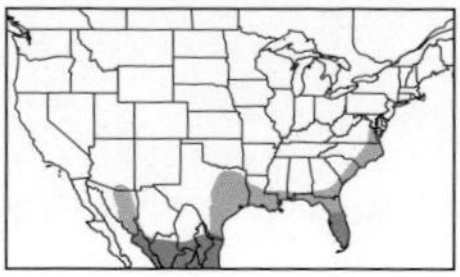

**1. Brazilian Skipper** *Calpodes ethlius*
**Translucent; white spots at rear**; head light brown or black. *Habitat* Gardens with cannas; wetlands. *Found* All year in the South; northward May–Oct. *Host* **Cannas** and relatives. *Garden Tip* Golden canna, ornamental cannas. *Comments* The folded nests of this beast are easily seen when plants are "infested."

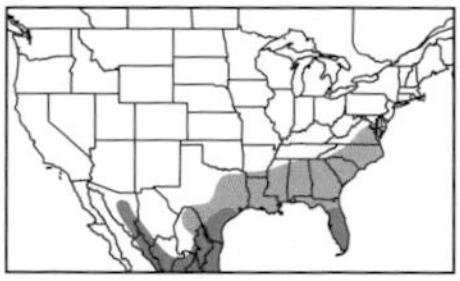

**3. Ocola Skipper** *Panoquina ocola*
Green with thin sublateral white stripe; body tapered toward green head. *Habitat* Open moist areas. *Found* All year in the South; northward late Aug–Sept. *Host* Aquatic and semiaquatic grasses.

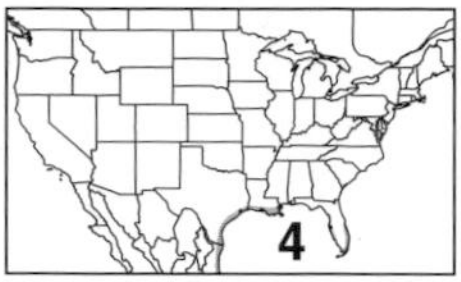

**4. Obscure Skipper** *Panoquina panoquinoides*
**5. Salt Marsh Skipper** *Panoquina panoquin* Not shown.
**6. Wandering Skipper** *Panoquina errans* Not shown.
Green with thin sublateral white stripe; body conspicuously tapered toward head; head green. *Habitat* Salt marshes and adjacent areas; Wandering Skipper mainly in brackish marshes in river mouths. *Found* All year, except Wandering Skipper, Aug–Nov. *Host* Salt grass.

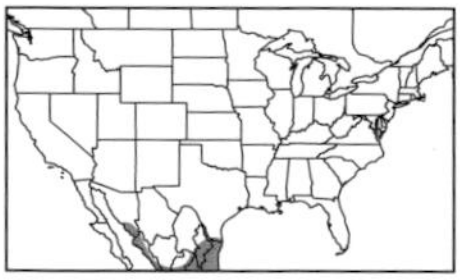

**7. Purple-washed Skipper** *Panoquina sylvicola*
Rare and essentially identical to Ocola Skipper, which is common and co-occurs. *Habitat* Subtropical open areas and open woodlands. *Found* All year. *Host* Grasses, often thin-leafed. *Comments* One of few skippers that does not construct a nest.

See pages 173–76 for more information about giant-skippers

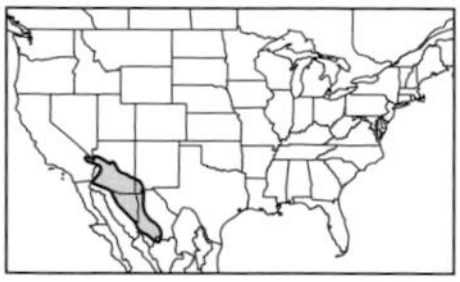

**8. Arizona Giant-Skipper** *Agathymus aryxna*
Greenish or bluish-white; head brown. Uses different agave than Orange Giant-Skipper. *Habitat* Open, arid grasslands and rocky slopes near the host. *Found* Most of year, especially May–Sept. *Host* Mainly Palmer's agave. *Comments* Trap door is smooth, brown, and on underside of leaf.

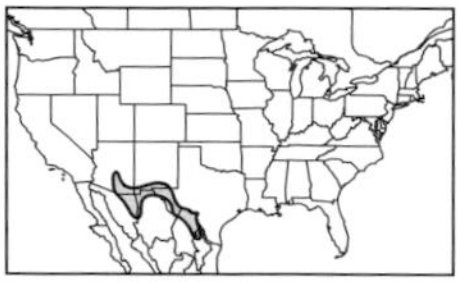

**9. Ursine Giant-Skipper** *Megathymus ursus*
Creamy-white; head black. Sometimes found in same habitats as Yucca Giant-Skipper. Usually lags behind that species in development and has smaller tents. *Habitat* Yucca grassland/oak woodland interface. *Found* Most of year, especially April–June. *Host* Schott's, Torrey's, banana yuccas, and others. *Comments* Much easier to locate early stages of this rarity than adults.

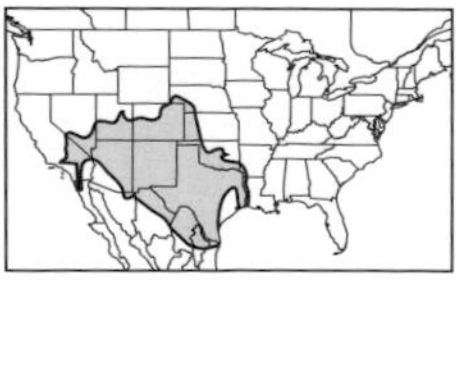

**10. Yucca Giant-Skipper** *Megathymus yuccae*
Whitish; head black. *First Instar* Red with black head. *Habitat* Yucca grassland, high desert, pine flats, and sand dunes with yuccas. *Found* Most of year, especially May–Nov. *Host* Many yuccas. *Comments* This is the giant skipper caterpillar that is most likely to be encountered.

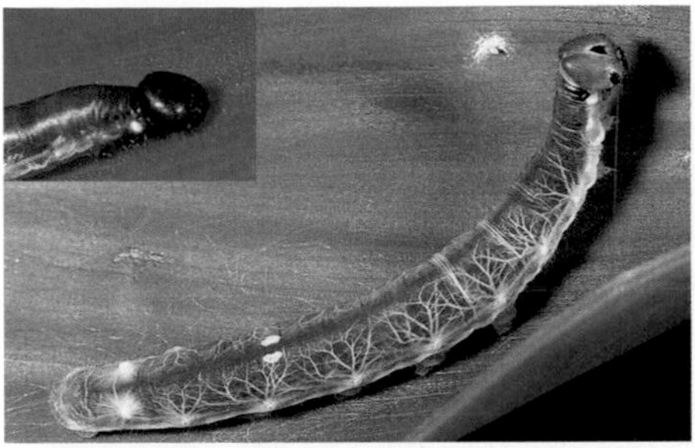

1. Brazilian Skipper

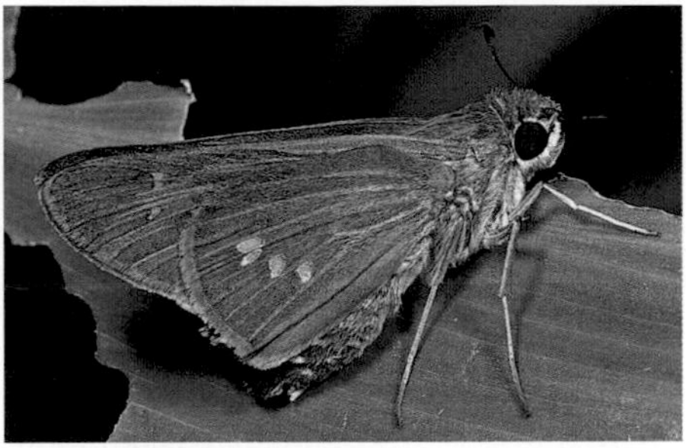

2. Brazilian Skipper adult

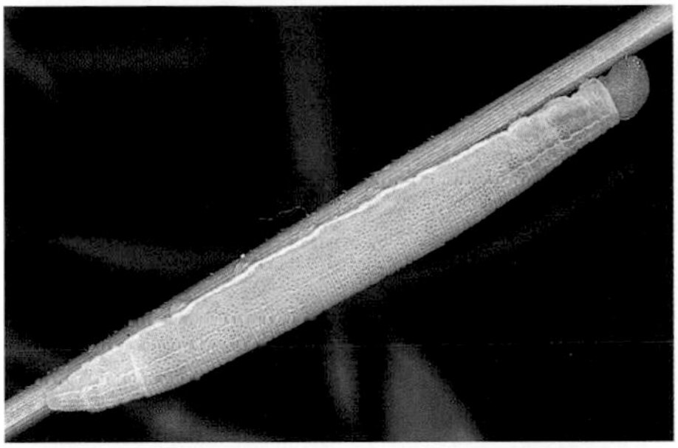

3. Ocola Skipper

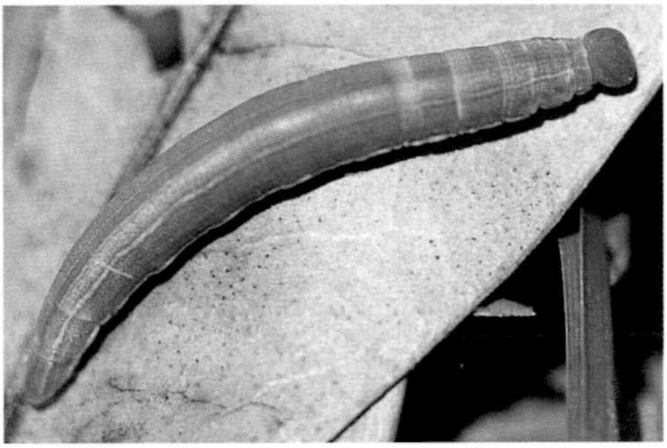

4. Obscure Skipper

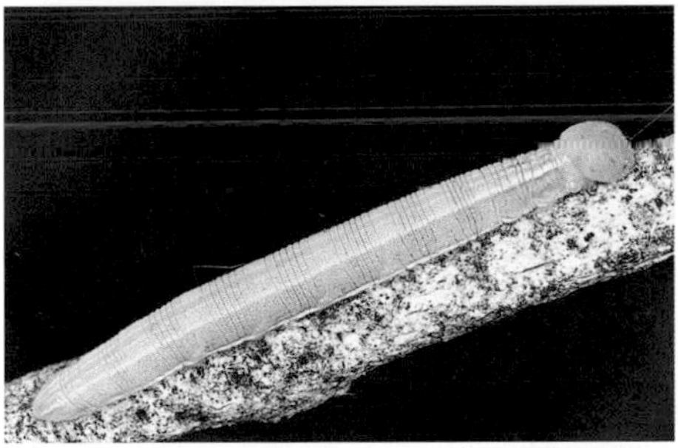

7. Purple-washed Skipper

8. Arizona Giant-Skipper

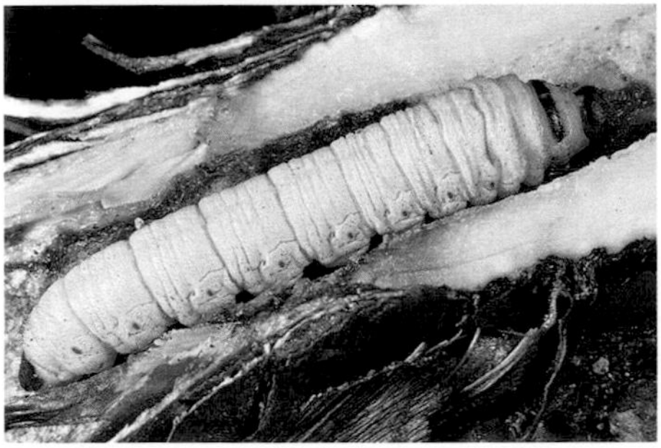

9. Ursine Giant-Skipper

10. Yucca Giant-Skipper (first instar)

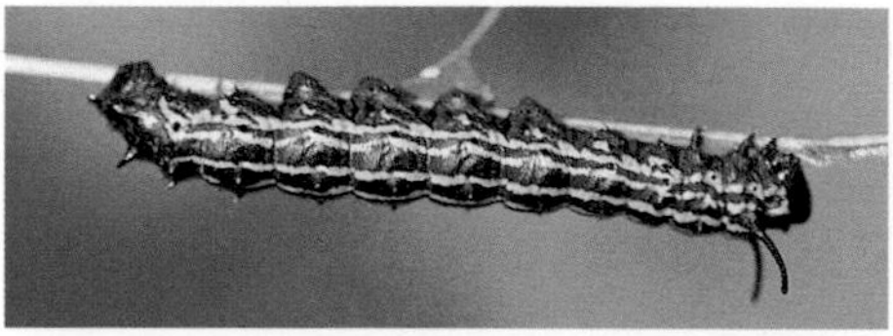

Orange-striped Oakworm *Anisota senatoria*

Eight-spotted Forester *Alypia octomaculata*

Saddle-back Caterpillar *Sibine stimulea*

Unicorn Caterpillar *Schizura unicornis*

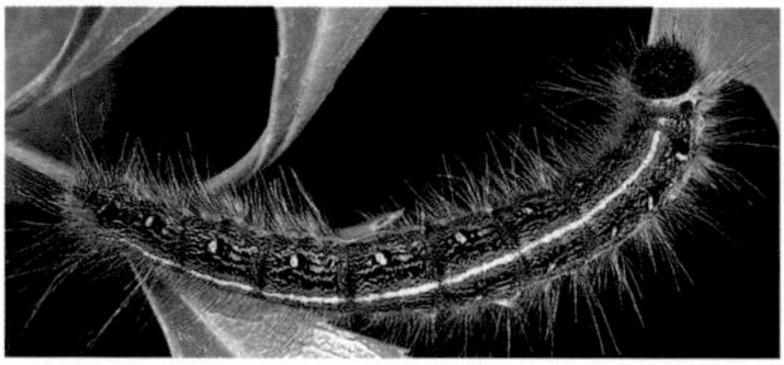

Eastern Tent Caterpillar *Malacosoma americanum*

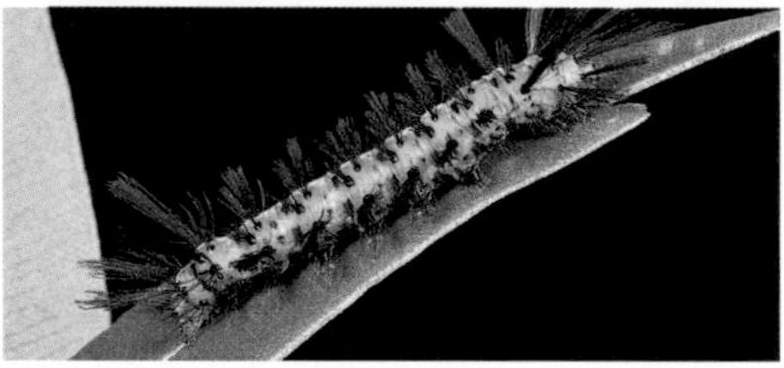

Polka-dot Wasp Moth *Syntomeida epilais*

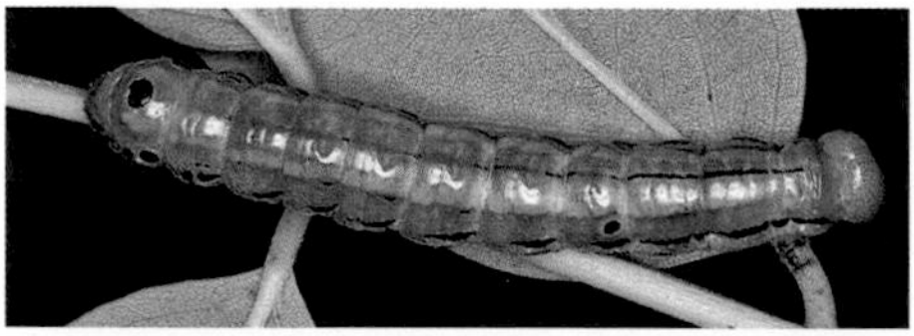

Black-spotted Prominent *Dasylophia anguina*

Double-toothed Prominent *Nerice bidentata*

Pepper and Salt Geometer *Biston betularia*

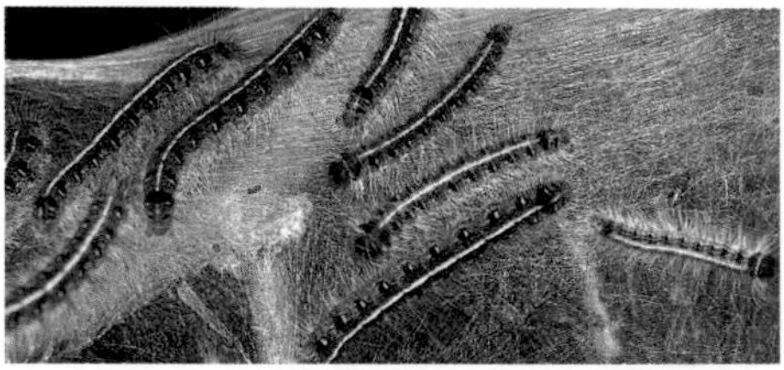

Eastern Tent Caterpillar *Malacosoma americanum*

Tulip-tree Silkmoth *Callosaimia angulifera*

Yellow-shoulder Slug *Lithacodes fasciola*

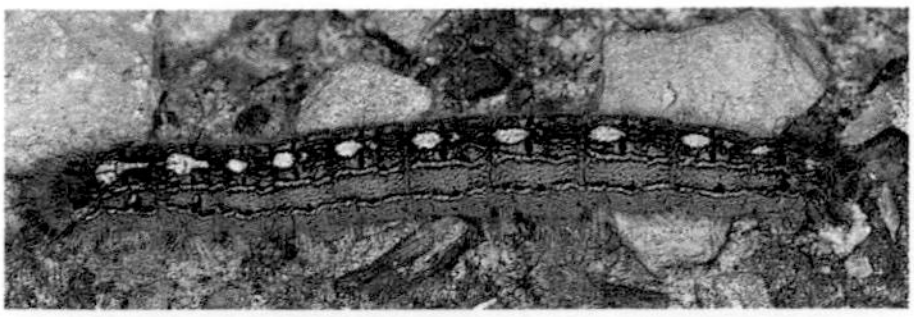

Forest Tent Caterpillar *Matacosoma disstria*

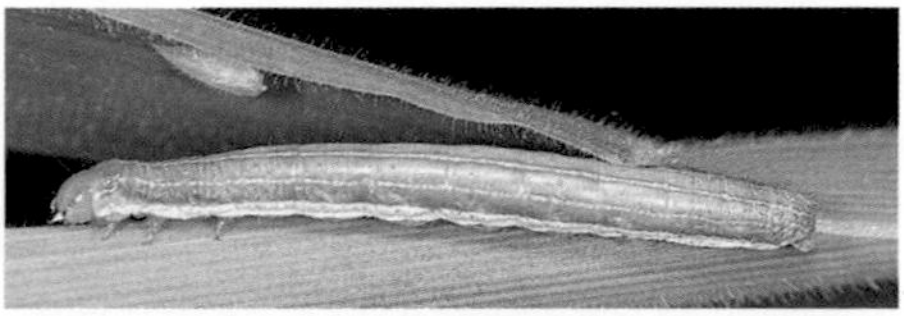

Veiled Ear Moth *Amphipoea velata*

Hitched Arches *Melanchra adjuncta*

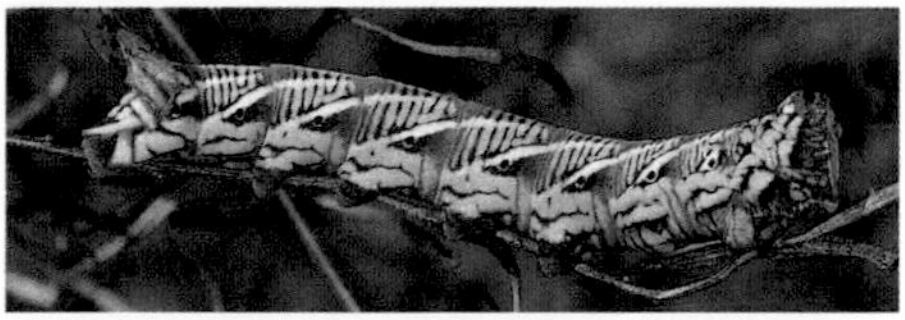

Banded Sphinx *Eumorpha fasciata*

Cecrops-eyed Silkmoth *Automeris cecrops*

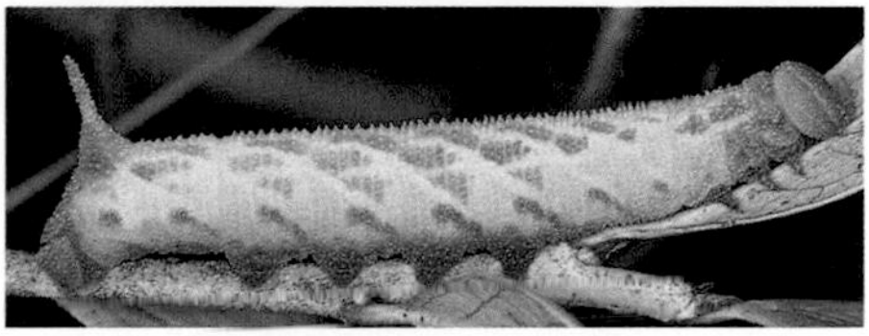

Huckleberry Sphinx *Paonias astylus*

Hickory Tussock Moth *Lophocampa caryae*

Milkweed Tussock Moth *Euchaetes egle*

Wooly Bear *Pyrrharctia isabella*

Giant Leopard Moth *Ecpantheria scribonia*

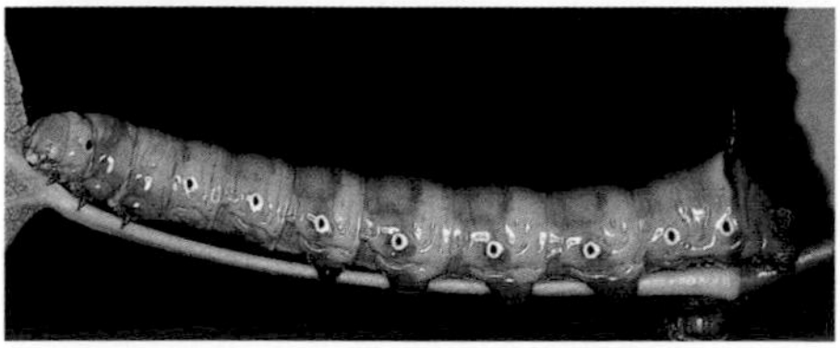

Black-rimmed Prominent *Pheosia rimosa*

Blinded Sphinx *Paoias exaecatus*

Hubbard's Small Silkmoth *Sphingicampa hubbardi*

Montana Small Silkmoth *Sphingicampa montana*

Imperial Moth *Eacles imperialis*

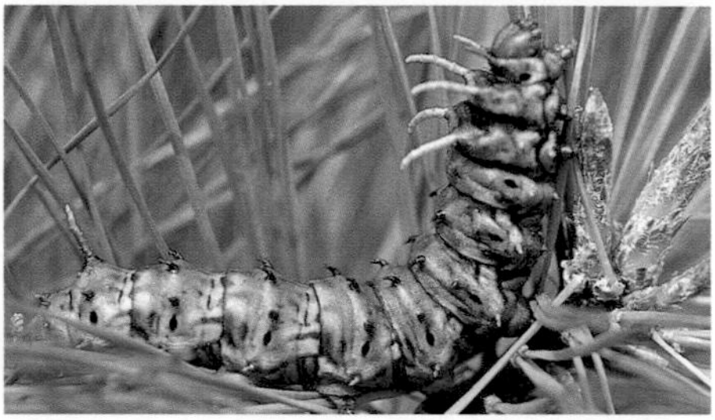

Pine-devil Moth *Citheronia sepulcralis*

Oslar's Oakworm *Anisota oslari*

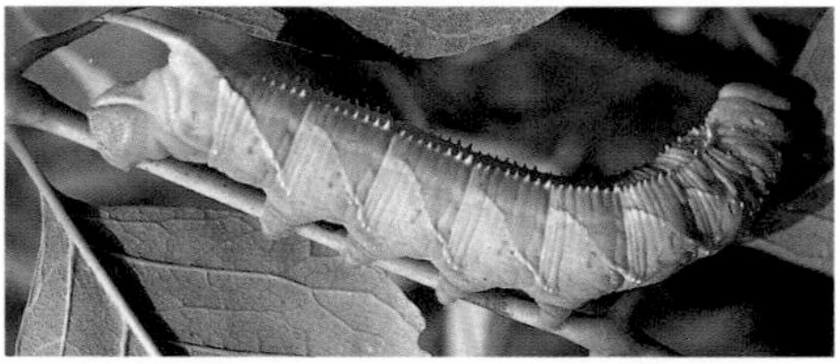

Franck's Sphinx *Sphinx franckii*

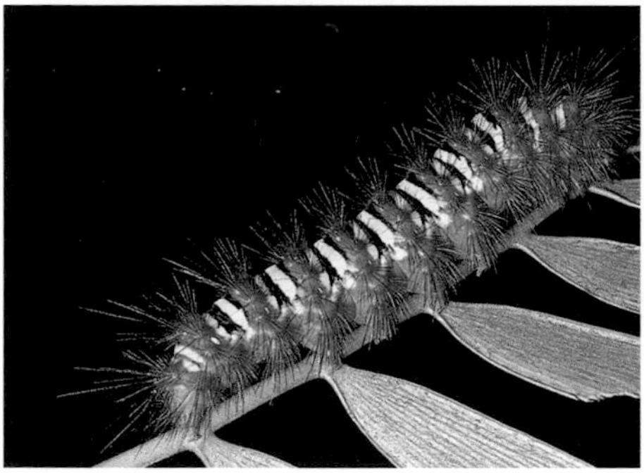

Echo Moth *Seirarctia echo*

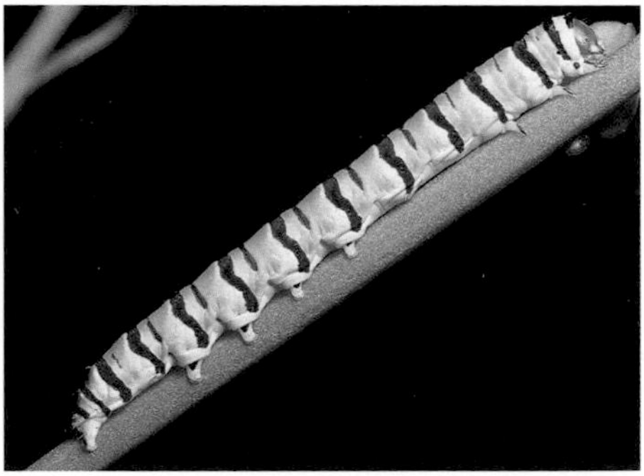

Clouded Crimson *Schinia gaurae*

Copper Underwing *Amphipyra pyramidoides*

The Brother *Raphia frater*

# Supplementary Text

### **Colias Sulphurs** (genus *Colias*) (Page 42)

The basic color and pattern for most caterpillars in the *Colias* group of sulphurs is green with a white or cream lateral stripe often dotted with yellow, orange, or red, rendering them difficult to distinguish from one another. In addition, we have not seen some of the species, and we have not seen enough examples of others to know the range of variation. Sometimes host plants will be of use for identification. For instance, Pink-edged Sulphur uses blueberries, and Orange and Clouded Sulphurs prefer alfalfa and clovers. Locality and habitat may also be useful for identification at times. High-altitude species are exceptionally difficult to locate in the field, often spending the winter under snow in difficult-to-reach habitats. This leaves a short window of time to find feeding caterpillars on hard-to-recognize hosts between snowmelt and their flights.

### **Dina Yellow** *Eurema dina* (Page 46)

Bright to dark green with a thin white lateral stripe. Similar to the Barred Yellow.
*Habitat* Tropical woodlands and woodland edges.
*Found* Most of the year following flights.
*Food* Mexican alveradoa tree and bitterbush.
*Comments* Eggs are laid on young leaves which the caterpillars prefer throughout growth.

### **Barred Yellow** *Eurema daira* (Page 46)

Green or blue-green with white, cream, or yellow lateral stripe. Probably not distinguishable from other yellows. It does, however, prefer different hosts.
*Habitat* Disturbed open situations; roadsides, vacant fields, gardens, etc.
*Found* All year in the South.
*Host* Pencil flowers and joint vetches.

### **Great Copper** *Lycaena xanthoides* (Page 48)

Green, or sometimes maroon, usually with a maroon dorsal stripe. Lilac-bordered and Edith's Coppers occur at higher altitudes.
*Habitat* Chaparral, dry grasslands, grassy hillsides, and open fields.
*Found* April–May in lowlands; May, June or July, at higher elevations.
*Host* Docks.
*Comments* Tended by ants.

### **Edith's Copper** *Lycaena editha* (Page 48)

Green or sometimes maroon, usually with a maroon dorsal stripe. Great Copper lives at lower altitudes. Lilac-bordered Copper (page 50) feeds on knotweeds.
*Habitat* Varied, including moist mountain meadows, openings in coniferous woodland, and moist areas in sagebrush steppes.
*Found* May–July.
*Host* Docks.
*Comments* Tended by ants.

### **Gray Copper** *Lycaena dione*

May not be separable from Bronze Copper (Page 48) where the two co-occur.
*Habitat* Areas with an abundance of dock, including moist meadows, fields and grasslands, roadside ditches, edges of ponds and streams.
*Found* May–July.
*Host* Broad dock and other docks.

### **Lustrous Copper** *Lycaena cupreus* (Page 50)

Green, sometimes with red lateral stripe. Probably not distinguishable from American Copper in high-altitude Western habitats.
*Habitat* Mainly moderate to high-elevation forest openings and meadows in California and western Oregon; mainly rockslides and high mountain meadows elsewhere.
*Found* May–June, early instars after single flight, August.
*Host* Docks and sorrels.
*Comments* Partially grown caterpillars probably over winter.

### **Mariposa Copper** *Lycaena mariposa* (Page 50)

Bluish-green, may have yellow lateral line. Hostplant is unique for coppers.
*Habitat* Moist openings in coniferous forests.
*Found* June–July.
*Host* Blueberries.

### **Colorado Hairstreak** *Hypaurotis crysalus* (Page 52)

Green with wavy diagonal dashes on sides and long white hairs below a white sublateral line. Overall somewhat flattened and elongated like a copper. Golden Hairstreak occupies a different range, except in Arizona where it is extremely rare.
*Habitat* Canyons and oak-covered hillsides near the host.
*Found* May–July, prior to flight in most of range; Aug–Sept, between flights in southern Arizona, on leaves.
*Host* Gambel oak and probably also silverleaf oak, in southern Arizona.
*Comments* This and the following species are related to Old World hairstreaks. They lack honey glands, and their adults rarely attend flowers.

### **Golden Hairstreak** *Habrodais grunus* (Page 52)

Light bluish-green with long white hairs below sublateral region. Overall flattened and elongated like a copper. Colorado Hairstreak occupies a different range. Goldhunter's Hairstreak more robust with white lateral stripe.

*Habitat* Dry slopes in chaparral; foothill and lower mountain canyons.
*Found* May–July, prior to single flight, on leaves.
*Host* Canyon oak, chinquapin (in Oregon), and others.
*Comments* Adults are often found swarming around the host, but the caterpillars are much more secretive and difficult to find.

**Sooty Hairstreak** *Satyrium fulginosum* (Page 58)
Although a hairstreak, both the adults and the caterpillars resemble blues! The caterpillars can most easily be confused with Boisduval's, Silvery and Arrowhead Blues. They are light green with whitish dorsal chevrons and are tended by ants. They are not reliably separable from blues that use the same hosts and that are also tended by ants.
*Habitat* High sagebrush-covered slopes and plateaus; high mountain meadows.
*Found* May–July, prior to single flight, on leaves.
*Host* Lupines.
*Comments* Feeds at night; found resting at the base of host during the day just at or slightly below soil surface with attendant ants.

**Lacey's Scrub-Hairstreak** *Strymon alea* (Page 64)
White with green circles dorsally and extra green marks laterally on thorax and toward the rear.
*Habitat* Thorn scrub and rocky canyons with host plant.
*Found* Probably most of year, whenever buds and flowers are available.
*Host* Southwestern bernardia.

**Bartram's Scrub-Hairstreak** *Strymon acis* (Page 64)
Reportedly whitish or yellowish-green to tan with pale dorsal and lateral bumps and a lateral line. Probably the only species on croton besides the ever-present Gray Hairstreak.
*Habitat* Pine rockland with the host.
*Found* All year.
*Host* Pineland croton.

**Fulvous Hairstreak** *Electrostrymon angelia* (Page 64)
Reddish-green with two dorsal rows of pale spots. Host is rather unique. The similar Red-banded Hairstreak usually prefers rotting leaves to living plant material.
*Habitat* Gardens, disturbed shrubby areas, and margins of tropical hammocks.
*Found* Most of year.
*Host* Brazilian pepper, Jamacian dogwood.

**Leda Ministreak** *Ministrymon leda* (Page 64)
Olive green, with white dorsal markings, the back deeply lobed or saw-toothed. Other species using the same host lack saw-toothed profile.
*Habitat* Arid region washes and canyons with mesquite.
*Found* April–Sept, prior to flights.
*Host* Mesquites, cat-claw acacia.

*Comments* Prefers flowers, thus flights are synchronized to blooming of the host, which can be sporadic.

### **Clytie Ministreak** *Ministrymon clytie* (Page 64)

Caterpillar Unreported.
*Habitat* Thorn scrub.
*Found* Unreported, but probably synchronized to blooming of the host.
*Host* Creeping mesquite.

### **Gray Ministreak** *Ministrymon azia* (Page 64)

Pale green with red and white markings, the back deeply lobed or sawtoothed.
*Habitat* Open scrub and disturbed areas.
*Found* Mainly May–Dec.
*Host* Lead tree.

### **Reakirt's Blue** *Hemiargus isola* (Page 66)

Variable. Yellow, green, or red with chevron markings dorsally. Range of variation is not well known. It may be that the small dorsal band on the thorax is distinctive. Ceraunus and Marine Blues are similar.
*Habitat* Many types of open situations, including prairies and weedy fields.
*Found* All year in southern regions.
*Host* A variety of legumes (Acacia, Dalea, Desmanthus, etc.), on buds and flowers.
*Garden Tips* Prairie-mimosa in Texas, indigo bushes (*Dalea frutescens, Dalea pulchra*) in the Southwest.
*Comments* Does not hibernate in warmer southern regions. One of us (J.B.) has caterpillars in the yard in December and January on a regular basis.

### **Ceraunus Blue** *Hemiargus ceraunus* (Page 66)

Variable. From green or yellow-green to red, usually with oblique dorsal markings (sometimes without) and a dorsal stripe. Marine Blue, Cassius Blue, and Gray Hairstreak are similar.
*Habitat* A variety of habitat at low to middle elevations.
*Found* All year in southern areas, March–Nov, elsewhere, on buds and flowers.
*Host* Buckwheats, mesquite, wild bean and other legumes.
*Comments* Not *as* variable as Marine Blue but well-marked forms are tough to identify as either one.

### **Miami Blue** *Hemiargus thomasi* (Page 66)

Similar Species: Silver-banded Hairstreak, Gray Hairstreak, Nickerbean Blue. Variable. Green to brown? with mid-dorsal stripe. Restricted in range. Silver-banded Hairstreak and Gray Hairstreak also burrow into balloons of the host and may be difficult to separate from each other. Caterpillars found on nickerbean may be either this species, Nickerbean Blue, Ceraunus Blue, or Gray Hairstreak.
*Habitat* Beach scrub and hammock edges.
*Found* All year. Inside balloon pods of the host if on balloon-vine, or on new leaf growth or flowers if on nickerbean.

*Host* Balloon-vines and nickerbean.
*Garden Tips* Balloon-vine or nickerbean.
*Comments* An endangered species that could use an assist from butterfly gardeners in south Florida.

### Nickerbean Blue *Hemiargus ammon* (Page 66)

Dark green or brown, with either white, green, or yellow lateral stripe, bordered with a narrower stripe that is dark green or pink.
*Habitat* Coastal hammocks and pine rockland.
*Found* Probably all year.
*Host* Nickerbean and pineland acacia.

### Buckwheat Blues (genus *Euphilotes*) (Page 70)

Challenging to identify as adults and no less so as caterpillars. The taxonomy for this western group is still in a state of flux. Patterns and colors vary. All feed on flowers, buds, or seeds, and overwinter as chrysalids. Several different buckwheats may be used by one species but usually only one at any given locality. Rarely do two species share a host at the same spot, much less at the same time. Thus, host, flight time, and locality will help in separating Euphilotes to species. However, other blues, some coppers, and some hairstreaks also use buckwheats.
*Host* Square-spotted Blue uses California, sulphur, and other buckwheats. Dotted Blue uses nude sulphur, reniform, and other buckwheats. Rita Blue uses loose, Wright's and other buckwheats. Spalding's Blue uses only red root buckwheat.

### Shasta Blue *Plebejus shasta* (Page 72)

Reported to be brown to maroon with yellow dorsal chevrons. Occurs at high altitudes.
*Habitat* Northward in high meadows and forest openings or sagebrush steppes. Southward more restricted to high-elevation rocky hilltops and ridges.
*Found* May–July, prior to single flight.
*Host* Vetches, clovers, and lupines.
*Comments* Probably biennial at highest elevations, overwintering as eggs first winter and as nearly full-grown caterpillars the next.

### Veined Blue *Plebejus neurona* (Page 72)

Gray-green to dull pinkish-gray, mostly unmarked. Lupine Blue uses different buckwheat. An unmarked form of Acmon Blue would be indistinguishable.
*Habitat* Openings in coniferous forest, usually over 5000′, rarely chaparral.
*Found* April–July, prior to flights; early instars follow flights then overwinter.
*Host* Kennedy's and Wright's buckwheats.

### Scintillant Metalmarks (genus *Calephelis*) (Page 74)

A group of tiny caterpillars with long, white or gray hair tufts, easily mistaken for some of the many moth species that also have long hair tufts. There are two rows of long dorsal hair tufts (on either side of the mid-dorsal line) and a lateral row of long hair tufts (below the spiracles). Most scintillant metalmarks have three small, dorsal, silver blotches on the first three segments behind the head. These caterpillars are all very similar to one another, and so they are best identified to species by range and host. Younger caterpillars also have long white hairs. Most feed on leaves or stems of their host. They overwinter partially grown.

### Edwards' Fritillary *Speyeria edwardsii*

Black with pale yellow dorsal and subdorsal bands (sometimes more yellow in appearance overall), the upper four rows of spines are gray at the base; the lower two rows of spines are orange at the base. Most similar to Great Basin Fritillary.
*Habitat* Prairies and foothills.
*Found* April–July.
*Host* Violets.

### Coronis Fritillary *Speyeria coronis*

Mottled brown or pale yellow and black; the upper four rows of spines black, lighter at the base; the lower two rows of spines yellow.
*Habitat* Moist openings in mountain forest, chaparral, and sagebrush and other brushy habitats.
*Found* May–July.
*Host* Violets.

### Great Basin Fritillary *Speyeria egleis*

Variable. Brown or black and yellow or orange with pale yellow or orange dorsal stripe; the top four rows of spines black or yellow; the lower two rows of spines yellow. Most similar to Edwards' Fritillary.
*Habitat* Varied, but usually at fairly high elevations in openings of mixed or coniferous forest.
*Found* May–June.
*Host* Violets.

### Unsilvered Fritillary *Speyeria adiaste*

Reported to be similar to Callippe Fritillary but with lighter gray sides.
*Habitat* Central coastal California only. Openings in redwood forest in San Mateo and Santa Cruz Counties. High mountain meadows in Monterey County. Chaparral/oak woodland in San Luis Obispo County.
*Found* May.
*Host* Violets.
*Comments* Losing ground from the south end of its range northward. In-depth life history field studies sorely needed.

### Hydaspe Fritillary *Speyeria hydaspe*

Black with pale yellow twinned mid-dorsal stripes; upper two rows of spines black; lower four rows of spines, orange-brown to yellow. The mid-dorsal yellow stripes are much lighter than on Zerene Fritillary.
*Habitat* Moist, dense woodlands.
*Found* April–June.
*Host* Violets.

### **Mormon Fritillary** *Speyeria mormonia*

Reported to be similar to other fritillaries. Brown to gray with pale mid-dorsal stripe; spines are short and have pale bases.
*Habitat* High-elevation mountain meadows.
*Found* May–June.
*Host* Violets.

### **Lesser Fritillaries** (genus *Boloria*) (Page 82)

Smaller than greater fritillaries with spines not quite half as long. Other than our more common eastern species, little is known about their life histories (especially of those in the high mountain of the West), so comparisons are impossible at this point. Many species are biennial, requiring 2 years for development, spending the first year as an early instar and the second year as a nearly full-grown caterpillar. Thus, the caterpillars of most species are probably present throughout the summer, but are extremely difficult to find.

### **Alberta Fritillary** *Boloria alberta* (Page 82)

Caterpillar Unreported.
*Habitat* Windswept scree slopes and ridges above treeline in the high Rocky Mountains.
*Found* May–June.
*Host* Mountain avens.

### **Astarte Fritillary** *Boloria astarte* (Page 82)

Caterpillar Unreported.
*Habitat* High, rocky windswept ridges above treeline.
*Found* Unreported.
*Host* Spotted saxifrage.

### **Checkerspots and relatives** (Page 84)

Mainly black caterpillars with varying amounts of orange or white, with black or orange branching spines. All overwinter or spend extended dry periods as partially grown caterpillars in a special "thicker" skin with a more dense covering of spines. Eggs are usually laid in clusters (sometimes of more than 100) on the host. Young caterpillars feed together (sometimes in a silken web), becoming solitary in later instars. Leaves are the primary diet for most, although some paintbrush feeders feed on the flowers in later instars. The best time to look for checkerspot caterpillars is 2–3 weeks prior to flights. Young caterpillars should be sought within 2–3 weeks after flights. Chrysalids hang suspended.

### **Crescents** (genus *Phyciodes*) (Page 92)

Mostly black to brown, with stripes of various widths; the branching spines are short, similar to lesser fritillary caterpillars. Eggs are laid in clutches. Caterpillars eat leaves. Partially grown caterpillars overwinter, so it is best to search prior to flights (especially for those with 1 brood), although early stages are present for about a month or so after flights. Most are difficult to identify because we have not seen enough samples to determine the extent of variation among individual species. In some instances range, habitat, and host preference will be aid in identification.

**Vesta Crescent** *Phyciodes vesta* (Page 92)
Similar to other crescents; dark gray with cream dorso-lateral stripe; spines tan.
*Habitat* Thorn scrub, roadsides, grasslands, and other dry open situations.
*Found* March–Oct, before and after flights.
*Host* Hairy tubetongue and *Dischoriste decumbens*.

**Painted Crescent** *Phyciodes picta* (Page 92)
Similar to Field Crescent but lighter and prefers different host in the field.
*Habitat* Moist areas in arid grasslands, prairies, and deserts.
*Found* Feb–Oct in southern regions; April–Aug northward.
*Host* Bindweed and slimleaf bursage.

**California Crescent** *Phyciodes orseis* (Page 92)
Reported to be purple-black with a paler, wide lateral stripe.
*Habitat* Mountain creeksides.
*Found* April–July, June–Aug.
*Host* Thistles.

**Anglewings** (genus *Polygonia*) (Page 94)
In addition to branching spines on the body, Commas have a pair of short branching spines on the head. The Eastern and Satyr Commas conceal themselves in tied leaves of their host. Overwintering is as an adult, thus caterpillars should be sought after the appearance of these individuals in the spring, prior to the flight of their offspring.

**Cloudywings** (genus *Thorybes*)(in part) (Page 118)
All are light green turning pinkish with dark green speckling and a thin, pale orange lateral stripe; head black, without facial spots.

**Southern Cloudywing** *Thorybes bathyllus* (Page 118)
*Habitat* A wide variety of open situations, especially dry fields with low brushy areas.
*Found* May–Oct.
*Host* Beggar weeds.

**Northern Cloudywing** *Thorybes pylades* (Page 118)
*Habitat* Widespread in open situations, such as power-line cuts, moist meadows, and pine-oak woodland.
*Found* June–Sept.
*Host* Various legumes, beggar weeds, bush clovers, alfalfa, vetch, clovers, narrow-leafed cologania, and others.

**Western Cloudywing** *Thorybes diversus* (Page 118)
*Habitat* Openings in moist coniferous forests.
*Found* June–July, perhaps into August.
*Host* Clovers and probably others.

### **Mexican Cloudywing** *Thorybes mexicanus* (Page 118)

*Habitat* Mountain openings and clearings.
*Found* June–Aug.
*Host* Clovers and other legumes.

### **Confused Cloudywing** *Thorybes confusis* (Page 118)

*Habitat* Dry open situations, such as dry prairies, hillside fields, and sand barrens.
*Found* Late May–July, Aug–Sept in North; March–Dec in South.
*Host* Beggar weeds and bush clovers.

### **Drusius Cloudywing** *Thorybes drusius* (Page 118)

*Habitat* Oak woodlands.
*Found* Aug–Sept.
*Host* So far only known to use narrow-leafed cologania.
*Comments* Half-grown caterpillars are unable to conceal themselves in the small leaves of the host, so a nest is created by silking together debris at the base of the plant. Older caterpillars begin feeding just after dark.

### **Duskywings** (genus *Erynnis*) (Page 122)

Duskywing caterpillars, like their adults, present a challenge in terms of identification. Head patterns and body colors are similar among the many species that occur together. There may be considerable variation among individual species, further complicating identification. Most are green or bluish-green (sometimes yellow-green) and up to an inch long when fully grown for larger species. The top of the head forms a small point on either side, and there are various markings on the faces. It remains to be seen whether these markings will aid in identification or not. More work is needed. All duskywing caterpillars construct nests on the host. They cut two slots in the edge of a young host leaf and fold the center portion onto the top of the leaf in a flap, which is tied down with silk. Duskywings can often be found quickly with a search for these small flaps cut in the host leaf. Most duskywings hibernate as full-grown caterpillars and should be searched for after flights.

### **Zarucco Duskywing** *Erynnis zarucco* (Page 122)

Green or bluish-green, covered with tiny white dots, with pale yellow lateral stripe; head brown, rimmed on either side with three yellow-orange patches. Funereal Duskywing is identical and uses same hosts.
*Habitat* Hot, sandy situations: sandy pineland, power-line cuts, etc.
*Found* All year in south Florida, May–Sept northward.
*Host* Black locust, milk-peas, indigos, orange sebania, and many other legumes.
*Comments* The facial markings of this species and Funereal Duskywing resemble a number of oak-feeding duskywing species.

### **Funereal Duskywing** *Erynnis funeralis* (Page 122)

See Zarucco Duskywing
*Habitat* A wide variety, including desert, woodland edges, and spruce forest, but preference is for hot, dry situations.

*Found* Most of year in southern regions.
*Host* A wide variety of legumes including alfalfa, indigos, New Mexico locust, ironwood, and deer vetches; recently found on *Janusia* (in the Malpighia family) in southern Arizona.
*Garden Tips* Babybonnets, orange sesbania.

### **Wild Indigo Duskywing** *Erynnis baptisiae* (Page 122)

Green or bluish-green, covered with tiny white dots, with pale yellow lateral stripe; head brown, mottled with lighter markings.
*Habitat* Widespread in open areas, especially along roadsides and railroad embankments with plantings of crown vetch.
*Found* May–June, July–Aug, Sept–Oct.
*Host* Legumes: crown vetch, wild indigo, wild blue indigo and lupines.

### **Afranius Duskywing** *Erynnis afranius* (Page 122)

Green or bluish-green, covered with tiny white dots, with pale yellow lateral stripe; head brown, mottled with orange. Persius Duskywing uses similar hosts but has tan facial mottling, not orange.
*Habitat* High prairie, badlands, canyons and chaparral.
*Found* May–Sept.
*Host* Locoweed, deer vetches, lupines and other legumes.

### **Persius Duskywing** *Erynnis persius* (Page 122)

Green or bluish-green, covered with tiny white dots, with pale yellow lateral stripe; head brown, mottled tan and dark brown. Afranius Duskywing similar but has orange mottling on face.
*Habitat* Mainly mountain meadows and forest openings, but also in prairie.
*Found* June–Sept.
*Host* Golden banners, locoweeds, lupines, and other legumes.

### **Sleepy Duskywing** *Erynnis brizo* (Page 122)

Green or bluish-green, covered with tiny white dots, with pale yellow lateral stripe; head brown, rimmed on either side with three orange patches, the two on either side of the mouth boldest. Other oak-feeding Duskywings are similar enough to be indistinguishable.
*Habitat* Dry habitat with oaks, especially pine-oak woodlands and barrens.
*Found* May–Oct.
*Host* Oaks.
*Comments* Development is slow, hence there is a long search period. Propensity of this species to use 2–3 foot seedlings instead of mature trees aids in identification.

### **Juvenal's Duskywing** *Erynnis juvenalis* (Page 122)

Green or bluish-green, covered with tiny white dots, with pale yellow lateral stripe; head brown, rimmed on either side with three yellow-orange patches. Other oak-feeding duskywings are similar enough to be indistinguishable.
*Habitat* Open oak woodlands and oak barrens.

*Found* May–Sept.
*Host* Oaks.

### **Rocky Mountain Duskywing** *Erynnis telemachus* (Page 122)
Green or bluish-green, covered with tiny white dots, with pale yellow lateral stripe; head brown, rimmed on either side with three yellow-orange patches. Other oak-feeding duskywings are similar enough to be indistinguishable.
*Habitat* Oak and pine-oak woodlands.
*Found* May–Aug.
*Host* prefers Gambel's oak.
*Comments* Usually found higher (8000´ and above) than other oak-feeding duskywings, but there is some overlap in some regions.

### **Propertius Duskywing** *Erynnis propertius* (Page 122)
Caterpillar is unreported but is probably similar to other oak-feeding duskywings.
*Habitat* Oak, and mixed oak, woodlands.
*Found* May–Aug.
*Host* Oaks, Garry oak in the Northwest, coast live oak farther south.

### **Meridian Duskywing** *Erynnis meridianus* (Page 122)
Green or bluish-green, covered with tiny white dots, with pale yellow lateral stripe; head brown, rimmed on either side with three yellow-orange patches. Other oak-feeding duskywings in its range are similar enough to be indistinguishable.
*Habitat* Oak woodlands.
*Found* April–Sept.
*Host* Emory oak, Arizona white oak, and perhaps others.

### **Scudder's Duskywing** *Erynnis scudderi* (Page 122)
Unreported but probably similar to other oak-feeding species.
*Habitat* Oak and pine-oak woodlands above 6000´
*Found* April–Sept.
*Host* Unreported, probably oaks.
*Comments* A rarity and one of the few spread-wing skippers for which we have no life history information.

### **Horace's Duskywing** *Erynnis horatius* (Page 122)
Green or bluish-green, covered with tiny white dots, with pale yellow lateral stripe; head brown, rimmed on either side with three yellow-orange patches. Other oak-feeding duskywings in its range are similar enough to be indistinguishable.
*Habitat* Open woodlands, especially those on poor soils.
*Found* May–June and Aug–Sept.
*Host* Red oaks favored, also scrub oak, post oak, laurel oak and others.

### **Mournful Duskywing** *Erynnis tristis* (Page 122)
Green or bluish-green, covered with tiny white dots, with pale yellow lateral stripe; head brown, rimmed on either side with three yellow-orange patches. Other oak-feeding duskywings are so similar as to be indistinguishable.

*Habitat* Oak woodlands.
*Found* Most of year.
*Host* Oaks.

### **Poweshiek Skipperling** *Oarisma poweshiek* (Page 130)

Green with numerous white longitudinal stripes; head pale green.
*Habitat* High-quality, tall-grass prairie.
*Found* Early instars August; late instars May–June, prior to flight.
*Host* Spikerush.
*Comments* Reported to make a nest only as a first instar; otherwise, no nests.

### **Garita Skipperling** *Oarisma garita* (Page 130)

Green form: green with darker mid-dorsal band divided by white and numerous green and white logitudinal stripes; head green, rounded, with two paler streaks on forehead. Tan form: tan to pinkish-tan with purple mid-dorsal stripe divided by a white line and numerous red and tan longitudinal stripes; head tan with two paler streaks on forehead. Both forms have a pointed tail with a small notch.
*Habitat* Mountain meadows, and short-grass or mixed-grass prairies.
*Found* Early instars July–Sept; late instars April–June, prior to flight.
*Host* Deer grass, squirrel-tail, blue grasses, and many others. Individual caterpillars often use fall grasses before overwintering then a different grass in the spring prior to the flight.
*Comments* Caterpillars shake (perhaps resembling a piece of grass blowing in the wind) while they walk. Overall behavior and appearance very satyr-like. No nests.

### **Edwards' Skipperling** *Oarisma edwardsii* (Page 130)

Caterpillar is unreported.
*Habitat* Open mountain woodland, usually pine-oak between 5000´ and 8000´.
*Found* Probably like Garita Skipperling, both before and after single flight.
*Host* Unreported, probably grasses.

### **Clouded Skipper** *Lerema accius* (Page 130)

Green with whitish bloom dorsally; head white, rimmed with blackish-brown stripe and bisected by a blackish-brown central stripe with a straight dark brown stripe on either side, not connected to the forehead and arising from the lower face. Delaware Skipper has the two stripes on either side slightly arched, not straight. Some roadside-skippers have two stripes on either side of the central stripe, but they arise from higher up on the face.
*Habitat* Open woodlands, preferring moist grassy areas in or near woods.
*Found* All year in South; Aug–Oct northward.
*Host* Large-leafed grasses, eastern gamma grass, Guinea grass, Johnson grass, and others.

### **Three-spotted Skipper** *Cymaenes tripunctus* (Page 130)

Blue-green with darker mid-dorsal stripe and gray-green lateral stripes; head white with brown vertical stripes.
*Habitat* Grassy disturbed woodlands and adjacent areas.

*Found* All year in southern Florida.
*Host* Grasses.

**Fawn-spotted Skipper** *Cymaenes odilia* (Page 130)
Unreported.
*Habitat* Shaded edges of tropical woodlands.
*Found* July–Dec in south Texas.
*Host* Knotgrass.

**Hesperia skippers** (genus *Hesperia*) (Page 132)
All Hesperia that we have seen arc similar enough and show enough specific variation to preclude reliable identification based on color and pattern.

**Juba Skipper** *Hesperia juba* (Page 132)
Tan; head dark brown with whitish stripes on forehead.
*Habitat* Mountain meadows, chaparral, and sagebrush grassland.
*Found* March–Oct, before, between, and after flights.
*Host* Blue grass, blue grama, bromegrass, and other grasses.

**Common Branded Skipper** *Hesperia comma* (Page 132)
Brown or dark brownish-black with black mid-dorsal stripe; head brown or black and may have two whitish vertical stripes on forehead.
*Habitat* Meadows, roadsides, open fields (East), rocky outcrops above treeline, open coniferous forest, sagebrush steppes, and prairies.
*Found* April–June; Aug–Sept, in double-brood areas.
*Host* Grasses.
*Comments* OverWinter as eggs, especially in U.S.

**Apache Skipper** *Hesperia woodgatei* (Page 132)
Unreported.
*Habitat* Openings in high-elevation pine and pine-oak forest.
*Found* Probably April–July or August, prior to single brood.
*Host* Probably grasses.

**Ottoe Skipper** *Hesperia ottoe* (Page 132)
Light greenish-brown; head dark brown.
*Habitat* Tall grass and short grass prairies, especially along ridges.
*Found* April–May, Sept.
*Host* Bluestems, side-oats grama, and other grasses.
*Comments* Females may lay eggs in flowers of coneflowers or on or near host grasses. The only Hesperia known to make aerial nests.

**Leonard's Skipper** *Hesperia leonardus* (Page 132)
Tan; head dark brown with whitish stripes on forehead.
*Habitat* Prairie (West), open fields with thick, low vegetation and nectar sources, often in moist open meadows (East).

*Found* May–July, prior to flight probably best.
*Host* Grama grasses, orchard grass, and others.

**Pahaska Skipper** *Hesperia pahaska* (Page 132)
Tan; head dark brown with whitish stripes on forehead.
*Habitat* A wide variety, from sparsely wooded grasslands in desert ranges to open pine forest.
*Found* Probably best between flights in areas with more than one brood May–July; after flight in single-brood areas June–Oct.
*Host* Blue grama grass and others.

**Columbian Skipper** *Hesperia columbia* (Page 132)
Tan; head dark brown with whitish stripes on forehead.
*Habitat* Chaparral and oak woodlands.
*Found* Not reported but probably between flights in double-brood areas.
*Host* Grasses.

**Cobweb Skipper** *Hesperia metea* (Page 132)
Brown or dark brownish-black; head brown or black and may or may not have two whitish vertical stripes on forehead.
*Habitat* Dry, open fields with bluestem grasses, often on hillsides, but also power-line cuts, etc.
*Found* July–Sept, into Oct.
*Host* Bluestem grasses, especially little bluestem.

**Green Skipper** *Hesperia viridis* (Page 132)
Brown or dark brownish-black; head brown or black and may have two whitish vertical stripes on forehead.
*Habitat* Low to mid-elevation canyons and gulches.
*Found* Both before and after flights, April–Oct.
*Host* Grasses.

**Dotted Skipper** *Hesperia attalus* (Page 132)
Brown or dark brownish-black with black mid-dorsal stripe; head brown or black and may have two whitish vertical stripes on forehead.
*Habitat* Atlantic Coast: sandy barrens, including pine barrens and associated railroad tracks, airport runways, and power-line cuts. Kansas–Texas; short grass prairies.
*Found* April–June, July–Aug.
*Host* Grasses.

**Meske's Skipper** *Hesperia meskei* (Page 132)
Grayish-brown; head brown or black.
*Habitat* Open dry or sandy pinelands and adjacent areas.
*Found* April–May, July–Aug; May–Oct in Florida.
*Host* Grasses.

### **Dakota Skipper** *Hesperia dacotae* (Page 132)

Light brown; head dark brown and pitted (unlike most other Hesperia).
*Habitat* Moist or dry, ungrazed, calcareous (alkaline) prairies, often with purple coneflower.
*Found* April–May, Aug–Sept.
*Host* Grasses.
*Comments* Females lay eggs haphazardly on or near grasses or on leaves of flowering plants.

### **Lindsey's Skipper** *Hesperia lindseyi* (Page 132)

Tan; head dark brown with whitish stripes on forehead.
*Habitat* Open grassy areas within chaparral or open oak woodland in foothills.
*Found* Feb–May, prior to flight. Eggs overwinter.
*Host* Grasses.
*Comments* Females lay their eggs on bizarre substrates (tree trunks, lichens, etc.) leaving the first instar caterpillars to fend for themselves.

### **Indian Skipper** *Hesperia sassacus* (Page 132)

Brown or dark brownish-black with black mid-dorsal stripe; head brown or black and may have two whitish vertical stripes on forehead.
*Habitat* Brushy fields and open meadows near woodlands, both dry and moist.
*Found* June–Oct.
*Host* Grasses.

### **Polites Skippers** (genus *Polites*) (Page 136)

Like Hesperia skippers, these species are all very similar and difficult to distinguish from one another in most cases. They differ only slightly from Hesperia in that they have a more speckled overall appearance because they have numerous dots dorsally and laterally. Nests are constructed in or near the base of grass clumps (sometimes partially subterranean), making them nearly impossible to locate in the field. Some species must make horizontal nests to survive on mowed lawns. Partially grown caterpillars overwinter.

### **Dusted Skipper** *Atrytonopsis hianna* (Page 144)

Dull green to tan, sometimes with a pinkish hue, covered with fine short hairs creating a velvety appearance; the rear segments more tan than green; head dark brown, pubescent.
*Habitat* Dry fields, prairies, barrens, and power-line cuts where bluestem grasses are found.
*Found* March–April, June–Aug where double-brooded; July–Oct, where single-brooded.
*Host* Bluestem grasses.
*Comments* Constructs nests and overwinters at base of host.

### **Deva Skipper** *Atrytonopsis deva* (Page 144)

Unreported, but undoubtedly similar to other dusted-skippers.
*Habitat* Openings in mid-elevation mountain oak woodland, especially in canyons.

*Found* Unreported, but probably March–April, July, or Aug–Oct.
*Host* Unreported, grasses suspected.

**Moon-marked Skipper** *Atrytonopsis lunus* (Page 144)
Dull green to tan, sometimes with a pinkish hue, covered with fine short hairs creating a velvety appearance; the rear segments more tan than green; head brown, pubescent.
*Habitat* Openings in mid-elevation mountain oak woodland, especially in canyons.
*Found* Prior to flight, April–June, and after flight late Aug–early Oct.
*Host* Bull grass.
*Comments* Early instars make aerial nests. Overwinters in a tube nest (partially subterranean) at base of the host, with only the front of the head exposed at the top of the leaf tube.

**Viereck's Skipper** *Atrytonopsis vierecki* (Page 144)
Unreported, but undoubtedly similar to other Dusted Skippers.
*Habitat* Washes through sparse oak-juniper-pinyon woodland and dry gulches on high prairies.
*Found* Unreported, but probably March–April, July, or Aug–Oct.
*Host* Unreported; grasses suspected.

**White-barred Skipper** *Atrytonopsis pittacus* (Page 144)
Dull green to tan, sometimes with a pinkish hue, covered with fine short hairs creating a velvety appearance; the rear segments more tan than green; head yellow-brown, pubescent. Flies with Sheep Skipper and the two may or may not feed concurrently on the same grasses.
*Habitat* Mid-elevation grassy oak woodlands and open grasslands.
*Found* Mostly June–Aug where single-brooded.
*Host* Side-oats grama and probably other grasses.
*Comments* After devouring the eggshell, first instar caterpillars wait 3–4 weeks before feeding, perhaps holding out for fresh host leaves after the onset of summer rains in July.

**Python Skipper** *Atrytonopsis python* (Page 144)
Dull green to tan, sometimes with a pinkish hue, covered with fine short hairs creating a velvety appearance; the rear segments more tan than green; head brown, pubescent. Flies with Deva Skipper in parts of its range.
*Habitat* Openings in mid- to high-elevation mountain oak woodlands.
*Found* Unreported, but probably March–April, July, or Aug–Oct.
*Host* Unreported; grasses suspected.

**Cestus Skipper** *Atrytonopsis cestus* (Page 144)
Dull green to tan, sometimes with a pinkish hue, covered with fine short hairs creating a velvety appearance; the rear segments more tan than green; head brown, pubescent. Both Sheep Skipper and White-barred Skipper use different grasses.
*Habitat* Rocky mid-elevation canyons with the host.

*Found* March–April, May–Aug, Sept–Oct, before, between, and after flights.
*Host* Bamboo muhly.
*Comments* Nests often found near terminal ends of host stems, often 5–7 feet above ground.

### **Sheep Skipper** *Atrytonopsis edwardsii* (Page 144)

Dull green to tan, sometimes with a pinkish hue, covered with fine short hairs creating a velvety appearance; the rear segments more tan than green; head brown, pubescent. Cestus Skipper uses different host. Flies with White-barred Skipper and the two may or may not feed concurrently on the same grasses.
*Habitat* Openings in low to mid-elevation oak or mesquite grassland.
*Found* March–April, May–Aug, Sept–Oct, before, between, and after flights.
*Host* Side-oats grama and probably other grasses.
*Comments* Last instar nests are vertical and can be found in the center of the grass at the base.

### **Large Roadside-Skipper** *Amblyscirtes exoteria* (Page 146)

Bluish-green; head white with dark red-brown vertical stripe, distinctly lighter on top of forehead. No other roadside skipper is known to use same host.
*Habitat* Openings in mid- to high-elevation oak and coniferous woodlands.
*Found* Aug–Sept.
*Host* Bull grass.
*Comments* Later instar nests consist of numerous grass blades tied together.

### **Cassus Roadside-Skipper** *Amblyscirtes cassus* (Page 146)

Yellow-green; head white with brown vertical stripe and two shorter vertical stripes extending to near top of forehead. Bronze Roadside-Skipper is bluish-green and has two shorter vertical stripes extending only partway to top of forehead.
*Habitat* Moist, open areas in mid- to high-elevation woodlands.
*Found* Aug–Sept.
*Host* Bulb panic grass and probably other grasses.

### **Bronze Roadside-Skipper** *Amblyscirtes aenus* (Page 146)

Light bluish-green; head white with brown vertical stripe and two shorter vertical stripes extending to near top of forehead. May not be distinguishable from Nysa Roadside-Skipper. Cassus Roadside-Skipper is yellow-green and has two vertical stripes extending closer to the forehead.
*Habitat* Woodland openings, rocky canyons, and prairie gulches.
*Found* June–Sept, after flights.
*Host* A wide variety of grasses. Usually shying away from those used by other roadside-skippers in any given habitat.

### **Linda's Roadside-Skipper** *Amblyscirtes linda* (Page 146)

Light bluish-green; head white with brown vertical stripe and two shorter vertical stripes extending to near top of forehead. May not be distinguishable from Nysa Roadside-Skipper and Bronze Roadside-Skipper.

*Habitat* Rich, moist woodlands.
*Found* June–Oct.
*Host* Broad-leaved uniola.

### **Oslar's Roadside-Skipper** *Amblyscirtes oslari* (Page 146)

Light yellowish-green; head white with central brown vertical stripe. Similar to Slaty and Elissa Roadside-Skippers.
*Habitat* Most often found in prairie ravines, badlands, streambeds, canyons, and gulches in arid regions.
*Found* July–Sept.
*Host* Side-oats grama.

### **Pepper and Salt Skipper** *Amblyscirtes hegon* (Page 146)

Whitish-green; head tan with darker, wide, central vertical stripe tapering at both ends.
*Habitat* Edges and openings within northern or mountain woodlands, especially along grass-lined watercourses.
*Found* June–mid-Aug.
*Host* Fowl manna grass, Indian grass, broad-leaved uniola, and others.

### **Elissa Roadside-Skipper** *Amblyscirtes elissa* (Page 146)

Bluish-green; head white with brown central vertical stripe, fairly narrow just above inverted V mark on the face. Oslar's Roadside Skipper uses the same host, usually in different habitat, and has central vertical brown stripe not narrowing above the V mark on the face.
*Habitat* Edges of desert canyons, washes and oak woodlands.
*Found* Mid-Aug–Sept.
*Host* Side-oats grama.
*Comments* Many-spotted Skipperling is found on the same host, but facial markings are quite different.

### **Texas Roadside-Skipper** *Amblyscirtes texanae* (Page 146)

Green; head white with brown central vertical stripe. Cassus and Slaty Roadside-Skippers are similar.
*Habitat* Canyons and washes (often limestone) in low-elevation arid regions.
*Found* May–Sept.
*Host* Bulb panic grass in Arizona. Unreported for Texas.

### **Slaty Roadside-Skipper** *Amblyscirtes nereus* (Page 146)

Green; head white with brown central vertical stripe. Similar to Toltec Roadside-Skipper, which inhabits lower elevations.
*Habitat* Grassy openings in mid-elevation oak-juniper and pine woodlands.
*Found* Aug–Sept in Arizona; May–June and Sept–Oct in west Texas.
*Host* Common beardgrass.
*Comments* The host grass for this species is much more widespread and common than the skipper. No other roadside-skipper appears to use the same host in its range.

### **Nysa Roadside-Skipper** *Amblyscirtes nysa* (Page 146)

Light bluish-green; head white with brown vertical stripe and two shorter vertical stripes extending to near top of forehead. May not be distinguishable from Bronze and Linda's Roadside-Skippers.

*Habitat* Thorn scrub and canyons and washes in sparsely wooded arid regions; also lawns and gardens and roadsides in disturbed areas eastward.

*Found* April–Oct.

*Host* Many different grasses.

### **Celia's Roadside-Skipper** *Amblyscirtes celia* (Page 148)

Green; head white with brown central vertical stripe. Bell's Roadside-Skipper is almost identical.

*Habitat* Woodland trails, creek bottoms, moist open woodland, suburban lawns and gardens.

*Found* May–Nov.

*Host* Guinea grass and other grasses.

### **Bell's Roadside-Skipper** *Amblyscirtes belli* (Page 148)

Green; head white with brown central vertical stripe. Celia's Roadside-Skipper is almost identical.

*Habitat* Trails through moist, rich woodlands, woodland creeksides, suburban lawns and gardens.

*Found* May–Oct.

*Host* Broad-leaved uniola, Johnson grass, and others.

### **Dusky Roadside-Skipper** *Amblyscirtes alternata* (Page 148)

Whitish green; head whitish with pale brown lateral stripes. The face has 3 pale stripes; the center stripe is short and near the top.

*Habitat* Open pine flats with low grasses.

*Found* June–July, Sept–Oct.

*Host* Bluestem grasses.

### **Violet-clouded Skipper** *Lerodea arabus*

Essentially identical to Eufala Skipper.

*Habitat* Arid gullies and canyons, suburban and urban gardens.

*Found* All year.

*Host* Bermuda grass and other grasses.

### **Olive-clouded Skipper** *Lerodea dysaules*

Caterpillar unreported, but probably essentially identical to Eufala Skipper.

*Habitat* Thorn scrub, tropical woodland edges.

*Found* Probably most of year.

*Host* Probably weedy grasses.

### **Giant Skippers** (subfamily Megathyminae)

Grub-like caterpillars that bore into leaves, stems, and rootstocks of agaves, yuccas, and related plants.

### **Agave-feeding giant-skippers** (genus *Agathymus*)

A small group of skippers found in the deserts of the Southwest. Caterpillars bore into the fleshy leaves of century plants or agaves. Eggs are dropped onto the hosts. Young caterpillars bore into the leaves and overwinter in leaf tips. Feeding is resumed in the spring, usually at or near the base of a leaf, and finished in early summer. The caterpillars remain inactive during the summer. Before pupating in early fall prior to the flight, the caterpillars cover themselves with a heavy white bloom and cover their tunnel with silk while making a silken trap door in the side of the leaf, through which the adults will emerge. *Agathymus* are all similar in color and pattern and are best identified by the host agave and, in some cases, locality.

#### **Orange Giant-Skipper** *Agathymus neumoegeni*

Greenish or bluish-white; head dark brown. Arizona Giant-Skipper prefers a different host.
*Habitat* Arid grassland/open woodland usually at mid-elevations.
*Found* Central Arizona east to west Texas. Most of year, especially May–Sept. For map, see *Butterflies through Binoculars: The West.*
*Host* Agaves, especially Parry's agave.

#### **Huachuca Giant-Skipper** *Agathymus evansi*

Reddish or bluish-green; head reddish-brown. Arizona Giant-Skipper uses a different host.
*Habitat* Gaps in oak-juniper-pine forests near host.
*Found* Huachuca Mountains and vicinity of southeastern Arizona only. Most of year, especially June–Sept.
*Host* Huachuca agave.
*Comments* Trap door is rough, black, and usually on the underside of the leaf.

#### **Mary's Giant-Skipper** *Agathymus mariae*

Pale bluish; head black. Uses the same host as Coahuila Giant-Skipper and may not be reliably separable from it.
*Habitat* Lecheguilla-covered, rocky Chihuahuan desert.
*Found* Southeastern New Mexico, west Texas and adjacent Mexico. Most of year, especially April–Sept. For map, see *Butterflies through Binoculars: The West.*
*Host* Lecheguilla.
*Comments* Older caterpillars live in a tunnel through several leaves connected to the stem. Trap door tan, usually on the upper side of the leaf.

#### **California Giant-Skipper** *Agathymus stephensi*

White to bluish-green; head brown. No other agave-feeding giant-skipper occurs in its range.
*Habitat* Desert hillsides and canyons along the western edge of the Colorado desert.
*Found* Extreme southern California only. Most of year, especially April–Sept.
*Host* Desert agave.
*Comments* Older caterpillars live in a tunnel through several leaves and the stem. Trap door is tan, usually on upper side of the leaf.

### **Coahuila Giant-Skipper** *Agathymus remingtoni*

Light tan; head reddish-brown. Uses same host as Mary's Giant-Skipper and may not be reliably separable from it.

*Habitat* Rocky slopes in Chihuahuan desert with stands of the host.

*Found* Del Rio, Texas, and vicinity only. Most of year, especially April–Aug. For map, see *Butterflies through Binoculars: The West*.

*Host* Lecheguilla.

*Comments* Older caterpillars live in a tunnel through several leaves connected to the stem. Trap door is tan, usually on the upper side of the leaf.

### **Poling's Giant-Skipper** *Agathymus polingi*

Caterpillar is unreported. Host is unique to this skipper.

*Habitat* Dry shindagger-covered slopes.

*Found* Southeastern Arizona only. Most of year, especially June–Sept. For map, see *Butterflies through Binoculars: The West*.

*Host* Shindagger.

*Comments* Trap door is silky white, on the underside of leaf near the base.

### **Mojave Giant-Skipper** *Agathymus alliae*

Greenish or bluish-white; head brown. Host is unique to this species.

*Habitat* Mid-elevation dry rocky limestone hillsides with pinyon-juniper.

*Found* Extreme southern Nevada and adjacent areas in surrounding states only. Most of year, especially May–Sept. For map, see *Butterflies through Binoculars: The West*.

*Host* Utah agave.

*Comments* Trap door thin, white, and usually on underside of leaf near base.

### **Yucca-feeding giant-skippers** (genus *Megathymus*)

This group includes our largest skipper caterpillars. These grubby guys feed in the roots and stems of yuccas. Eggs are laid on the leaves of the host. Early instar caterpillars of most eat the top surfaces of yucca leaves before burrowing into the stem and eventually down into the root. As they grow a tube-like, chalk-lined burrow into the root is created with a small exit hole at the top (often in the center of the plant) through which droppings are expelled. This exit eventually transforms into a frass-covered thumb-like tent and aids in locating caterpillars or pupas. The caterpillar lives and feeds within the root, eventually pupating there. Pupas can move freely up and down the burrow by wiggling the abdomen. Caterpillars overwinter partially to full-grown and are present most of the year except just prior to and during flights.

### **Yucca Giant-Skipper** *Megathymus yuccae* (Page 150)

### **Cofaqui Giant-Skipper** *Megathymus cofaqui*

Creamy-white; head brownish-black. Much less likely to encounter this rarity than the Yucca Giant-Skipper.

*Habitat* Open situations with yuccas, especially pine flats and coastal dunes.

*Found* Central Florida north to North Carolina. For map, see *Butterflies through Binoculars: The East.* Most of year, between flights.
*Host* Spanish bayonet and others.
*Comments* First instars burrow directly into stem and root and do not eat leaves; nor do they construct a tent. Prior to pupation a tent is constructed at base of stem or soil surface.

### Strecker's Giant-Skipper *Megathymus streckeri*

Yellowish-white; head reddish-brown.
*Habitat* Rolling yucca grassland.
*Found* Northern Arizona, New Mexico, and Texas, and adjacent southern Utah and Colorado. Also Montana to Nebraska. Most of year. For map, see *Butterflies through Binoculars: The West.*
*Host* Bailey's yucca and others.
*Comments* First instars burrow directly into stem and root and do not eat leaves; nor do they construct a tent. Prior to pupation a tent is constructed at base of stem or soil surface.

### Ursine Giant-Skipper *Megathymus ursus* (Page 150)

### Manfreda Giant-Skipper *Stallingsia maculosa*

White with tan hairs; head tan. The host is unique for North American skippers.
*Habitat* Thorn scrub.
*Found* Only in extreme southern Texas and adjacent Mexico. Probably most of year.
*Host* Manfreda.

# Photo Locations and Credits

29.1: Sierra Nevada, Nevada Co., CA. Contributor: Ken Hansen. Photo: JPB
29.1: (inset) Sierra Nevada, Nevada Co., CA. Contrib: Ken Hansen. Photo: JPB
29.2: Elkins, Randolph Co., WV. Photo: TJA
29.3: Siskiyou Co., CA. Contrib: Ken Hansen. Photo: JPB
29.3: (inset) Clear Creek Co., CO. Contrib: Ken Hansen. Photo: JPB
29.4: Gainesville, Alachua Co., FL. Contrib: Jeff Slotten. Photo: TJA
29.5: Upper Tract, Pendelton Co., WV. Photo: TJA
29.6: Tucson, Pima Co., AZ. Photo: JPB
29.6: (inset) Tucson, Pima Co., AZ. Photo: Rob & Eve Gill
29.7: Dry Branch, Cabin Creek, Kanawha Co., WV. Photo: TJA
29.7: (inset) Dry Branch, Cabin Creek, Kanawha Co., WV. Photo: TJA
29.8: Santa Rosa, Sonora, MX. Photo: JPB

31.1: Elkins, Randolph Co., WV. Photo: TJA
31.2: San Diego Co., CA. Photo: JPB
31.3: Santa Ana NWR, Hidalgo Co., TX. Photo: JG
31.3: (inset) Bathurst, Gloucester Co., NB, Can. Contributor: Reginald Webster. Photo: TJA
31.4: Southern CA. July 1991. Photo: TJA
31.4: (inset) Southern CA. July 1991. Photo: TJA
31.5: Warsaw, MO. Photo: Richard Heitzman
31.6: Alaska. Contrib: Jack Harry. Photo: JPB
31.7: Eager, Apache Co., AZ. Photo: JPB
31.8: Sorrento Valley, San Diego Co., CA. Aug. 2000. Contrib: Chris Conlan. Photo: TJA
31.8: (inset) Kennedy Meadows, Tulare Co., CA. Photo: JPB

33.1: Beatty, Nye Co, NV. Photo: JPB
33.2: Chaparral WMA, LaSalle Co., TX. Photo: JG
33.3: San Rafael Swell, Emery Co., UT, Elev. 5200'. Aug. 1999. Contributor: Todd Stout. Photo: TJA
33.4: Spring Mountains, Clark Co., NV. June 2000. Photo: JPB
33.4: (inset) Cannon Canyon, Dugway Range, Toole Co., UT. Contrib: Todd Stout. Photo: TJA
33.5: Butterfly farm livestock. July 24, 1993. Photo: Keith Wolfe
33.6: Santa Ana NWR, Hidalgo Co, TX. Photo: JG

33.7: Elliott Key, Dade Co., FL. Aug. 1987. Photo: Thomas C. Emmel
33.8: Elliott Key, Dade Co., FL. Aug. 1987. Photo: Thomas C. Emmel

35.1: Elkins, Randolph Co., WV. July 1998. Photo: TJA
35.2: Beverly, Randolph Co., WV. Sept. 1990. Photo: TJA
35.3: Elkins, Randolph Co., WV. Sept. 1990. Photo: TJA
35.5: Black River, Apache Co., AZ. Photo: JPB
35.5: (inset) Ditch Camp, Apache Co., AZ. Photo: JPB
35.6: Tucson, Pima Co., AZ. Photo: JPB
35.6: (inset) Tucson, Pima Co., AZ. Photo: JPB
35.7: Mendocino Co., CA. Contributor: Ken Hansen. Photo: JPB
35.8: Beverly, Randolph Co., WV. Aug. 1986. Photo: TJA
35.8: (insets): Beverly, Randolph Co., WV. Aug. 1986. Photo: TJA
35.9: Ormond Beach, Volusia Co., FL. Sept. 1991. Photo: TJA

37.1: Files Creek, Beverly, Randolph Co., WV. May 2002. Photo: TJA
37.2: Morristown, Morris Co., NJ. Photo: JG
37.3: Broward Co., FL. Photo: Marc Minno
37.4: Elkins, Randolph Co., WV. June 1999. Photo: TJA
37.5: nr. Alpine Dam, Marin Co., CA. Apr. 1991. Photo: Keith Wolfe
37.6: Files Creek, Beverly, Randolph Co., WV. Apr. 2002. Photo: TJA
37.7: Round Spring Draw, Sevier Co., UT. Contributor: Todd Stout. Photo: JPB
37.8: Eco Pond, Everglades Nat. Park, Dade Co., FL. Jan. 1998. Photo: Jeff Fengler

39.1: Cedar Ridge, Coconino Co., AZ. Photo: JPB
39.2: Jefferson Co., CO. Photo: JG
39.3: Bobs Gap, Los Angeles County, CA. March 26, 1988. Photo: Keith Wolfe
39.5: Dickinson Park, Wind River Mt., Fremont Co., WY. July 2000. Photo: TJA
39.6: Black Rock Station, Tulare Co., CA. Photo: JPB
39.7: Romney, Hampshire Co., WV. Apr. 1988. Photo: TJA
39.8: Sabino Canyon, Pima Co., AZ. Photo: JPB
39.9: Black Rock Station, Tulare Co., CA. Photo: JPB

41.1: Tucson, Pima Co., AZ. Photo: JPB
41.1: (inset) Tucson, Pima Co., AZ. Photo: JPB
41.2: Los Angeles Co., CA. Photo: JG
41.3: Photo: TJA
41.4: Upper Tract, Pendleton Co., WV. May 1988. Photo: TJA
41.5: Alamos, Sonora, Mex. Photo: JPB
41.5: (inset) Alamos, Sonora, Mex. Photo: JPB
41.6: Gomez Farias, Tamaulipas, Mex. Photo: JPB
41.7: Moctezuma, Sonora, Mex. Photo: JPB
41.8: Tucson, Pima Co., AZ. Photo: JPB
41.8: (inset) Bates Canyon, Santa Barbara Co., CA. Photo: JPB

43.1: Elkins, Randolph Co., WV. July 1997. Photo: TJA
43.2: Troy Meadows, Morris Co., NJ. Photo: JG
43.3: Elkins, Randolph Co., WV. July 1997. Photo: TJA
43.4: Horse Ridge, Wasatch Co., UT, Elev. 7900'. Contributor: Steve Sommerfeld. Photo: TJA
43.5: Laguna Mtn. Recreation Area, Cleveland Nat. For., San Diego, CA. Aug. 1999. Photo: Keith Wolfe
43.6: Alberta, Can. Contrib: Jack Harry. Photo: JPB
43.8: Spruce Knob Lake, Randolph Co., WV. May 1993. Photo: TJA
43.10: Slopes of Gaylor Peak, Mono/Tuolumne counties, CA. Aug. 19, 2001. Photo: Keith Wolfe

45.1: Cochise Co., AZ. Photo: JG
45.2: Santa Ana NWR, Hidalgo Co, TX. Photo: JG
45.3: Ormond Beach, Volusia Co., FL. Aug. 1998. Photo: TJA
45.4: Tucson, Pima Co., AZ. Photo: JPB
45.5: Coral Springs, Broward Co., FL. Jan. 1998. Photo: Jeff Fengler
45.6: Tucson, Pima Co., AZ. Photo: JPB
45.8: Tucson Mtn. Park, Pima Co., AZ. Photo: JPB
45.9: Santa Ana Wildlife NWR, Hidalgo Co., TX. Photo: JPB

47.1: Wardensville, Hardy Co., WV. Sept. 1998. Photo: TJA
47.2: Loma Alta, Cameron Co., TX. Photo: JG
47.3: Slate Creek Preserve, Cayo District, Belize. Oct. 10, 2001. Photo: Keith Wolfe.
47.4: Wardensville, Hardy Co., WV. Sept. 1998. Photo. TJA
47.5: Tucson, Pima Co., AZ. Photo: JPB
47.6: Montezuma Canyon, Huachuca Mts., Cochise Co., AZ. Photo: JPB
47.7: Gainesville, Alachua Co., FL. Photo: TJA
47.9: California Gulch, Santa Cruz Co., AZ. Photo: JPB
47.9: (inset) California Gulch, Santa Cruz Co., AZ. Photo: JPB

49.1: Elkins, Randolph Co., WV. Aug. 1994. Photo: TJA
49.2: Kenvil, Morris Co., NJ. Photo: JG
49.3: Elkins, Randolph Co., WV. Aug. 1994. Photo: TJA
49.4: Bathurst, Gloucester Co., NB, Can. May 1999. Contrib: Reginald Webster. Photo: TJA
49.5: Salt Lake City, Salt Lake Co., UT. Oct. 2000. Photo: TJA
49.6: Upper Tract, Pendleton Co., WV, June 1990. Photo: TJA
49.7: Greenbottom WMA, Cabel Co., WV, June 1995. Photo: TJA
49.8: Forge Pond, nr. Pleasant Mills, Atlantic Co., NJ. June 1982. Photo: David Wright

51.1: Greer, Apache Co., AZ. Photo: JPB
51.2: Harts Pass, Okanogan Co., WA. Photo: JG
51.3: Lake of the Woods, Ventura Co., CA. Photo: JPB

51.4: Road's End, Tulare Co., CA. Photo: JPB
51.5: Bodfish, Kern Co., CA. Photo: JPB
51.6: San Diego Co., CA. Photo: John Emmel
51.7: Siskiyou Co., CA. Contributor: Ken Hansen. Photo: JPB
51.10: Upper Tract, Pendleton Co., WV. June 1990. Photo: TJA

53.1: Delray Beach, Palm Beach Co., FL. Photo: JG
53.2: Riverside Co., CA. Photo: JG
53.3: Mt. Lemmon, nr. Tucson, Pima Co., AZ. June 1987. Photo: Greg Ballmer
53.4: Humbolt Co., CA. Contributor: Pete Haggard. Photo: JPB
53.5: Riverside, CA. Apr. 1986. Photo: Greg Ballmer
53.6: Llera de Canales, Tamaulipas, Mex. Photo: Rick and Nora Bowers
53.7: Molino Basin, Pima Co., AZ. Photo: JPB
53.8: Dade Co., FL. July 1984. Photo: Marc Minno

55.1: South of Whitmer, Randolph Co., WV. Apr. 1995. Photo: TJA
55.2: Upper Tract, Pendleton Co., WV. July 1995. Photo: TJA
55.3: Bennett Run, Criders, VA. May 1994. Photo: TJA
55.4: Hualapai Mts., Mojave Co., AZ. June 1987. Photo: Greg Ballmer
55.6: West Rock SP, Hamden, New Haven Co., CT. June 2000. Photo: Jeff Fengler
55.7: Birmingham, AL. May 2000. Contributor: Paulette Haywood. Photo: TJA
55.8: Elkins, Randolph Co., WV. June 2000. Photo: TJA
55.8: (inset) Litchfield Co., CT. July 1999. Contrib: David Wagner. Photo: TJA
55.9: Bear Mt., Litchfield Co., CT. July 1999. Contrib: David Wagner. Photo: TJA

57.1: Road's End, Tulare Co., CA. Photo: JPB
57.2: Sisters, Deschutes Co., OR. Photo: JG
57.3: Road's End, Tulare Co., CA. Photo: JPB
57.4: Hamden, New Haven CO., CT. May 1998. Photo: Jeff Fengler
57.5: Road's End, Tulare Co., CA. Photo: JPB
57.6: Westgard Pass, Inyo Co., CA. Photo: JPB
57.7: Hualapai Mtn. Park, Mohave Co., CA. Photo: JPB
57.7: (inset) Hualapai Mtn. Park, Mohave Co., CA. Photo: JPB
57.8: Democrat Hot Springs, Kern Co., CA. Photo: JPB

59.1: Kelso Valley, Kern Co., CA. Photo: JPB
59.2: Beatty, Nye Co., NV. Photo: JPB
59.3: Mendocino Co. Contributor: Ken Hansen. CA. Photo: JPB
59.3: (inset) Mendocino Co., CA. Contrib: Ken Hansen. Photo: JPB
59.4: N. of Yellowstone National Park, WY. Photo: TJA
59.5: Pine Canyon Trail, Big Bend NP, TX. Photo: JPB
59.7: Yavapai Co., AZ. Photo: Rob and Eve Gill
59.9: Beatty, Nye Co., NV. Photo: JPB
59.10: Devil's Gap, San Luis Obispo Co., CA. Photo: JPB

61.1: Larenim Park, N. of Burlington, Mineral Co., WV. May 1992. Photo: TJA
61.2: Spikebuck Meadow, Humboldt Co., CA. Photo: JG

61.3: Larenim Park, N. of Burlington, Mineral Co., WV. May 1992. Photo: TJA
61.4: Brushy Run, Upper Tract, Pendelton Co., WV. May 1994. Photo: TJA
61.5: King's Run, Elkins, Randolph Co., WV. June 1993. Photo: TJA
61.6: Chatsworth, Burlington Co., NJ. July 1997. Photo: David Wright
61.7: Del Norte Co., CA. Contributor: Ken Hansen. Photo: JPB
61.8: Fredericton, NB, Can. June 1999. Photo: TJA

63.1: Petersburg, Grant Co., WV. May 1992. Photo: TJA
63.2: Larenim Park, N. of Burlington, Mineral Co., WV. May 1994. Photo: TJA
63.3: Madera Canyon, Santa Cruz Co., AZ. Photo: JPB
63.4: Dismal Swamp, Norfolk, VA. June 1999. Photo: TJA
63.5: King's Run, Elkins, Randolph Co., WV. June 1999. Photo: TJA
63.6: King's Run, Elkins, Randolph Co., WV. June 1999. Photo: TJA
63.7: Deer Creek, Greenbank, Pocahontas Co., WV. July 1993. Photo: TJA
63.8: Key West, Monroe Co., FL. Photo: JG

65.1: Shelby Co., AL. Apr. 1999. Contributor: Paulette Haywood. Photo: TJA
65.2: Stonecoal Road, S. of Anmoore, Barbour Co.,WV. Photo: TJA
65.3: Gainesville, Alachua Co., FL. May 2000. Contrib: Jeff Slotten. Photo: TJA
65.4: New Braunfels, Comal Co., TX. Contrib: Jeff Slotten. Photo: JPB
65.6: Shelby C., AL. Apr. 1999. Contrib: Paulette Haywood. Photo: TJA
65.9: Tucson, Pima Co., AZ. Photo: JPB
65.12: Rose Peak, Greenlee Co., AZ. Photo: JPB
65.13: 5 miles no. of Spruce Knob Lake, Randolph Co., WV. June 1994. Photo: TJA

67.1: Buttonwillow, Kern Co., CA. Photo: JPB
67.1: (inset) Ft. Lauderdale, Broward Co., FL. 1983. Photo: Minno, Butler and Hall
67.2: Reston, Fairfax Co.,VA. Photo: JG
67.4: Tucson, Pima Co., AZ. Photo: JPB
67.4: (inset) Tucson, Pima Co., AZ. Photo: JPB
67.6: Tucson, Pima Co., AZ. Photo: JPB
67.7: Tucson, Pima Co., AZ. Photo: JPB
67.7: (inset) Tucson, Pima Co., AZ. Photo: JPB
67.8: North Key Largo, Monroe Co., FL. Photo: Marc Minno
67.10: Moorefield, Hardy Co., WV. May 1993. Photo: TJA
67.11: Ramsey Brook, Restigouche Co., NB, Can. July 2000. Contrib: Richard Boscoe. Photo: TJA
67.11: (inset) Westgard Pass, Inyo Co., CA. Photo: JPB

69.1: Dutch Run, Greenbrier Mt., Grenbrier Co., WV. May 2001. Photo: TJA
69.2: Fork Creek WMA, Nellis, Boone Co., WV. May 1997. Photo: TJA
69.3: State Game Lands #157, Bucks Co., PA. June 1987. Photo: David Wright
69.3: (inset) Bowden, Randolph Co., WV. June 1999. Photo: TJA
69.4: NJ Pine Barrens, Chatsworth, Burlington Co., NJ. May 1983. Photo: David Wright

69.5: State Game Lands #196, Bucks Co., PA. Sept. 1994. Photo: David Wright
69.6: Bowden, Randolph Co., WV. June 1999. Photo: TJA
69.7: NJ Pine Barrens, Chatsworth, Burlington Co., NJ. June 2000. Photo: David Wright
69.8: State Game Lands #127, Pocono Mts., Monroe Co., PA. June 1993. Photo: David Wright
69.8: (inset) State Game Lands #127, Pocono Mts., Monroe Co., PA. June 1993. Photo: David Wright

71.1: Cockscomb, Kane Co., UT. Photo: JPB
71.2: Plum Canyon, Anza-Borrego Desert SP, San Diego Co., CA. Photo: JG
71.3: Cockscomb, Kane Co., UT. Photo: JPB
71.7: Beatty, Nye Co., NV. Photo: JPB
71.7: (inset) Beatty, Nye Co., NV. Photo: JPB
71.8: Chimney Creek, Tulare Co., CO. Photo: JPB
71.9: Tioga Pass, Mono Co., CA. Photo: John Emmel
71.10: Above Upper Angora Lake, Dorado Co., CA. Photo: John Emmel
71.11: Weldon, Kern Co., CA. Photo: JPB

73.1: Sageland, Kern Co., CA. Photo: JPB
73.2: Great Basin NP, NV. Photo: JG
73.3: N. of Moorefield, Hardy Co., WV. May 1994. Photo: TJA
73.4: Shippagan, Gloucester Co., NB, Can. July 2000. Contrib: Richard Boscoe. Photo: TJA
73.5: Green's Peak Road, nr. US Hwy. 60, Apache Co., AZ. Photo: JPB
73.5: (inset) Green's Peak Road, nr. US Hwy. 60, Apache Co., AZ. Photo: JPB
73.6: Gaspe-Quest Co., Quebec, Can. July 2000. Contributor: Richard Boscoe. Photo: TJA
73.7: Aurora, Arapahoe Co., CO. Photo: JG
73.8: Brush Creek, Tulare Co., CA. Photo: JPB

75.1: Canaan, Litchfield Co., CT. May 1998. Photo: Jeff Fengler
75.2: Sussex Co., NJ. Photo: JG
75.3: Houston Co., AL. Aug. 2000. Contributor: Paulette Haywood. Photo: TJA
75.4: Tucson, Pima Co., AZ. Photo: JPB
75.5: Laguna Atascosa (butterfly garden), Cameron Co., TX. Photo: JPB
75.6: nr. Palm Desert, Riverside Co., CA. Apr. 1980. Photo: Greg Ballmer
75.7: Molino Basin, Pima Co., AZ. Photo: JPB
75.8: WI. Photo: Susan S. Borkin

77.1: Picachos Mts., Nuevo Leon, Mex. Photo: JPB
77.2: Catemaco, Veracruz, Mex. Photo: JG
77.3: Garden Canyon, Huachuca Mts., Cochise Co., AZ. Photo: JPB
77.4: Pinery Canyon, Chiricachua Mts., Cochise Co., AZ. Photo: JPB
77.5: Cottonwood Pass, San Luis Obispo Co., CA. Photo: JPB
77.6: Kitt Peak, Pima Co., AZ. Photo: JPB

77.7: Rose Peak, Greenlee Co., AZ. Photo: JPB
77.8: McAllen, Hidalgo Co., TX. Contrib: Shawn and Joe Patterson. Photo: JPB

79.1: Bennetts Run, Crider, VA. July 1994. Photo: TJA
79.2: N. of Del Rio, Val Verde Co., TX. Photo: JG
79.3: Chipinque Nat. Park, Nuevo Leon, Mex. Photo: JPB
79.4: Ormond Beach, Volusia Co., FL. Apr. 1995. Photo: TJA
79.5: Weslaco, Hidalgo Co., TX. Photo: JG
79.6: South of Anmoore, Barbour Co., WV. Sept. 2000. Photo: TJA
79.8: Butterfly farm livestock. Aug. 1, 1988. Photo: Keith Wolfe
79.8: (inset) Turner River, Big Cypress Nat. Pres., Collier Co., FL. April 1997. Photo: Jeff Fengler
79.9: Cape Coral, Lee Co., FL. May 2003. Photo: TJA

81.1: Fort Martin Power Station, Fort Martin, Monongalia Co., WV. Apr. 1993. Photo: TJA
81.2: Mullens, Wyoming Co., WV. Aug. 1987. Photo: TJA
81.3: Pimentel Meadows, Mono County, CA. May 17, 1994. Photo: Keith Wolfe
81.4: Nellis, Boone Co., WV. Oct. 1986. Photo: TJA
81.5: Spruce Knob, Randolph Co., WV. Oct. 1991. Photo: TJA
81.6: Elkins, Randolph Co., WV. Sept. 1988. Photo: TJA
81.7: Hondo, Medina County, TX. July 29, 1995. Photo: Keith Wolfe
81.8: Humbolt Co., CA. Contributor: Ken Hansen. Photo: JPB

83.1: Spruce Knob, Randolph Co., WV. Sept. 1994. Photo: TJA
83.2: Helmetta, Middlesex Co., NJ. Photo: JG
83.3: Elkins, Randolph Co., WV. Aug. 1988. Photo: TJA
83.4: Magalloway, Oxford Co., ME. June 1996. Contrib: Richard Boscoe. Photo: TJA
83.5: Beverly, Randolph Co., WV. Aug. 1988. Photo: TJA
83.6: Elkins, Randolph Co., WV. Aug. 1992. Photo: TJA
83.6: (inset) Canaan Valley, Tucker Co., WV. June 1996. Photo: TJA
83.7: Goldstream, AK. Contributor: Jack Harry. Photo: JPB
83.8: Gaspe-Quest Co., Quebec, Can. Apr. 2000. Contrib: Richard Boscoe. Photo: TJA

85.1: Alamos, Sonora, Mex. Photo: JPB
85.2: La Gloria, Starr Co., TX. Photo: JG
85.3: Santa Catalina Mts., Pima Co., AZ. Photo: JPB
85.4: Santa Catalina Mts., Pima Co., AZ. Photo: JPB
85.5: Tucson, Pima Co., AZ. Photo: JPB
85.5: (inset) Aduana, Sonora, Mex. Photo: JPB
85.6: Tucson, Pima Co., AZ. Photo: JPB
85.8: Hondo, Medina Co., TX. July 1995. Photo: Keith Wolfe
85.9: Gomez Farias, Tamaulipas, Mex. Photo: JPB
85.9: (inset) Gomez Farias, Tamaulipas, Mex. Photo: JPB

87.1: Fork Creek, Boone Co., WV. June 1990. Photo: TJA
87.2: Lucerne Valley, San Bernardino Co., CA. Photo: JG
87.3: Upper Tract, Pendleton Co., WV. June 1993. Photo: TJA
87.4: Fork Creek, Boone Co., WV. June 1990. Photo: TJA
87.5: Powell Co, MT. Contributor: Jacque Wolfe. Photo: JPB
87.7: West of Ellsworth, Sheridan Co., NE. Feb. 2001. Contrib: Troy Pabst. Photo: TJA
87.8: Molino Basin, Pima Co., AZ. Photo: JPB
87.9: Fork Creek WMA, Nellis, Boone Co., WV. June 1990. Photo: TJA

89.1: Del Puerto Canyon, Stanislaus Co., CA. Contrib: Mike Smith. Photo: JPB
89.2: San Ygnacio, Zapata Co., TX. Photo: JG
89.3: Redington Pass, Pima Co., AZ. Photo: JPB
89.4: Redington Pass, Pima Co., AZ. Photo: JPB
89.5: Garden Canyon, Huachuca Mts., Cochise Co., AZ. Photo: Steve Prchal
89.6: 10 miles N. of Van Horn, Culberson Co., TX. Photo: JPB
89.7: Tollgate Pass, Union Co., NM. Photo: JPB
89.8: Dugway Mts., Dugway Pass, Tooele Co., UT. Contrib: Jacque Wolfe. Photo: JPB

91.1: Taylor Canyon , Weber Co., UT, 5300'. Apr. 2002. Contrib: Todd Stout. Photo: TJA
91.2: Vernon, Sussex Co., NJ. Photo: JG
91.3: Dugway Mts., Tooele Co., UT, 5000'. June 2001. Contrib: Jacque Wolfe. Photo: TJA
91.4: Yellowstone Nat. Park, S. of Mammoth Hot Springs, WY. May 1999. Photo: TJA
91.5: Kern Canyon, Kern Co., CA. Photo: JPB
91.6: 3 miles N. of Cecilville, Siskiyou Co., CA. Contrib: Jacque Wolfe. Photo: TJA
91.6: (inset) Yellowstone Nat. Park, WY. May 1999. Photo: TJA
91.7: E. of Alpine, Grey's River, Lincoln Co., WY, 5800'. Apr. 2001. Contrib: Jacque Wolfe. Photo: TJA
91.8: Elkins, Randolph Co., WV. June 1989. Photo: TJA

93.1: Elkins, Randolph Co., WV. Oct. 1991. Photo: TJA
93.2: Loma Alta, Cameron Co., TX. Photo: JG
93.4: Macon Co., NC. Photo: TJA
93.5: Collier Co., FL. Photo: Dave Wagner
93.6: Salt Lake City, Salt Lake Co., UT, 4200'. June 2002. Contributor: Todd Stout. Photo: TJA
93.6: (inset) Roadside Meadow, Mariposa County, CA. Aug. 10, 2001. Photo: Keith Wolfe
93.6: Junct. Hwy. & Blueberry Rd., N. Umberland Co., NB, Can. Contrib: Reginald Webster. Photo: TJA
93.7: Pittsburg, Contra Costa Co., CA. June 1997. Photo: TJA

93.9: Elliott Key, Dade Co., FL. Photo: Marc Minno
93.10: Tucson, Pima Co., AZ. Photo: JPB

95.1: Sierra Picachos, Nuevo Leon, Mex. Photo: JG
95.2: Ophir Canyon, Toiyabe Range, Nye Co., NV. Photo: JPB
95.3: Fork Creek WMA, Nellis, Boone Co., WV. June 1994. Photo: TJA
95.4: 20 miles N. of Yellowstone Nat. Park, Hwy. 89, WY. May 1999. Photo: TJA
95.4: (inset) Photo: TJA
95.5: Comfort Spring, Huachuca Mts., Cochise Co., AZ. Photo: JPB
95.6: Junct. Hwy. 8 & Blueberry Rd., N. Umberland Co., NB, Can. Contrib: Reginald Webster. Photo: TJA.
95.9: nr. Ludlowville, Tompkins Co., NY. 1978. Photo: J. Rawlins
95.10: East Turkey Creek, Chiricauhua Mts., Cochise Co., AZ. Photo: JPB
95.10: (insert): Photo: JG

97.1: Dolly Sods, Tucker Co., WV. June 1980. Photo: TJA
97.2: Santa Ana NWR, Hidalgo Co., TX. Photo: JG
97.3: Byron Center, MI. July 1994. Photo: TJA
97.4: Slate Creek Preserve, Cayo District, Belize. Oct. 28, 1999. Photo. Keith Wolfe
97.5: Neola, Greenbrier Co., WV. Aug. 1991. Photo: TJA
97.7: Photo: Dave Wagner
97.8: El Nacimiento, Tamaulipas, Mex. Sept. 29, 1995. Photo: Keith Wolfe
97.9: Naples, Collier County, FL. Oct. 17, 1995. Photo: Keith Wolfe

99.1: Moorefield, Hardy Co., WV. July 1989. Photo: TJA
99.2: Anza-Borrego Desert SP, San Diego Co., CA. Photo: JG
99.3: Moorefield, Hardy Co., WV. July 1989. Photo: TJA
99.4: Larenim Park, N. of Burlington, Mineral Co., WV. Photo: TJA
99.5: Greenbrier Mt., White Sulphur Springs, Greenbrier Co., WV. Photo: TJA
99.6: Elkins, Randolph Co., WV. June 1995. Photo: TJA
99.7: Salt Lake City, Salt Lake Co., UT. Contributor: Todd Stout. Photo: JPB
99.8: Elkins, Randolph Co., WV. June 1995. Photo: TJA

101.1: Elkins, Randolph Co., WV. June 1991. Photo: TJA
101.1: (inset) Blackwater Falls St. Park, Davis, Tucker Co., WV. June 1997. Photo: TJA
101.2: Sawmill Canyon, Huachuca Mts., Cochise Co., AZ. Photo: JPB
101.3: Molino Basin, Pima Co., AZ. Photo: JPB
101.4: Patagonia Lake St. Park, Santa Cruz Co., AZ. Photo: JPB
101.5: Blackwater Falls St. Park, Davis, Tucker Co., WV. June 1997. Photo: TJA
101.6: Brush Creek, Tulare Co., CA. Photo: JPB
101.6: (inset) San Diego, San Diego County, CA. July 27, 1985. Photo: Keith Wolfe
101.7: Alpine, Apache Co., AZ. Photo: JPB
101.8: Democrat Hot Springs, Kern Co., CA. Photo: JPB

103.1: Naples, Collier County, FL. Nov. 12, 1994. Photo: Keith Wolfe
103.1: (inset) Naples, Collier County, FL. Nov. 12, 1994. Photo: Keith Wolfe
103.2: Castello Hammock, Miami-Dade Co., FL. Contribs: Minno, Hall & Butler. Photo: Jerry Butler
103.3: La Estanzuela, Nuevo Leon, Mex. Photo: JPB
103.4: Horsetail Falls, Nuevo Leon, Mex. Photo: JPB
103.5: Xilitla, San Luis Potosi, Mex. Sept. 21, 1996. Photo: Keith Wolfe
103.6: Slate Creek Preserve, Cayo District, Belize. Sept. 1, 1997. Photo: Keith Wolfe
103.7: Costello Hammock, Dade Co., FL. Contribs: Minno, Hall & Butler. Photo: Jerry Butler
103.8: North Key Largo, Monroe Co., FL. Contribs: Minno, Hall & Butler. Photo: Jerry Butler

105.1: Aduana, Sonora, Mex. Photo: JPB
105.2: Santa Ana NWR, Hidalgo Co., TX. Photo: JG
105.3: Big Pine Key, Monroe Co., FL. Photo: Jane Ruffin
105.3: (inset) Big Pine Key, Monroe Co., FL. Photo: Jane Ruffin
105.4: Kino Springs area, Santa Cruz County, AZ. Sept. 3, 1991. Photo: Keith Wolfe
105.5: White Sulphur Springs, Greenbrier Co., WV. June 1988. Photo: TJA
105.6: Tucson, Pima Co., AZ. Photo: JPB
105.7: Florida Canyon, Santa Rita Mts., Pime Co., AZ. Photo: JPB
105.8: Riverton, Pendleton Co., WV. June 1993. Photo: TJA

107.1: Suffolk, VA. Aug. 1999. Contributor: Richard Boscoe. Photo: TJA
107.2: Canal Road, Sussex Co., NJ. Photo: JG
107.3: Rough Run, nr. Fort Seybert, Pendleton Co., WV. June 1995. Photo: TJA
107.4: Suffolk, VA. Aug. 1999. Contrib: Richard Boscoe. Photo: TJA
107.5: Robbins Swamp, Canaan, Litchfield Co., CT. Sept. 1999. Contrib: David Wagner. Photo: TJA
107.6: Robbins Swamp, Canaan, Litchfield Co., CT. Oct. 1995. Contrib: David Wagner. Photo: TJA
107.7: Deer Creek, Greenbank, Pocahontas Co., WV. June 1994. Photo: TJA
107.8: Fork Creek WMA, Nellis, Boone Co., WV. June 1995. Photo: TJA
107.8: (inset) Garden Canyon, Huachuca Mts., Cochise Co., AZ. Photo: JPB

109.1: Houston Co., AL. Sept. 1999. Contributor: Paulette Hayward. Photo: TJA
109.2: Helmetta, Middlesex Co., NJ. Photo: JG
109.4: Meadow Valley, Plumas County, CA. July 24, 1993. Photo: Keith Wolfe
109.5: Spruce Knob, Randolph Co., WV. May 1998. Photo: TJA
109.7: Ormond Beach, Volusia Co., FL. June 1995. Photo: TJA
109.8: Rustler Park, Chiricahua Mts., Cochise Co., AZ. Contrib: Tom Carr. Photo: JPB
109.9: Molino Basin, Santa Catalina Mts., Pima Co., AZ. Photo: JPB
109.10: Elkins, Randolph Co., WV. Apr. 1994. Photo: TJA
109.10: (inset) Elkins, Randolph Co., WV. May 1994. Photo: TJA

111.1: Elkins, Randolph Co., WV. Aug. 1998. Photo: TJA
111.2: Canal Rd., Sussex Co., NJ. Photo: JG
111.3: Sevier Co., UT. Contrib.: Todd Stout. Photo: JPB
111.6: Montana. June 2001. Contributor: Steve Koehler. Photo: TJA
111.7: NNE. of Hatch, Garfield Co., UT. June 2000. Contrib: Killian Roever. Photo: TJA
111.8: Sawmill Canyon, Huachuca Mts., Cochise Co., AZ. Photo: JPB
111.9: Humbolt Co., CA. Contrib: Ken Hansen. Photo: JPB
111.9: (insert): Humbolt Co., CA. Contrib: Ken Hansen. Photo: JPB
111.10: Magalloway, Oxford Co., ME. June 1999. Contrib: Richard Boscoe. Photo: TJA
111.10: (inset) Goldstream, AK. Contrib: Jack Harry. Photo: JPB

113.1: Troy Meadows, Morris Co., NJ. Photo: JG
113.2: Weslaco, Hidalgo Co., TX. Photo: JG
113.3: Elkins, Randolph Co., WV. Sept. 1991. Photo: TJA
113.4: Ormond Beach, Volusia Co., FL. Sept. 1991. Photo: TJA
113.5: Cape Coral, Lee Co., FL. June 2003. Photo: TJA
113.6: Elkins, Randolph Co., WV. Sept. 2000. Photo: TJA
113.7: Elkins, Randolph Co., WV. Sept. 2000. Photo: TJA
113.8: Ormond Beach, Volusia Co., FL. Sept. 1991. Photo: TJA
113.9: Cape Coral, Lee Co., FL. June 2003. Photo: TJA

115.1: Garden Canyon, Huachuca Mts., Cochise Co., AZ. Photo: JPB
115.2: Santa Ana NWR, Hidalgo Co., TX. Photo: JG
115.3: Weslaco, Hidalgo Co., TX. Photo: JG
115.3: (inset) Weslaco, Hidalgo Co., TX. Photo: JG
115.4: Key Largo, Monroe Co., FL. May 2001. Photo: TJA
115.4: (inset) Key Largo, Monroe Co., FL. May 2001. Photo: TJA
115.5: Aduana, Sonora, Mex. Photo: JPB
115.6: Bahia Honda SP, Monroe Co., FL. May 2001. Photo: TJA
115.6: (inset) Bahia Honda SP, Monroe Co., FL. Photo: TJA
115.7: Beverly, Randolph Co., WV. Mar. 1998. Photo: TJA
115.8: Stock Island, Monroe Co., FL. Photo: Marc Minno

117.1: Llera de Canales, Tamaulipas, Mex. Photo: JPB
117.2: Miami-Dade Co., FL. Photo: JG
117.3: Alamos, Sonora, Mex. Photo: JPB
117.4: Garden Canyon, Huachuca Mts., Cochise Co., AZ. Photo: JPB
117.5: Coronado Nat. Memorial, Cochise Co., AZ. Photo: JPB
117.6: 6 miles S. of Patagonia, Santa Cruz Co., AZ. Photo: JPB
117.7: Fork Creek WMA, Nellis, Boone Co., WV. Aug. 1998. Photo: TJA
117.8: Ormond Beach, Volusia Co., FL. Sept. 1994. Photo: TJA

119.1: Alamos, Sonora, Mex. Photo: JPB
119.2: Eager, Apache Co., AZ. Photo: JG

119.4: Stone Coal Lake WMA, Upshure Co., WV. Aug. 1992. Photo: TJA
119.5: 4 miles S. of Patagonia, Santa Cruz Co., AZ. Photo: JPB
119.6: Elkins, Randolph Co., WV. Photo: TJA
119.8: 4 miles S. of Patagonia, Santa Cruz Co., AZ. Photo: JPB
119.10: Photo: JPB
119.11: Eastside, Uinta Co., UT. July 1999. Contrib: Todd Stout. Photo: TJA

121.1: Santa Ana NWR, Hidalgo Co., TX. Photo: JPB
121.2: Santa Ana NWR, Hidalgo Co., TX. Photo: JG
121.3: Santa Ana NWR, Hidalgo Co., TX. Photo: JPB
121.4: Santa Catalina Mts., Pima Co., AZ. Photo: JPB
121.5: Tucson, Pima Co., AZ. Photo: JPB
121.8: Molino Basin, Santa Catalina Mts., Pima Co., AZ. Photo: JPB
121.9: Alamos, Sonora, Mex. Photo: JPB
121.10: Monterrey, Nuevo Leon, Mex. Photo: JPB

123.1: Big Pine Key, Monroe Co., FL. May 2000. Contrib: Richard Boscoe. Photo: TJA
123.2: Parsippany, Morris Co., NJ. Photo: JG
123.3: Powers Hollow, Cabins, Grant Co., WV. May 1998. Photo: TJA
123.4: Upper Tract, Pendleton Co., WV. July 1994. Photo: TJA
123.5: Taylor Canyon, Ogdon, Weber Co., UT, 5400'. June 2002. Contrib: Todd Stout. Photo: TJA
123.5: (inset) Black River, Apache Co., AZ, UT, 5000'. July 2002. Contrib: Todd Stout. Photo: TJA
123.18: Moorefield, Hardy Co., WV. July 1994. Photo: TJA

125.1: Greenbrier Mt., White Sulphur Springs, Greenbrier Co., WV. Aug. 1997. Photo: TJA
125.1: (inset) MI. Photo: TJA
125.2: Photo: JG
125.3: Petersburg, Grant Co., WV. Aug. 1994. Photo: TJA
125.4: Walker Canyon, Santa Cruz Co., AZ. Photo: JPB
125.7: Green's Peak Rd., US Hwy. 60, Apache Co., AZ. Photo: JPB
125.9: Jordan R. St. Park., Modelport, Salt Lake/Dans Co., UT, 4200'. June 2002. Contrib: Todd Stout. Photo: TJA
125.10: Patagonia, Santa Cruz Co., AZ. Photo: JPB
125.11: Harpers Ferry, Jefferson Co., WV. Aug. 1986. Photo: TJA
125.11: (inset) Port Norris, Cumberland Co., NJ. Contrib: Dale Schweitzer. Photo: Dave Wagner

127.1: Alamos, Sonora, Mex. Photo: JPB
127.2: Higbees Beach, Cape May Co., NJ. Photo: JG
127.3: St. George, UT. Contributor: Todd Stout. Photo: JPB
127.4: Santa Ana NWR, Hidalgo Co., TX. Photo: JPB
127.4: (inset) NWR, Hidalgo Co., TX. Photo: JPB

127.6: Molino Basin, Santa Catalina Mts., Pima Co., AZ. Photo: JPB
127.8: Along So. Branch Poyomac R., Upper Tract, Pendleton Co., WV. Aug. 1998. Photo: TJA
127.10: Buckhorn Wash, UT. Aug. 1999. Contributor: Todd Stout. Photo: TJA
127.11: Overton, Clark Co., NV. Photo: JPB

129.1: Wentworth Location, Coos Co., NH. Aug. 2001. Contrib: Richard Boscoe. Photo: TJA
129.2: Sussex Co., NJ. Photo: JG
129.3: South Fork, Apache Co., AZ. Photo: JPB
129.4: Pena Blanca Lake, Santa Cruz Co., AZ. Photo: JPB
129.5: Salt Lake Co., UT. Contributor: Todd Stout. Photo: JPB
129.6: Fort Martin Power Station, Fort Martin, Monongalia Co., WV. July 1982. Photo: TJA
129.9: Elkins, Randolph Co., WV. July 1994. Photo: TJA
129.10: Sabino Canyon, Santa Catalina Mts., Pima Co., AZ. Photo: JPB

131.1: Green's Peak Road at Hwy. 60, Apache Co., AZ. Photo: JPB
131.2: Santa Ana NWR, Hidalgo Co., TX. Photo: JG
131.3: Affleck Peak, Mountain Dell Canyon, Salt Lake Co., UT. Contrib: Todd Stout. Photo: TJA
131.6: Molino Basin, Santa Catalina Mts., Pima Co., AZ. Photo: JPB
131.7: Imuris, Sonora, Mex. Photo: JPB
131.8: Canelo, Santa Cruz Co., AZ. Photo: JPB
131.9: Elkins, Randolph Co., WV. May 1994. Photo: TJA
131.10: Ormond Beach, Volusia Co., FL. April 1995. Photo: TJA
131.11: Ohio Key, Monroe Co., FL. May 2002. Contrib: Richard Boscoe. Photo: TJA

133.1: Utah. Photo: JPB
133.2: Colliers Mills WMA, Ocean Co., NJ. Photo: JG
133.3: Gaspe-Quest Co., Quebec, Can. July 2000. Contrib: Richard Boscoe. Photo: TJA
133.4: Chimney Creek, Tulare Co., CA. Photo: JPB
133.11: Greenbrier Mt., White Sulphur Spr., Greenbrier Co., WV. May 2000. Photo: TJA
133.12: Ocean Co., NJ. June 2000. Contrib: Richard Boscoe. Photo: TJA
133.13: Newport Richie, Pasco Co., FL. Oct. 1999. Contrib: Richard Boscoe. Photo: TJA
133.16: Rich Mt., Job, Pendleton Co., WV, June 1994. Photo: TJA

135.1: Bishop, Inyo Co., CA. Photo: TJA
135.2: Morristown, Morris Co., NJ. Photo: JG
135.3: Coonskin Park, Charleston, Kanawha Co., WV. Oct. 1994. Photo: TJA
135.4: Green's Peak Road at Hwy. 60, Apache Co., AZ. Photo: JPB
135.6: Reddish Knob, Sugar Grove, Pendleton Co., WV. Oct. 1998. Photo: TJA
135.7: Hamden, New Haven Co., CT. July 1998. Photo: Jeff Fengler

135.8: Deland, Volusia Co., FL, Sept. 2000. Contributor: Richard Boscoe. Photo: TJA

137.1: Ormond Beach, Volusia Co., FL. Nov. 1994. Photo: TJA
137.2: Chappaqua, Westchester Co., NY. Photo: JG
137.3: Elkins, Randolph Co., WV. May 1994, Photo: TJA. Aug. 1995, Photo: TJA
137.4: Spruce Knob Lake, Randolph Co., WV. June 1998. Photo: TJA
137.6: Elkins, Randolph Co., WV. May 1994, Photo: TJA. Aug. 1995, Photo: TJA
137.7: Elkins, Randolph Co., WV. May 1994, Photo: TJA. Aug. 1995, Photo: TJA
137.8: Homestead, Dade Co., FL. June 2000. Contributor: Richard Boscoe. Photo: TJA
137.9: Salt Lake City, Salt Lake Co., UT. Aug. 1999. Contrib: Todd Stout. Photo: TJA
137.10: Del Norte Co., CA. Contrib: Ken Hansen. Photo: JPB

139.1: Green's Peak Road at US Hwy. 60, Apache Co., AZ. Photo: JPB
139.2: Chappaqua, Westchester Co., NY. Photo: JG
139.4: Sunnytown, Montgomery Co., PA. July 2000. Contributor: Richard Boscoe. Photo: TJA
139.5: Elkins, Randolph Co., WV. May 1994, Photo: TJA. July 1994, Photo: TJA
139.6: Elkins, Randolph Co., WV. May 1994, Photo: TJA. July 1995, Photo: TJA
139.8: Monck's Corner, Berkeley Co., SC. May 2000. Contrib: Richard Boscoe. Photo: TJA
139.9: Milford, New Haven Co., CT. Contrib: D. Wagner. Photo: TJA
139.10: Deland, Volusia Co., FL. Apr. 2002. Contrib: Richard Boscoe. Photo: TJA

141.1: Deer Creek, Greenbank, Pocahontas Co., WV. June 1998. Photo: TJA
141.1: (inset) N. Attleboro, MA. June 1999. Photo: TJA
141.2: Wasatch Race Pk., Linden, Utah Co., UT, 4600'. May 2002. Contrib: Todd Stout. Photo: TJA
141.3: Bull Creek WMA, Osceola Co., FL. Photo: Marc Minno
141.4: Monck's Corner, Berkeley Co., SC. June 2001. Contributor: Richard Boscoe. Photo: TJA
141.5: Atlantic Co., NJ. July 2000. Contributor: Richard Boscoe. Photo: TJA
141.6: W. of Salt Lake City, Salt Lake Co., UT, 4600'. May 2002. Contrib: Todd Stout. Photo: TJA
141.7: Bodfish, Kern Co., CA. Photo: JPB
141.8: Wasatch Race Pk., Linden, Utah Co., UT, 4600'. May 2002. Contrib: Todd Stout. Photo: TJA

143.1: Newport Richie, Pasco Co., FL. Oct. 2001. Contrib: Richard Boscoe. Photo: TJA
143.2. Chappaqua, Westchester Co., NY. Photo: JG
143.3: Dade Co., FL. Contributor: Todd Stout. Photo: JPB
143.6: Cranesville, Preston Co., WV. Oct. 2000. Photo: TJA
143.7: Deland, Volusia Co., FL. Apr. 2002. Contrib: Richard Boscoe. Photo: TJA

143.8: Deer Creek, Greenbank, Pocahontas Co., WV. June 1994. Photo: TJA
143.9: Oklawaha River, Putnam Co., FL. Sept. 2000. Contrib: Richard Boscoe. Photo: TJA
143.10: Spruce Knob Lake, Randolph Co., WV. June 1998. Photo: TJA

145.1: Monck's Corner, Berkeley Co., SC. May 2000. Contrib: Richard Boscoe. Photo: TJA
145.2: Mountainside Park, Morris Co., NJ. Photo: JG
145.3: Ventana Canyon, Santa Catalina Mts., Pima Co., AZ. Photo: JPB
145.4: Willington, Tolland Co, CT. June 1999. Contrib: David Wagner. Photo: TJA
145.6: Tucson, Pima Co., AZ. Photo: JPB
145.9: Post Canyon, Santa Cruz Co., AZ. Photo: JPB
145.11: Sabino Canyon, Santa Cruz Co., AZ. Photo: JPB
145.12: Clark Co., NV. Sept. 1999. Contrib: Todd Stout. Photo: TJA

147.1: 4 miles S. of Patagonia, Santa Cruz Co., AZ. Photo: JPB
147.2: Garden Canyon, Cochise Co., AZ. Photo: JG
147.3: Garden Canyon, Huachuca Mts., Cochise Co., AZ. Photo: JPB
147.4: Molino Basin, Santa Catalina Mts., Pima Co., AZ. Photo: JPB
147.6: Pena Blanca Canyon, Santa Cruz Co., AZ. Photo: JPB
147.7: Glady Creek, Glady, Randoph Co., WV. July 1995. Photo: TJA
147.8: French Joe Canyon, Whetstone Mts., Cochise Co., AZ. Photo: JPB
147.9: Garden Canyon, Huachuca Mts., Cochise Co., AZ. Photo: JPB

149.1: Grant Mere Dunes St. Pk., Burrian Co., MI. Aug. 98. Contrib: David Wagner. Photo: TJA
149.2: Northwest River Park, Chesapeake, VA. Photo: JG
149.3: Suffolk, VA. Aug. 1999. Contributor: Richard Boscoe. Photo: TJA
149.4: nr. Foxfire Village, Moore Co., NC. Sept. 1999. Contrib: Richard Boscoe. Photo: TJA
149.6: Alamos, Sonora, Mexico. Photo: JPB
149.7: Box Canyon, Santa Rita Mts., Pima Co., AZ. Photo: JPB
149.8: Comfort Spring, Huachuca Mts., Cochise Co., AZ. Photo: JPB
149.9: Green's Peak Rd. at US Hyw. 60, Apache Co., AZ. Photo: JPB

151.1: Ormond Beach, Volusia Co., FL. Aug. 1998. Photo: TJA
151.2: Ormond Beach, Volusia Co., FL. July 2002. Photo: TJA
151.3: Elkins, Randolph Co., WV. Oct. 1994. Photo: TJA
151.4: Elliott Key, Monroe Co., FL. Photo: Marc Minno
151.7: Gomez Farias, Tamaulipas, Mex. Photo: JPB
151.8: nr. Canelo, Santa Cruz Co., AZ. July 1999. Contrib: Kilian Roever. Photo: TJA
151.9: Garden Canyon, Huachuca Mts., Cochise Co., AZ. Contrib: Kilian Roever. Photo: TJA
151.10: 9 miles SE. of Agua Prieta, Sonora, Mex. Photo: JPB

152.1: Neola, Greenbrier Co., WV. Photo: TJA
152.2: Elkins, Randolph Co., WV. Photo: TJA
152.3: Elkins, Randolph Co., WV. Photo: TJA
152.4: Brushy Fork, Barbour Co., WV. Photo: TJA
152.5: Brushy Fork, Barbour Co., WV. Photo: TJA
152.6: Ormond Beach, Volusia Co., FL. Photo: TJA
152.7: Pendleton Co., WV. Photo: TJA
152.8: Pendleton Co., WV. Photo: TJA
152.9: Elkins, Randolph Co., WV. Photo: TJA
152.10: Worthington SF, Warren Co., NJ. Photo: JG
152.11: Neola, Greenbrier Co., WV. Photo: TJA
152.12: Barbour Co., WV. Photo: TJA

153.1: Hardy Co., WV. Photo: TJA
153.2: Randolph Co., WV. Photo: TJA
153.3: Pendleton Co., WV. Photo: TJA
153.4: Ormond Beach, Volusia Co., FL. Photo: TJA
153.5: Photo: TJA
153.6: Elkins, Randolph Co., WV. Photo: TJA
153.7: Randolph Co., WV. Photo: TJA
153.8: Randolph Co., WV. Photo: TJA
153.9: Barbour Co., WV. Photo: TJA
153.10: Barbour Co., WV. Photo: TJA
153.11: Pendleton Co., WV. Photo: TJA
153.12: Elkins, Randolph Co., WV. Photo: TJA

154.1: Photo: TJA
154.2: Photo: TJA
154.3: Neola, Greenbrier Co., WV. Photo: TJA
154.4: Neola, Greenbrier Co., WV. Photo: TJA
154.5: Photo: TJA
154.6: Madison, Boone Co., WV. Photo: TJA
154.7: Ormond Beach, Volusia Co., FL. Photo: TJA
154.8: Cape Coral, Lee Co., FL. Photo: TJA
154.9: Pendleton Co., WV. Photo: TJA
154.10: Pendleton Co., WV. Photo: TJA

# Organizations Concerned with Butterflies

**The North American Butterfly Association (NABA)** promotes public enjoyment, awareness, and conservation of butterflies and all aspects of recreational, non-consumptive butterflying, including field identification, butterfly gardening, and photography. NABA publishes a full-color magazine, *American Butterflies*; a newsletter *Butterfly Gardening News*; has chapters throughout North America and runs the annual NABA 4th of July Butterfly Counts. These one-day counts, held mainly in June-July (centered on the 4th of July period) are growing rapidly. Currently almost 400 counts are conducted each year, at sites across North America. They are a fun-filled way to help monitor butterfly populations, to learn about butterfly identification, and to meet other butterfliers.

NABA
4 Delaware Rd.
Morristown, NJ 07960
Web site: http://www.naba.org

**The Lepidopterists' Society** is an international organization devoted to the scientific study of all lepidoptera. The Society publishes the *Journal of the Lepidopterists' Society* as well as the *News of the Lepidopterists Society*.

Lepidopterists' Society
1608 Presidio Way
Roseville, CA 95661
Web site: http://www.furman.edu

**The Xerces Society** is an international organization dedicated to the global protection of habitats for all invertebrates, including butterfflies. The Society publishes *Wings*.

Xerces Society
4828 Southeast Hawthorne Blvd
Portland, OR 97215

# Glossary

**Anterior.** Toward the front, or the head end.

**Band.** A line of color (of varying thickness) that runs from the dorsal (back) side of the caterpillar to the ventral (belly) side of the caterpillar – the entire distance or only a portion of the distance.

**Biennial.** Having a two-year cycle.

**Bristle.** Stiff hair-like structure.

**Caudal.** Referring to the hind end; toward the tail.

**Chevron**. A "V" shaped mark on a caterpillar.

**Chrysalis.** The hard case in which a caterpillar transforms into an adult. Also referred to as a pupa.

**Club.** A knob or enlarged tip of a horn or antenna.

**Collar.** The dorsal plate behind the head of a caterpillar, usually dark in color.

**Cremaster**. The posterior end of a pupa (chrysalis) which contains hooks that fasten it to a pad of silk spun by the caterpillar.

**Diapause.** A period of reduced or halted development during any lifestage; usually to pass a period of unfavorable environmental conditions.

**Dorsal (Dorsum).** The upper portion or back surface.

**Dorsolateral.** The upper portion of the side.

**Eye spots.** Refers to the colored rings on a caterpillar that resemble eyes.

**Filaments.** Fleshy appendages on caterpillars.

**Form.** A color or seasonal variation of a species.

**Frass.** The excrement or droppings of a caterpillar.

**Hibernaculum.** The winter nest or shelter of a caterpillar.

**Instar.** Any stage in the development of a caterpillar. After a caterpillar molts it becomes a new "instar," until finally it pupates and becomes a pupa. Caterpillars go through 4–8 instars depending upon the species.

**Larva.** The immature stage of insects.

**Lateral.** Along the side.

**Longitudinal.** Running lengthwise along the body from head to tail.

**Middorsal.** The middle of the back or upper surface.

**Molt.** To shed the skin or exoskeleton.

**Oblique.** Forming a diagonal line or dash.

**Osmeterium.** A fleshy orange, forked gland which can be everted from the back of a caterpillar's head when alarmed. It gives off a pungent odor that some find foul-smelling. Found in species in the swallowtail family.

**Ovipositor.** The egg laying structure of an insect.

**Powdery bloom.** A white powder which covers portions of a pupa in some species.

**Proleg.** One of the first pair of legs closest to the head, foreleg.

**Prothorax.** The anterior segment of the thorax which bears the first pair of legs.

**Pubescence.** A dense covering of short hairs creating a soft appearance.

**Pupa.** The development stage of a butterfly between the caterpillar and the adult. Also called a chrysalis.

**Pupate.** The transformation from a caterpillar to a pupa.

**Segment.** One of several ring-like sections of a caterpillar's abdomen.

**Seta(ae).** The hairs of a caterpillar.

**Setose.** Hair-like bristle.

**Spiracle.** An oval breathing vent along the side of a caterpillar.

**Stripe.** A line of color (of varying thickness) that runs from the head to the rear of the caterpillar —the entire distance or only a portion of the distance.

**Subdorsal.** The area of a caterpillar between the back and side.

**Sublateral.** An area of the body below the sides.

**Suture.** The dividing line or joint between 2 segments of a caterpillar's body.

**Tubercle.** Any small, rounded projection on the body of a caterpillar or pupa.

**Ventral.** The bottom or lower surface.

# Selected Bibliography


Allen, T. J. 1997. *The Butterflies of West Virginia and Their Caterpillars*. Pittsburgh: University of Pittsburgh Press.

Bailowitz, R. A., and Brock, J. P. 1991. *Butterflies of Southeastern Arizona*. Tucson: Sonoran Arthropod Studies, Inc.

Brock, J. P., and Kaufman, K. 2003. *Butterflies of North America*. Boston: Houghton Mifflin.

Glassberg, J. 1993. *Butterflies through Binoculars: A Field Guide to the Butterflies of the Boston-New York-Washington Region*. New York: Oxford University Press

Glassberg, J. 1999. *Butterflies through Binoculars: The East*. New York: Oxford University Press.

Glassberg, J. 2001. *Butterflies through Binoculars: The West*. New York: Oxford University Press.

Glassberg, J. 2002. *Butterflies of North America*. New York: Barnes and Noble.

Glassberg, J., Minno, M. C., and Calhoun, J. V. 2000. *Butterflies through Binoculars: Florida*. New York: Oxford University Press.

Minno, M. C., Butler, J. F., and Hall, D. W. 2004. *A Field Guide to Florida Butterfly Caterpillars and Their Host Plants*. Gainesville: University Press of Florida. In press.

Stamp, N. E., and Casey, T. M., eds. 1993. *Caterpillars. Ecological and Evolutionary Constraints on Foraging*. New York: Chapman and Hall.

Sutton, P. T., and Sutton, C. 1999. *How to Spot a Butterfly*. Boston: Houghton Mifflin.

Tuskes, P. M., Tuttle, J. P., and Collins, M. M. 1996. *The Wild Silk Moths of North America*. Ithaca, N.Y.: Cornell University Press.

Wright, A.B. 1993. *Peterson First Guide to Caterpillars of North America*. Boston: Houghton Mifflin.

# Foodplant Scientific Name Index

Acacias *(Acacia spp.)*
Mimosa Yellow, Reakirt's Blue
Acanthus *(Acanthaceae* [family]*)*
Elf, Crimson Patch, Elada Checkerspot, Texas Crescent, White Peacock
Adelias *(Adelia spp.)*
Mexican Bluewing
Agaves *(Agave spp.)*
Orange Giant-Skipper
Alder *(Alnus spp.)*
Alfalfa *(Medicago sativa)*
Clouded Sulphur, Funereal Duskywing, Northern Duskywing, Orange Sulphur
Alkali sacaton *(Sporobolus airoides)*
Sandhill Skipper
American beech *(Fagus grandifolia)*
Early Hairstreak
American Holly *(Ilex opaca)*
Henry's Elfin
Angelica *(Angelica spp.)*
Short-tailed Swallowtail
Anise, wild *(Pimpinella spp.)*
Anise Swallowtail
Antelope horned milkweed *(Aclepias viridis)*
Monarch
Arizona water willow *(Jacobinia [Justica] candicans)*
Texas Crescent
Arizona white oak *(Quercus arizonica)*
Dull Firetip, Meridian Duskywing, Short-tailed Skipper
Arugula *(Eruca vesicaria var. sativa)*
Great Southern White
Ash *(Fraxinus spp.)*
Two-tailed Swallowtail
Aspens
Canadian Tiger Swallowtail, Lorquin's Admiral, Weidemeyer's Admiral, Western Tiger Swallowtail

Asters *(Aster spp.)*
Bordered Patch, Field Crescent, Gabb's Checkerspot, Hoffmann's Checkerspot, Northern Checkerspot, Northern Crescent, Pearl Crescent, Silvery Checkerspot, Tawny Crescent
Atlantic white cedar *(Chamaecyparis thyoides)*
Hessel's Hairstreak
Ayenia *(Ayenia spp.)*
Common Streaky-Skipper, Scarce Streaky-Skipper
Babybonnets *(Coursetia spp.)*
Funereal Duskywing
Bahama senna *(Senna mexicana Var. C.)*
Little Yellow
Bailey's yucca *(Yucca baileyi)*
Strecker's Giant-Skipper
Balloonvine *(Cardiospermun halicacabum)*
Miami Blue, Silver-banded Hairstreak
Bamboo muhly *(Muhlenbergia dumosa)*
Cestus Skipper
Banana (yucca) *(Yucca baccata)*
Ursine Giant-Skipper
Barbados cherry *(Malpighia emarginata)*
White-patched Skipper
Bay cedar *(Suriana maritima)*
Mallow Scrub-Hairstreak, Martial's Scrub-Hairstreak
Beach aster *(Corethrogyne filaginifolia)*
Gabb's Checkerspot
Beans, Garden Beans *(Phaseolus spp.)*
Long-tailed Skipper
Bearberry *(Arctostaphylos uvaursi)*
Freija Fritillary, Hoary Elfin
Beardtongues *(Penstemon spp.)*
Arachne Checkerspot, Dotted Checkerspot, Variable Checkerspot
Beggar ticks *(Bidens spp.)*
Arizona Metalmark, Dainty Sulphur
Beggar weeds *(Desmodium spp.)*
Confused Cloudywing, Desert Cloudywing, Dorantes Longtail, Hoary Edge, Northern Cloudywing, Siilver-spotted Skipper, Southern Cloudywing
Bent grass *(Agrostis spp.)*
Four-spotted Skipperling
Bermuda grass *(Cynodon dactylon)*
Baracoa Skipper, Eufala Skipper, Fiery Skipper, Julia's Skipper, Orange Skipperling, Sandhill Skipper, Satchem, Southern Skipperling, Violet-clouded Skipper, Whirlabout
Big bluestem *(Andropogon geradii)*
Arogos Skipper, Crossline Skipper, Orange-headed Roadside-Skipper

Bindweed *(Convolvulus spp.)*
Painted Crescent
Birches *(Betula spp.)*
Canadian Tiger Swallowtail, Compton Tortoiseshell, Early Hairstreak, Mourning Cloak
Birchleaf mountain mahogany *(Cercocarpus betuloides)*
California Hairstreak, Mountain Mahogany Hairstreak
Bird's beak *(Cordylanthus spp.)*
Fulvia Checkerspot
Bitterbrush *(Purshia spp.)*
Behr's Hairstreak
Bitterbush *(Picramnia Pentandra)*
Dina Yellow
Bittercress *(Cardamine spp.)*
Falcate Orangetip
Black cherry *(Prunus serotina)*
Canadian Tiger Swallowtail, Coral Hairstreak, Eastern Tiger Swallowtail, Spring Azure
Black cohosh *(Cimicifuga racemosa)*
Appalachian Azure, Spring Azure
Black dalea *(Dalea frutescens)*
Reakirt's Blue
Black locust *(Robinia pseudo acacia)*
Silver-spotted Skipper, Zarucco Skipper
Black mangrove *(Avicennia germinans)*
Mangrove Buckeye
Black spruce *(Picea mariana)*
Bog Elfin
Blackbead *(Pithecellobium keyense)*
Large Orange Sulphur
Bladderpod *(Isomeris arborea)*
Becker's White
Bleeding hearts *(Dicentra spp.)*
Clodius Parnassius
Bloodleaf *(Iresine spp.)*
Hayhurst's Scallopwing
Blue grama grass *(Bouteloua gracilis)*
Juba Skipper, Pahaska Skipper, Rhesus Skipper, Riding's Satyr, Simius Roadside-Skipper, Uncus Skipper
Blue grasses *(Poa pratense)*
Garita Skipperling, Juba Skipper, Least Skipper, Long Dash
Blueberries *(Vaccinium spp.)*
Brown Elfin, Freija Fritillary, Henry's Elfin, Pelidne Sulphur, Pink-edged Sulphur, Mariposa Copper, Striped Hairstreak, Spring Azure

California encelia *(Encelia californica)*
Fatal Metalmark
California scrub oak *(Quercus dumosa)*
Gold-hunter's Hairstreak
Candle plant *(Senna alata)*
Orange-barred Sulphur
Canc *(Arundinaria spp.)*
Carolina Roadside-Skipper, Creole Pearly-eye, Lace-winged Roadside-Skipper, Southern Pearly-eye, Reversed Roadside-Skipper
Cannas (ornamental cana) *(Canna spp.)*
Brazilian Skipper
Canyon oak *(Quercus dumosa)*
Golden Hairstreak
Capers *(Capparis spp.)*
Great Southern White, Florida White
Carpetweed *(Phyla nodiflora)*
Phaon Crescent
Carrots *(Daucus carota)*
Black Swallowtail, Ozark Swallowtail
Catclaw acacia *(Acacia greggii)*
Leda Ministreak
Cauliflower *(Brassica oleraceae var. botroytis)*
Cabbage White
Cedars
Juniper Hairstreak
Cherries, wild
Coral Hairstreak, Red-spotted Admiral, Spring Azure, Striped Hairstreak, Two-tailed Swallowtail, Western Tiger Swallowtail
Chinese house *(Collinsia heterophylla)*
Edith's Checkerspot
Chinese wisteria *(Wisteria sinensis)*
Silver-spotted Skipper
Chinquapin *(Castanea spp.)*
Golden Hairstreak
Chock cherry *(Prunus virginiana)*
Spring Azure
Christmas palm *(Veitchia merrillii)*
Monk Skipper
Christmas senna *(Senna bicapsularis)*
Little Yellow, Orange-barred Sulphur, Sleepy Orange
Chuparosa *(Beloperone californica)*
Tiny Checkerspot
Cinquefoil *(Potentila spp.)*
Grizzled Skipper, Mountain Checkered-Skipper, Two-banded Checkered-Skipper

Citrus *(Citrus spp.)*
Anise Swallowtail, Bahamian Swallowtail, Giant Swallowtail, Ornythion Swallowtail, Ruby-spotted Swallowtail
Cliff rose *(Cowania mexicana)*
Desert Elfin
Cloudberry *(Rubus chamaemorus)*
Grizzled Skipper
Clovers *(Trifolium spp.)*
Eastern Tailed-Blue, Greenish Blue, Mexican Cloudywing, Northern Cloudywing, Queen Alexandra's Sulphur, Shasta Blue, Southern Cloudywing, Western Cloudywing, Western Sulphur
Coast live oak *(Quercus agrifolia)*
Propertius Duskywing
Coconut palm *(Cocos nucifera)*
Monk Skipper
Coinvine *(Dalbergia spp.)*
Large Orange Sulphur
Common beardgrass *(Bothriochloa barbinodis)*
Slaty Roadside-Skipper
Common reed grass *(Phragmites communis)*
Broad-winged Skipper
Coontie *(Zamia pumila)*
Atala
Corkystemmed passionvine *(Passiflora suberosa)*
Julia Heliconian, Zebra Heliconian
Cow parsnip *(Heracleum maximum)*
Short-tailed Swallowtail
Coyotillo *(Karwinskia spp.)*
Two-barred Flasher
Crab grasses *(Digitaria spp.)*
Sachem, Southern Broken-Dash, Whirlabout
Crabwood *(Ateramnus lucidus)*
Florida Purplewing
Cranberries
Bog Copper
Creeping mesquite *(Prosopis reptans)*
Blue Metalmark, Clytie Ministreak
Crimson dicliptera *(Dicliptera assurgeris)*
Cuban Crescent
Crotons *(Croton spp.)*
Goatweed Leafwing, Tropical Leafwing
Crown vetch *(Coronilla varia)*
Wild Indigo Duskywing
Cudweeds *(Gnaphalium spp.)*
American Lady

Curly dock *(Rumex crispus)*
Bronze Copper
Cycads *(Zamiaceae* [family]*)*
Atala
Cypresses *(Cypressus spp.)*
Juniper Hairstreak
Cynanchums *(Cynanchum spp.)*
Soldier
Daleas *(Dalea spp.)*
Reakirt's Blue
Deer grass *(Muhlenbergia rigeris)*
Garita Skipperling
Deer vetches *(Lotus spp.)*
Acmon Blue, Afranius Duskywing, Funereal Duskywing, Melissa Blue, Northern Blue, Silvery Blue
Deerweed *(Lotus scoparius)*
Avalon Scrub-Hairstreak, Bramble Hairstreak
Desert agave *(Agave deserti)*
California Giant-Skipper
Desert aster *(Machaeranthera tortifolia)*
Sagebrush Checkerspot
Desert hackberry *(Celtis pallida)*
Empress Leilia
Desert hollyhock *(Sphaeralcea ambigua)*
Northern White Skipper
Desert scrub oak *(Quercus turbinella)*
Ilavia Hairstreak
Desert senna *(Senna covesii)*
Sleepy Orange
Desert sunflower *(Viquiera deltoides)*
California Patch
Deserthoneysuckle *(Anisacanthus thurberi)*
Elada Checkerspot
Desmanthus *(Desmanthus spp.)*
Reakirt's Blue
Dill *(Anethum graveolens)*
Black Swallowtail
Docks *(Rumex spp.)*
American Copper, Edith's Copper, Gray Copper, Great Copper, Lustrous Copper, Purplish Copper, Ruddy Copper
Dodder *(Cuscuta spp.)*
Brown Elfin
Dogweeds *(Dyssodia spp.)*
Dainty Sulphur
Dogwoods *(Cornus spp.)*
Spring Azure

Dragonwort *(Artemisia spp.)*
Old World Swallowtail
Dudleyas *(Dudleya spp.)*
Sonoran Blue
Dwarf bilberry *(Vaccinium caespitosum)*
Sierra Sulphur
Dwarf birch *(Betula [glandulosa/minor])*
Frigga Fritillary
Dwarf mistletoe *(Arceuthobium spp.)*
Johnson'sHairstreak, Thicket Hairstreak
Eastern gama grass *(Tripsacum dactyloides)*
Byssus Skipper
Elms *(Ulmus spp.)*
Eastern Comma, Question Mark
Elongate buckwheat *(Eriogonum elongatum)*
Gorgon Checkerspot
Emory oak *(Quercus emoryi)*
Ares Metalmark, Dull Firetip, Meridian Duskywing, Short-tailed Skipper
English plantain *(Plantago lanceolata)*
Baltimore Checkerspot
European buckthorn
Henry's Elfin
False indigo *(Amorpha californica)*
California Dogface, Southern Dogface
False nettle *(Boehmeria cylindrica)*
Red Admiral
Feather tree *(Lyciloma microphylla)*
Large Orange Sulphur
Fendler's buckbrush *(Ceanothus fenleri)*
Nais Metalmark
Fennel *(Foeniculum vulgare)*
Anise Swallowtail, Black Swallowtail, Ozark Swallowtail
Fern acacia *(Acacia angustissima)*
Acacia Sulphur, Gold-costa Sulphur, Mexican Yellow
Figs *(Ficus spp.)*
Ruddy Daggerwing
Firs *(Abes spp.)*
Western Pine Elfin
Flattopped white aster *(Aster umbellatus)*
Harris' Checkerspot
Flaxes *(Linum spp.)*
Mexican Fritillary, Variegated Fritillary
Fleabanes *(Erigeron spp.)*
Rockslide Skipper
Florida trema *(Trema micranthum)*
Martial Scrub-Hairstreak

Foldwing *(Dicliptera spp.)*

Texas Crescent

Fourwing saltbush

Mojave Sootywing, Saltbush Sootywing, San Emigdio Blue

Fowl manna grass *(Glyceria striata)*

Pepper and Salt Skipper

Frogfruits *(Phyla spp.)*

Common Buckeye, Phaon Crescent, White Peacock

Gambel's oak *(Quercus gambelii)*

Colorado Hairstreak, Rocky Mountain Duskywing

Garden beans

Dorantes Skipper, Long-tailed Skipper

Garry oak *(Quercus garryana)*

Propertius Duskywing

Giant bent grass *(Agrostis gigantea)*

Long Dash

Giant buckwheat *(Eriogonum giganteum)*

Avalon Scrub-Hairstreak

Giant cane *(Arundinaria gigantea)*

Yehl Skipper

Giant reed grass *(Calamagrostis spp.)*

Yuma Skipper

Glasswort *(Salicornia perennis)*

Eastern Pigmy-Blue

Globe mallows *(Sphaeralcea spp.)*

Small Checkered Skipper, West Coast Lady

Goatsbeard *(Aruncus dioicus)*

Dusky Azure

Golden alexander *(Zizia aurea)*

Ozark Swallowtail

Golden banners *(Thermopsis spp.)*

Persius Duskywing

Golden canna *(Canna flaccida)*

Brazilian Skipper

Gooseberries *(Ribes spp.)*

Gray Comma, Hoary Comma, Oreas Comma, Tailed Copper

Grapefruit

Giant Swallowtail

Grasses *(Gramineae* [family]*)*

Alpines, Apache Skipper, Appalachian Brown, Arctic Satyr, Arctics, Bell's Roadside-Skipper, Bronze Roadside-Skipper, Brown Longtail, Canyonland Satyr, Carolina Satyr, Cassus Roadside-Skipper, Celia's Roadside-Skipper, Columbian Skipper, Common Branded Skipper, Common WoodNymph, Common Ringlet, Crossline Skipper, Dakota Skipper, Dotted Skipper, Edward's Skipperling, Fiery Skipper, Gemmed Satyr, Georgia Satyr, Great

Grasses *(Gramineae* [family]*) (continued)*
Arctic, Great Basin WoodNymph, Green Skipper, Hayden's Satyr, Indian Skipper, Juba Skipper, Jutta Arctic, Least Skipper, Leonard's Skipper, Lindsey's Skipper, Little Wood Satyr, Mead's WoodNymph, Meske's Skipper, Morrison's Skipper, Nabokov's Satyr, Nevada Skipper, Northern Pearlyeye, Nysa RoadsideSkipper, Olive-clouded Skipper, Ottoe Skipper, Pahaska Skipper, Pepper and Salt Skipper, Pine Satyr, Poweshiek Skipperling, Python Skipper, Red Satyr, Russet Skipperling, Sheep Skipper, Sierra Skipper, Small Wood-Nymph, Southern Broken-Dash, Three-spotted Skipper, Tropical Least Skipper, Viereck's Skipper, Violet-clouded Skipper White-barred Skipper, Whirlabout

Gray oak *(Quercus grisea)*
Dull Firetip

Green shrimp plant *(Blechum brownei)*
Malachite, White Peacock

Green violet *(Hybanthus concolor)*
Mexican Fritillary, Variegated Fritillary

Greenthread *(Thelesperma spp.)*
Dainty Sulphur

Guamuchil *(Manila tamarind) (Pithecellobium dulce)*
Large Orange Sulphur, Red-bordered Pixie

Guava *(Psidium spp.)*
Guava Skipper

Guinea grass *(Panicum maximum)*
Celia's Roadside-Skipper, Clouded Skipper

Gumbo limbo *(Bursera simaruba)*
Dingy Purplewing

Hackberries *(Celtis spp.)*
American Snout, Hackberry Emperor, Question Mark, Tawny Emperor

Hairy tubetongue *(Siphonoglossa pilosella)*
Vesta Crescent

Harvard's plum *(Prunus harvardii)*
Nais Metalmark

Hazelnut *(Corylus spp.)*
Early Hairstreak

Hickories *(Carya spp.)*
Banded Hairstreak, Hickory Hairstreak

Hog peanut *(Amphicarpa bracteata)*
Golden-banded Skipper, Silver-spotted Skipper

Hollies *(Ilex spp.)*
Spring Azure

Hollyhocks *(Althaea spp.)*
Painted Lady, West Coast Lady

Honey locust *(Gleditsia tracanthos)*
Silver-spotted Skipper

Honey mesquite *(Prosopis glandulosa* [also *juliflora])*
Palmer's Metalmark
Hop tree *(Ptelea trifoliata)*
Giant Swallowtail, Two-tailed Swallowtail
Hops *(Humulus spp.)*
Eastern Comma
Horkelias *(Horkelia spp.)*
Two-banded Checkered-Skipper
Huachuca agave *(Agave parryi var. huachucensis)*
Huachuca Giant-Skipper
Incense cedar *(Libocedrus decurrens)*
Juniper Hairstreak
Indian grass *(Sorghastrum nutans)*
Pepper and Salt Skipper
Indian mallow *(Abutilon spp.)*
Arizona Powdered Skipper, Erichson's White-Skipper, Texas Powdered-Skipper
Indigo bushes *(Dalea frutescens, pulchra)*
Reakirt's Blue, Southern Dogface
Indigos *(Baptisia spp.)*
False Duskywing, Funereal Duskywing, Silver-spotted Skipper, Zarucco Duskywing
Intertidal cordgrass *(Spartina spp.)*
Rare Skipper
Ironwood *(Olneya tesota)*
Funereal Duskywing
Jamaican caper *(Capparis cynophallophora)*
Florida White
Jamaican dogwood *(Piscidia piscipula)*
Fulvous Hairstreak, Hammock Skipper
Janusia *(Janusia gracilis)*
Funereal Duskywing, White-patched Skipper
Japanese lantern *(Bladder mallow) (Herrisantia crispa)*
Arizona Powdered-Skipper, Erichson's White-Skipper, Texas Powdered-Skipper
Johnson grass *(Sorphum halepense)*
Bell's Roadside-Skipper, Clouded Skipper, Eufala Skipper
Joint vetches *(Aeschynomene spp.)*
Barred Yellow
Junipers *(Juniperus spp., Thuja spp.)*
Juniper Hairstreak
Kennedys' buckwheat *(Eriogonum kennedyi)*
Veined Blue
Kidneywood *(Eysenhardtia spp.)*
Arizona Skipper

Knotgrass *(Paspalum spp.)*
Fawn-spotted Skipper, Sunrise Skipper, Tropical Least Skipper
Knotweeds *(Paspalum spp.)*
Lilac-bordered Copper, Purplish Copper
Lacepod *(Thysanocarpus)*
Desert Orangetip, Spring White
Lambsquarters *(Chenopodium spp.)*
Common Sootywing, Golden-headed Scallopwing, Hayhurst's Scallopwing, Mazans Scallopwing, Mexican Sootywing, Western Pigmy-Blue,
Lead tree *(Leucaena leucocephaia)*
Gray Ministreak
Leadplant *(Amorpha fruticosa)*
Southern Dogface
Leadworts *(Plumbago auriculata/scandens)*
Cassius Blue, Marine Blue
Lecheguilla *(Agave lecheguilla)*
Coahuila Giant-Skipper, Mary's Giant-Skipper
Legumes *(Leguminosae* [family]*)*
Afranius Duskywing, Cassius Blue, Ceraunus Blue, Christina's Sulphur, Clouded Sulphur, Dorantes Skipper, Eastern Tailed-Blue, Funereal Duskywing, Golden Banded-Skipper, Gray Hairstreak, Hoary Edge, Marine Blue, Meads Sulphur, Melissa Blue, Mexican Cloudywing, Northern Blue, Northern Blue, Northern Cloudywing, Orange Sulphur, Persius Duskywing, Queen Alexandra's Sulphur, Reakirt's Blue, Southern Cloudywing, Southern Dogface, Western Sulphur, Western Tailed-Blue, Zarucco Duskywing, Zestos Skipper
Lemon *(Citru limon)*
Giant Swallowtail
Lignum vitae *(Guajacum sanctum)*
Lyside Sulphur
Little bluestem *(Andropogon scoparius)*
Arogos Skipper, Cobweb Skipper, Common Roadside-Skipper, Neomanthla Skipper, Swathy Skipper
Live forever *(Dudleya cymosa)*
Sonoran Blue
Locoweeds *(Astragalus spp.)*
Afranius Duskywing, Persius Duskywing
Locust *(Robinia spp.)*
Mexican Yellow
Locustberry *(Byrsonima lucida)*
Florida Duskywing
Lopsided Indian grass *(Sorphastrum secundum)*
Arogos Skipper
Lousewort *(Pedicularis spp.)*
Edith's Checkerspot

Lupines *(Lupinus spp.)*
Afranius Duskywing, Arrowhead Blue, Boisduval's Blue, Frosted Elfin, Persius Duskywing, Shasta Blue, Silvery Blue, Sooty Hairstreak, Wild Indigo Duskywing
Mallows *(Fam. Malvaceae) (Malvaceae* [family]*)*
Arizona Powdered-Skipper, Common Checkered-Skipper, Common Streaky-Skipper, Desert Checkered-Skipper, Gray Hairstreak, Laviana White-Skipper, Mallow Scrub-Hairstreak, Red-crescent Scrub-Hairstreak, Scarce Streaky-Skipper, Small Checkered-Skipper, Texas Powdered-Skipper, Tropical Checkered-Skipper, Turks'-cap White-Skipper, West Coast Lady, White Checkered-Skipper
Manfreda *(Manfreda maculosa)*
Ursine Giant-Skipper
Manzanita *(Arctostaphylos spp.)*
Arizona Hairstreak, Brown Elfin
Maypop *(Passiflora incarnata)*
Gulf Fritillary, Julia Heliconian, Zebra Heliconian
Meadow parsnip *(Thaspium barbinode)*
Ozark Swallowtail
Melic grass *(Melica spp.)*
Rural Skipper
Mesquites *(Prosopis spp.)*
Ceraunus Blue, Leda Ministreak, Palmer's Metalmark
Mexican alvaradoa *(Alvaradoa amorphoides)*
Dina Yellow
Mexican buckeye *(Ungnadia speciosa)*
Henry's Elfin
Milk peas *(Galacia spp.)*
Long-tailed Skipper, Silver-spotted Skipper, Zarucco Duskywing
Milk vetches *(Astragalus canadenses)*
Acmon Blue, Melissa Blue, Northern Blue, Silvery Blue, Western tailed-Blue
Milkweed vines
Monarch, Queen, Soldier
Milkweeds *(Asclepias spp.)*
Monarch, Queen
Mimosas *(Albizia spp.)*
Mimosa Skipper, Mimosa Yellow, Outis Skipper
Mistflowers *(Eupatorium spp.)*
Rawson's Metalmark, Rounded Metalmark
Mistletoe *(Phoradendron flavescens)*
Great Purple Hairstreak
Monkeyflower *(Mimulus spp.)*
Common Buckeye, Tropical Buckeye, Variable Ceckerspot
Mountain avens *(Dryas octopetala)*
Alberta Fritillary

Mountain heather *(Cassiope mertensiana)*
  Heather Blue
Mountain mahogany *(Cercocarpus montanus)*
  Behr's Hairstreak
Mustards *(Brassica spp.)*
  Becker's White, Cabbage White, Checkered White, Desert Orangetip, Falcate Orangetip, Gray Marble, Great Southern White, Large Marble, Mustard White, Pearly Marble, Sara Orangetip, Spring White, Western White
Narrow-leafed cologania *(Cologania spp.)*
  Drusius Cloudywing, Northern Cloudywing
Narrow-leaved croton *(Croton linearis)*
  Bartram's Scrub Hairstreak
Narrowleaved milkweed *(Asclepias fascicularis)*
  Monarch
Nasturtium *(Tropaeolum majus)*
  Cabbage White, Great Southern White
Needlegrass *(Piptochaetium spp.)*
  Morrison's Skipper
Nettles *(Urtica spp.)*
  Eastern Comma, Milbert's Tortoiseshell, Question Mark, Red Admiral, Satyr Comma, West Coast Lady
New Jersey tea *(Ceanothus americana)*
  Mottled Duskywing
New Mexican locust *(Robinia neomexicana)*
  Funereal Duskywing, Silver-spotted Skipper
Nickerbean *(Caesaloinia spp.)*
  Miami Blue, Nickerbean Blue
Noseburns *(Tragia spp.)*
  Blue-eyed Sailor, Common Mestra, Gray Cracker, Red Rim
Nude buckwheat *(Eriogonum nudum)*
  Dotted Blue, Gorgon Copper
Oaks *(Quercus spp.)*
  Ares Metalmark, Arizona Hairstreak, Banded Hairstreak, California Hairstreak, California Sister, Dull Firetip, Edwards Hairstreak, Juvenile's Duskywing, Mournful Duskywing, Oak Hairstreak, Poling's Hairstreak, Propertius Duskywing, Scudder's Duskywing, Short-tailed Skipper, Sleepy Duskywing, White M Hairstreak, Zela Metalmark
Obtuse panic grass *(Panicum obtusum)*
  Dotted Roadside-Skipper
Orange
  Giant Swallowtail
Orange milkweed *(Asclepias tuberosa)*
  Monarch
Orange sesbania *(Sesbania exaltata)*
  Funereal Duskywing

Pineland croton *(Croton linearis)*
Bartram's Scrub-Hairstreak, Florida Leafwing
Pink purslane *(Portulaca pilosa)*
Mallow Scrub-Hairstreak
Pipevine (Mexico) *(Aristolochia watsoni)*
White-dotted Cattleheart
Pipevines *(Aristolochia spp.)*
Polydamus Swallowtail, White-dotted Cattleheart
Plantains *(Plantago spp.)*
Common Buckeye, Edith's Ckeckerspot, Tropical Buckeye
Plum, wild
Coral Hairstreak, Pale Swallowtail
Plume grasses *(Erianthus spp.)*
Byssus Skipper
Pond spice *(Litsea aestivalis)*
Palamedes Swallowtail
Ponderosa pine *(Pinus ponderosa)*
Chiricahua White, Pine White
Pongam *(Pongamia pinnata)*
Hammock Skipper
Poplars *(Populus spp.)*
Dreamy Duskywing, Mourning Cloak, Red-spotted Admiral
Post oak *(Quercus stellata)*
Horace's Duskywing
Prairie mimosa *(Desmanthus illinoensis)*
Reakirt's Blue
Prickly ash *(Zanthoxylum americanum)*
Giant Swallowtail
Prince's plume *(Stanleya pinnata)*
Becker's White
Prostrate apine willows *(Salix [arctica, vestita])*
Dingy Fritillary
Punctured bract *(Oxytheca perfoliata)*
Small Blue
Purpletop grass *(Troides flava)*
Little Glassywing
Pussytoes *(Antennaria spp.)*
American Lady
Quail saltbush *(MacNeill's) (Atriplex lentiformis)*
Saltbush Sootywing
Rabbitbrushes *(Chrysothamnus spp.)*
Gabb's Checkerspot, Northern Checkerspot, Sagebrush Checkerspot
Rabbitfoot grass *(Polypogon spp.)*
Tropical Least Skipper
Ratany *(Krameria spp.)*
Mormon Metalmark

Rattleweed *(Crotalaria retusa)*
Queen Alexandra's Sulphur, Western Sulphur
Red bay *(Persea borbonia)*
Palamedes Swallowtail, Spicebush Swallowtail
Red bud *(Cercis canadensis)*
Henry's Elfin
Red cedar *(Juniperus virginiana)*
Juniper Hairstreak
Red mangrove *(Rhizophora mangle)*
Mangrove Skipper
Red oaks *(Quercus [rubra])*
Horace's Duskywing
Redberry *(Rhamnus crocea)*
Hermes Copper
Red-root Buckwheat *(Eriogonum racemosum)*
Spalding Blue
Reniform buckwheat *(Eriogonum reniforme)*
Dotted Blue, Small Blue
Rice cutgrass *(Leersia oryzoides)*
Least Skipper
Rock cresses *(Arabis spp.)*
Falcate Orangetip, Gray Marble, Olympia Marble, Sara Orangetip, Spring White
Rock primroses *(Androsaceae spp.)*
Arctic Blue, Heather Blue
Roundleaved ragwort *(Senecio obovatus)*
Northern Metalmark
Ruellias *(Ruellia spp.)*
Banded Peacock, Texas Crescent
Russian thistle *(Salsola kali)*
Western Pigmy-Blue
Saltbushes *(Atriplex spp.)*
Western Pigmy-Blue
Saltgrass *(Distichlis spicata)*
Aaron's Skipper, Alkali Skipper, Obscure Skipper, Salt Marsh Skipper, Sandhill Skipper, Wandering Skipper
Saltwort *(Batis maritima)*
Eastern Pigmy-Blue, Great Southern White
Sassafras *(Sassafras spp.)*
Spicebush Swallowtail
Saw palmetto *(Serenoa repens)*
Monk Skipper, Palmetto Skipper
Sawgrass *(Cladium spp.)*
Palatka Skipper
Saxafrage *(Saxafraga spp.)*
Arctic Blue

Scarlet milkweed *(Asclepias curassavica)*
Monarch, Queen
Schott's (yucca) *(Agave schottii)*
Ursine Giant-Skipper
Scotchman's lovage *(Ligusticum scothicum)*
Short-tailed Swallowtail
Screwbean mesquite *(Prosopis pubescens)*
Palmer's Metalmark
Scrub oak *(Quercus turbinella* [west] *ilicifolia* [north] *inopina* [south]*)*
Edward's Hairstreak, Horace's Duskywing
Sea purslane *(Sesuvium spp.)*
Western Pigmy-Blue
Sea rocket *(Cakile lanceolata)*
Great Southern White
Sedges *(Cyperaceae* [family]*)*
Aaron's Skipper, Appalachian Brown, Bay Skipper, Berry's Skipper, Black Dash, Dion Skipper, Duke's Skipper, Dun Skipper, Eyed Brown, Georgia Satyr, Mitchell's Satyr, Mulberry Wing, Two-spotted Skipper
Seepwillow *(Baccharis glutimosa)*
Fatal Metalmark
Sennas, Tree sennas *(Senna spp.)*
Cloudless Sulphur, Little Yellow, Mottled Longtail, Orange-barred Sulphur, Sleepy Orange, White Angled-Skipper
Sensitive pea *(Chamaecrista spp.)*
Tailed Orange
Serviceberry *(Amelanchier spp.)*
Weidemeyer's Admiral
Shadscale Saltbush *(Atriplex canescens)*
Mojave Sootywing
Sheep sorrel *(Rumex acetosella)*
American Copper
Shindagger (Schott's yucca) *(Agave schottii)*
Poling's Giant-Skipper
Shooting star *(Dodecatheon media)*
Arctic Blue, Heather Blue
Shrimpflower *(Justicia brandegeana)*
Cuban Crescent
Shrubby cinquefoil *(Potentilla fruticosa)*
Dorcas Copper
Shrubby willows
Bog Fritillary, Frigga Fritillary
Sicklepod *(Senna obtuifolia)*
Sleepy Orange
Sideoats grama *(Bouteloua curtipendula)*
Elissa Roadside-Skipper, Many-spotted Skipperling, Morrison's Skipper, Orange Skipperling, Oslar's Roadside-Skipper, Ottoe Skipper, Sheep Skipper, White-barred Skipper

Silverleaf oak *(Quercus hypoleucoides)*
Colorado Hairstreak
Silverleafs *(Leucophyllum spp.)*
Chinati Checkerspot, Theona Checkerspot
Silver-leaved lotus *(Lotus argophyllum)*
Avalon Scrub-Hairstreak
Slimleaf bursage *(Ambrosia spp.)*
Painted Crescent
Slimpod senna *(Senna hirsuta)*
Cloudless Sulphur, Sleepy Orange
Small cranberry *(Vaccinium oxycoccos)*
Bog Fritillary
Small-leafed Nissolia vine *(Nissolia spp.)*
Zilpa Longtail
Smooth bromegrass *(Bromus inermis)*
Delaware Skipper, Four-spotted Skipperling
Smooth cordgrass *(Spartina alterniflora)*
Aaron's Skipper
Smooth water hyssop *(Bacopa monnieri)*
White Peacock
Snapdragons *(Antirrhinum spp.)*
Common Buckeye
Snow willow *(Salix nivalis)*
Dingy Fritillary
Snowberries *(Symphoricarpus spp.)*
Variable Checkerspot
Sorrels
Lustrous Copper
Southwestern bernardia *(Bernardia myricaefolia)*
Lacey's Scrub Hairstreak
Spanish bayonet *(Yucca aloifolia)*
Cofaqui Giant-Skipper
Spanish needles *(Bidens bipinnata)*
Dainty Sulphur
Speedwell *(Veronica spp.)*
Common Buckeye, Tropical Buckeye
Spicebush *(Lindera spp.)*
Spicebush Swallowtail
Spikerush *(Eleocharis spp.)*
Poweshiek Skipperling
Spotted saxifrage *(Saxifrage bronchialis)*
Astarte Fritillary
Spiny hackberry
Empress Leilia, Red-bordered Metalmark
Spreading buckwheat *(Eriogonum effusum)*
Rita Blue

Spreading snakeherb *(Dischoriste decubens)*
    Vesta Crescent
Spurge (Family) *(Euphobiacaea)*
    Blue-eyed Sailor
Squirrel-bananas *(Deeringothamnus spp.)*
    Zebra Swallowtail
Squirrel-tail *(Sitanion hystix)*
    Garita Skipperling
Stonecrops *(Sedum spp.)*
    Moss' Elfin, Phoebus Parnassius, Xamia Hairstreak
Strawberries *(Fragaria spp.)*
    Grizzled Skipper
Striated milkpea *(Galactia spicaformis)*
    Zestos Skipper
Sulphur buckwheat *(Eriogonum umbellatum)*
    Blue Copper, Dotted Blue, Square-spotted Blue
Sumac *(Rhus spp.)*
    Dusky-blue Groundstreak, Red-banded Hairstreak
Sunflowers *(Helianthus spp.)*
    Bordered Patch, Dainty Sulphur, Gorgone Checkerspot, Silvery Checkerspot
Swamp bay *(Persea palustris)*
    Palamedes Swallowtail, Spicebush Swallowtail
Swamp milkweed *(Asclepias perennis)*
    Monarch, Queen
Swamp thistle *(Circium muticum)*
    Swamp Metalmark
Sweet bay *(Magnolia virginiana)*
    Eastern Tiger Swallowtail
Sweet shaggytuft *(Stenandrium spp.)*
    Definite Patch
Sweetbush *(Bebbia juncea)*
    Wright's Metalmark
Sweetleaf *(Symplocus tintoria)*
    King's Hairstreak
Switch cane *(Arundinaria gigantea)*
    Creole Pearly-eye, Southern Pearly-eye
Sycamore *(Platanus occidentalis)*
    Western Tiger Swallowtail
Tamarind, wild *(Lysiloma latisiliquum)*
    Large Orange Sulphur
Tansy mustard *(Descurainia spp.)*
    Large Marble, Pearly Marble
Tarragon, wild
    Old World Swallowtail
Tephrosia *(Tephrosia leiocarpa)*
    Valeriana Skipper

Tetramerium *(Tetramerium spp.)*
Elf, Tiny Checkerspot
Texas beargrass *(Nolina texana)*
Sandia Hairstreak
Texas ebony *(Pithecellobium flexicaule)*
Large Orange Sulphur
Thistles *(Circium spp.)*
California Crescent, Mylitta Crescent, Painted Lady, Pale Crescent
Timothy *(Phleum pratense)*
European Skipper
Toothworts *(Dentaria spp.)*
West Virginia White
Torchwood *(Amyris elemifera)*
Bahamian Swallowtail, Giant Swallowtail, Ornythion Swallowtail, Schaus' Swallowtail
Torrey's yucca *(Yucca torreyi)*
Ursine Giant-Skipper
Tower mustard *(Arabis glabra)*
Large Marble
Trailing arbutus *(Epigaea repens)*
Hoary Elfin
Tree sennas
White Angled-Sulphur, Yellow Angled-Sulphur
Tubetongue *(Siphonogossa pilosella)*
Tiny Checkerspot
Tufted vetch *(Vicia cracca)*
Silvery Blue
Tulip tree *(Liriodendron tulipifera)*
Eastern Tiger Swallowtail
Turpentine broom *(Thamnosma montana)*
Desert Black Swallowtail
Turtlehead *(Chelone glabra)*
Baltimore Checkerspot
Twinberry honeysuckle *(Lonicera involucrata)*
Gillette's Checkerspot
Twinseed
Rosita Patch, Texan Crescent
Twist flower *(Streptanthella longirostris)*
Desert Orangetip, Pearly Marble
Uniolas *(Uniola spp.)*
Pepper and Salt Skipper
Utah agave *(Agave utahensis)*
Mojave Giant-Skipper
Velvet mesquite *(Prosopis velutina)*
Palmer's Metalmark

# Caterpillar Index